# Animal Welfare

Volume 15

**Series Editor**

Professor Clive Phillips
Foundation Chair of Animal Welfare
Centre for Animal Welfare and Ethics
School of Veterinary Science
University of Queensland
Gatton 4343, QLD
Australia

More information about this series at
http://www.springer.com/series/5675

David A.H. Wilson

# The Welfare of Performing Animals

## A Historical Perspective

Springer

David A.H. Wilson
School of History
University of Leicester
Leicester, United Kingdom

ISSN 1572-7408
ISBN 978-3-662-45833-4 ISBN 978-3-662-45834-1 (eBook)
DOI 10.1007/978-3-662-45834-1

Library of Congress Control Number: 2015931642

Springer Heidelberg New York Dordrecht London


Printed on acid-free paper

Springer-Verlag GmbH Berlin Heidelberg is part of Springer Science+Business Media (www.springer.com)

*To Judy, Dan and Kit*

# Series Preface

Animal welfare is attracting increasing interest worldwide, especially in developed countries where knowledge and resources are available to, at least potentially, provide better management systems for farm animals, as well as companion, zoo, laboratory and performance animals. The key requirements for adequate food, water, a suitable environment, companionship and health are important for animals kept for all of these purposes.

There has been increased attention given to animal welfare in the West in recent years. This derives largely from the fact that the relentless pursuit of financial reward and efficiency, to satisfy market demands, has led to the development of intensive animal management systems that challenge the conscience of many consumers in this part of the world, particularly in the farm and laboratory animal sectors. Livestock are the world's biggest land users (FAO 2002), and the farmed animal population is increasing rapidly to meet the needs of an expanding human population. This results in a tendency to allocate fewer resources to each animal and to value individual animals less, for example, in the case of farmed poultry where flocks of over 20,000 birds are not uncommon. In these circumstances, the importance of each individual's welfare is diminished.

In developing countries, human survival is still a daily uncertainty, so that provision for animal welfare has to be balanced against human welfare. Animal welfare is usually a priority only if it supports the output of the animal, be it food, work, clothing, sport or companionship. However, in many situations, the welfare of animals is synonymous with the welfare of the humans that look after them, because happy, healthy animals will be able to assist humans best in their struggle for survival. In principle, the welfare needs of both humans and animals can be provided for, in both developing and developed countries, if resources are properly husbanded. In reality, the inequitable division of the world's riches creates physical and psychological poverty for humans and animals alike in many parts of the world.

Increased attention to welfare issues is just as evident for zoo, companion, laboratory, sport and wild animals. Of growing importance is the ethical management of breeding programmes, since genetic manipulation is now technically advanced, but there is less public tolerance of the breeding of extreme animals if it comes at the expense of animal welfare. The quest for producing novel genotypes has fascinated breeders for centuries. Dog and cat breeders have produced a variety of deformities that have adverse effects on their welfare, but nowadays the breeders are just as active in the laboratory, where the mouse is genetically manipulated with equally profound effects.

The intimate connection between animals and humans that was once so essential for good animal welfare is rare nowadays, having been superseded by technologically efficient production systems where animals on farms and in laboratories are tended by increasingly few humans in the drive to enhance labour efficiency. With today's busy lifestyles, companion animals too may suffer from reduced contact with humans, although their value in providing companionship, particularly for certain groups such as the elderly, is beginning to be recognised. Animal consumers also rarely have any contact with the animals that are kept for their benefit.

In this estranged, efficient world, people struggle to find the moral imperatives to determine the level of welfare that they should afford to animals within their charge. A few people, and in particular many companion animal owners, strive for what they believe to be the highest levels of welfare provision, while others, deliberately or through ignorance, keep animals in impoverished conditions in which their health and well-being can be extremely poor. Today's multiple moral codes for animal care and use are derived from a broad range of cultural influences, including media reports of animal abuse, guidelines on ethical consumption and campaigning and lobbying groups.

This series has been designed to contribute towards a culture of respect for animals and their welfare by producing learned treatises about the provision for the welfare of the animal species that are managed and cared for by humans. The early species-focused books were not detailed management blueprints; rather they described and considered the major welfare concerns, often with reference to the behaviour of the wild progenitors of the managed animals. Welfare was specifically focused on animals' needs, concentrating on nutrition, behaviour, reproduction and the physical and social environment. Economic effects of animal welfare provision were also considered where relevant, as were key areas where further research is required.

In this volume the series again departs from the species focus to address the welfare of performance animals and, in particular, the controversy surrounding these animals that developed over the course of the twentieth century. Taking a historical perspective, Wilson charts the development of public sentiment and legislation during this period. What emerges is not just about performance animals but an insightful commentary on some of the major social issues of today. Following Darwin's shocking revelations in the mid-nineteenth century of humans' close relationship to other animals, especially primates, the public gained much pleasure from watching animals in circuses and theatres perform ridiculous tricks. And that appears to have been precisely their motivation: having been told that they were not any longer inherently superior to animals, the public delighted in shows where animals could be ridiculed. During the exploration, and colonisation, of foreign lands by European powers, returning with exotic animals such as lions and tigers for shows in Europe became popular. However, it was not just the exotic wild animals that people came to see. Domesticated animals were dressed in costumes, made to ride bicycles and engage in many other unnatural acts, all helping to reaffirm our mastery of the animal kingdom. During the course of the twentieth century, however, the cruelty involved in procuring, training and managing these animals was gradually revealed, and the battle was started to introduce suitable legislation to curb these practices. Much was revealed in the aftermath of the First World War, which ushered in a new sense of openness and moral rectitude in the British public.

Wilson narrates the story without passion, but with a keen eye for detail and historical accuracy. Nevertheless, the extent of the cruelty that is revealed is shocking and must be a lesson for anyone who still believes that animals, and in particular wild animals, have a place in circuses. Until humanity completely and collectively distances itself from such abuse of animals, we cannot truly discover the enormous benefits that close cooperation with animals can bring. On what may be the eve of a ban on using wild animals in circuses in England,[1] this book appropriately describes the depth of cruelty to which circus animals have been exposed, especially during training and confinement. It also serves to inform us of the long path to legislative change on animal welfare issues, even when there is major concern by the public. Although a ban on using wild animals in circuses already exists in a few countries, others should also follow their lead in curbing this important and long-standing animal welfare problem.

Professor of Animal Welfare and Director
Centre for Animal Welfare and Ethics
School of Veterinary Sciences
University of Queensland, Australia

Clive Phillips

## Reference

Food and Agriculture Organisation (2002). http://www.fao.org/ag/aga/index_en.htm

[1] See Wild Animals in Circuses Bill, 2014, expected at the time of writing to be enacted and have effect from 1 December 2015. http://services.parliament.uk/bills/2014-15/wildanimalsincircuses.html. Accessed 9/10/14; https://www.gov.uk/government/uploads/system/uploads/attachment_data/file/228862/8538.pdf. Accessed 7/7/14.

# Acknowledgements

I have pleasure in acknowledging Professor Clive Phillips for his generous advice and patience during the preparation of this book and Anette Lindqvist for support in its production. Access to primary source material was made possible through the kind cooperation of Michael Day of Equity (British Actors Equity Association, custodians of the Variety Artistes' Federation archives), Chris Reed and Pat Squire (RSPCA Archive) and Catherine Haill (Theatre Museum, London). I am especially grateful to Mrs Esther Denham for allowing me to study the papers of her father, Edmund MacMichael, and to his granddaughters Caroline MacMichael Dales and Juliet MacMichael for further information. Among others who have given their time and assistance are Will Travers of the Born Free Foundation, Malcolm Clay of the Association of Circus Proprietors of Great Britain, David Cheshire, Richard Baker, Peter Bailey, Roger Wilmut, Michael Kilgarriff, Kim Stallwood, Ann Rogers, Mike Matthews, Henry S. Salt Archive, Susan Furmage of the Vegetarian Society, and David Kenworthy, 11th Baron Strabolgi. The material from United Kingdom Parliament publications contains Parliamentary information licensed under the Open Parliament Licence v3.0 (http://www.parliament.uk/site-information/copyright/open-parliament-licence/). The University of Cumbria met the reproduction fees for many of the illustrations.

# Abbreviations

| | |
|---|---|
| ABE | Animal Behavior Enterprises |
| ACP | Association of Circus Proprietors of Great Britain |
| AHA | American Humane Association |
| ASPCA | American Society for the Prevention of Cruelty to Animals |
| BBC | British Broadcasting Corporation |
| BBFC | British Board of Film Censors |
| CAPS | Captive Animals' Protection Society |
| CWG | Circus Working Group |
| DEFRA | Department of Environment, Food and Rural Affairs |
| n.d. | Not dated |
| NCDL | National Canine Defence League |
| ODFL | Our Dumb Friends' League |
| PADL | Performing Animals' Defence League |
| PETA | People for the Ethical Treatment of Animals |
| RBBBC | Ringling Bros. and Barnum & Bailey Circus |
| RSPCA | Royal Society for the Prevention of Cruelty to Animals |
| SGGB | Showmen's Guild of Great Britain |
| SPCA | Society for the Prevention of Cruelty to Animals |
| USDA | United States Department of Agriculture |
| VAF | Variety Artistes' Federation |

Numerical references in brackets within the text indicate the numbered paragraphs of the minutes in the Reports from the Select Committee on Performing Animals (United Kingdom Parliament 1921 and 1922), those in italics referring to the Report of 1921 and those in roman to that of 1922.

# Contents

# Chapter 1
# Introduction

**Abstract** Historical perspectives on the welfare of performing animals have so far received little attention within the expanding field of human-animal studies. In contrast, because the exhibition of performing animals has remained a public activity, public opinion about its ethical implications has been highly visible; and the secrecy surrounding preparatory training, as well as concerns about conditions of captivity, fuelled arguments that also won the attention of the media, government and science. In Britain especially, these arguments intensified from the beginning of the twentieth century, encouraged by organized campaigns of opposition and defence and involving influences from the United States and Germany.

This book describes and analyses a neglected aspect of the history of concern for animal welfare in westernized societies. It examines the welfare of performing animals as debated in the public arena, especially over the past 100 years and when the term 'welfare' was often used on a subjective 'common sense' basis to apply to a variety of circumstances and disputed problems. With the rapid expansion of academic studies of animals in human society of the last four decades, neglect of this subject is all the more surprising. Furthermore, while there has been much published on the ethical implications of animal experimentation, intensive farming and hunting, and on theoretical examinations of animal interests, there has been very little relating to the use of trained animals for popular entertainment, even though this is an area where there has been the greatest opportunity to gauge the opinions (often based on personal experience) of the widest cross-section of society regarding the treatment of animals.

Ironically, as criticism began to form, the public felt confident about expressing negative views on an activity with which they had already been encouraged to become familiar. In this way, the public debate on animal performance differed from that on vivisection, which was an exclusive activity against which relatively few opponents could present scientific arguments based on personal acquaintance. Nevertheless, one stage in animal performance, the preparatory training, shared a key characteristic with vivisection in that it was carried out behind closed doors. It was this element of secrecy that fuelled suspicion, excited the imagination and helped to secure the longevity of opposition. Further concerns about standards of

 
D.A.H. Wilson, *The Welfare of Performing Animals*, Animal Welfare 15,
DOI 10.1007/978-3-662-45834-1_1

travel and accommodation for wild and domesticated animals and arguments about the significance and effects of capture and captivity would also have increasing relevance beyond the use of animals in entertainment and invited scientific and veterinary opinion together with proposals for new controlling legislation.

In Britain by the early 1920s, the public debate about animal performance even eclipsed that concerning vivisection and field sports. The intensity of feeling and the story surrounding it have been all but forgotten. A special emphasis is given here to the major public controversy in Britain at that time, when many of the continuing questions around the rights and wrongs of animal performance were first crystallized through a major parliamentary enquiry and by accompanying attention in the press. During 1921 and 1922, before the passage of the Performing Animals (Regulation) Act in 1925, a British parliamentary select committee undertook an exhaustive investigation into the degree to which animal performance in the circus and on the music-hall stage depended on cruelty. International exchange within the entertainment industry also enabled contemporary assessments, sometimes biased, to be made of practices abroad and by visiting artistes. Unresolved problems became the basis of agitation and calls for new legislation that persist today in Britain and beyond.

Following a survey of the long history of animal performance, this book records the emergence of opposition to it and analyses the arguments and evidence of critics and supporters to the present day, using Britain as the main example. It discusses the means and ends of capture and training and the role of pressure groups, politics, press and vested interests, so that the historical context of this aspect of animal welfare issues is revealed. At the same time, developments in the United States, Germany and elsewhere affected British reactions to animal performance: for example, a crucial influence in the escalation of the debate had been Jack London's novel, *Michael, Brother of Jerry*, and the establishment of the international Jack London Club. Both the Royal Society for the Prevention of Cruelty to Animals (RSPCA) and the more radical Performing Animals' Defence League (PADL) aligned themselves with the club's aims and tactics. The PADL was especially active from its foundation in 1914 until 1968 and under its eccentric and uncompromising director, Edmund MacMichael, who was an early proponent of direct-action tactics in animal welfare campaigns. His unpublished and uncatalogued personal papers represent primary source material of much interdisciplinary interest, revealing the League's motives, arguments and tactics. These papers were kindly made available to the author by MacMichael's daughter, Esther Denham, and in this work, they are referred to as the MacMichael Papers.

In order to make this study manageable, it has been necessary to confine it to developments in what might be called westernized societies. In the same way, it has been necessary to address the difficulty of deciding what activities should be included as animal performance. This difficulty had also been encountered in the British parliamentary debates. However, as then, in this work the emphasis is made on animals trained for use in popular, commercial entertainment mainly in the circus, fairground, music hall and variety theatre and, later, in film and advertising. The historic legitimate theatre has also been included, but only occasionally, when performing animals were the leading attraction of a given production. Training for obedience trials, military purposes, dressage and the like is not dealt with here. The flowering of the British controversy in parliament and in the press in the 1920s had meanwhile witnessed the advent of cinema and radio broadcasting, and so debates about the economic exploitation of animals for public entertainment were then linked to another issue particular of the time, the continued viability of the music hall and circus in the face of new competition.

# Chapter 2
# The Legacy of Animal Performance

**Abstract** Trained animals were among those slaughtered on a vast, organised scale in the arenas of the classical world. The abuse of animals as well as humans—as revealed by contemporary sources—reflected a political contract between those in power and common citizens, religious influence, and an oppressive view of the world beyond ordered Roman society. Trained animal displays then left the fixed arena to become itinerant, small-scale business operations, until eventually they returned to be part of the theatre, fairground and, more recently, the circus arena as well. Concern for the welfare of performing animals was almost non-existent until the post-medieval period, becoming more apparent in the Enlightenment and consolidating in the nineteenth century. Many of the spectacles and performances, together with contemporary critiques of them, are discussed in this chapter, leading to the point when misgivings about the commercial training of animals for entertainment became publicly aired and organised into campaigns in the early twentieth century. The high visibility of new campaigning organisations ensured press coverage in Britain which in turn led to political interest, investigation and legislation.

## 2.1 From the Classical World to the Nineteenth Century

The first comprehensive records of the exploitation of animals for entertainment exist in the works of the authors of the Greek and Roman world, complemented by the archaeological remains of many sites of the spectacles they describe. Roman civilisation in particular was notorious for the uncivilised slaughter of people and animals on a massive scale. Besides the mass execution of humans, the slaughter of wild animals, sometimes subjected to training before their death, was so extensive over so many centuries that wildlife populations across the empire were decimated.[1] Together with other entertainments, this means of satisfying and encouraging the bloodlust of Roman citizens came to be expected as a service from those in authority who might in turn expect the favour and obedience of those they governed. At the same time, it reflected an ethos of domination and control of

[1] See Kyle (1998, p. 78), who gives further references to the assessment of this problem.



D.A.H. Wilson, *The Welfare of Performing Animals*, Animal Welfare 15,
DOI 10.1007/978-3-662-45834-1_2

nature and of those who as enemies, lawbreakers or heretics threatened Roman authority and its religion. These unfortunates had no value beyond their role as victims, and the idea of welfare was therefore irrelevant and absent.[2] However, there seems to have been some unusual regard for elephants, not just as specially exotic and symbolic of foreign conquest, but also because they were sometimes credited with desirable human moral characteristics; for example, their reaction to mistreatment by Pompey at a show in 55 BC elicited a rare instance of compassion and anger in the crowd.[3]

The killings in these ancient spectacles have attracted most of the attention both of contemporary and later writers, but it is clear that many animals were also tamed and trained to entertain. In 1937, George Jennison drew on a range of classical sources that refer to the public use of animals in the Roman world, adding to them his commentaries as one 'who has spent a lifetime in practical contact with wild animals': he was superintendent of Belle Vue Zoological Gardens in Manchester, founded by his family in 1837 (Jennison 1937, p. xiii).

In a review of 1950, Harold Nicolson said that 'the book should not be read by those who have been taught to admire the Roman virtues, since it constitutes one of the most damning indictments of that repugnant race that I have ever read'. The Romans 'delighted in performing animals, always a sign of a low-grade sensibility ... It is wholly impossible to respect a nation which derives amusement from the sad artificial antics of animals ... No nation which could enjoy this mixture of the savage and the jocose can expect to be revered by posterity'. Animal performances in the Roman Empire were 'the beastly diversions of a beastly race' (Nicolson 1950). Jennison noted that there was actually little information on taming and training techniques in Graeco-Roman times, but that the traditions of the art probably came largely from the Greek world, particularly Alexandria, later a wild beast market in the Middle Ages and an international centre of the animal trade until the twentieth century (Jennison 1937, p. 168).

Wild animals were frequently trained to take part in processions at festivals. At one in Egypt for Dionysus, celebrated by Ptolemy II in the late third century BC, there were chariots drawn by elephants, goats, antelopes, oryx, hartebeest, ostriches and wild asses.[4] Antony appeared in public in a chariot drawn by lions in 48–47 BC, the first record of tamed and trained lions in Italy.[5] Suetonius described Caesar as going up to the Capitol in triumph in 46 BC with 40 highly trained elephants to the left and right of him and carrying torches with their trunks.[6] In the second century

[2] But concerning an animal victim, in Statius's unusually sympathetic account of the accidental death of a trained fighting lion in the arena, he asks of the lion what good has therefore been achieved by its taming and training, to endure commands and to obey a weaker master (Statius, *Silvae* 2, 5, cited by Jennison 1937, pp. 79–80).

[3] See Futrell (2006, pp. 7 and 27), citing Cicero, *Letters to His Friends* 7.1, and Pliny, *Natural History* 8.7. See also Kyle (1998, p. 4).

[4] Athenaeus, *The Deipnosophists* 5, 198d ff, cited by Jennison (1937, pp. 29–32).

[5] Pliny, *Natural History* 8.21 and Plutarch *Antony* 9.5, cited by Jennison (1937, p. 56).

[6] Suetonius, *Julius Caesar* 37, cited by Jennison (1937, p. 58).

AD, Pausanias saw a procession with a car drawn by stags, and Jennison observes that it would have been very difficult to train them to go into harness.[7] But Elagabalus used them in this way as well, alongside lions and tigers, and for his amusement turned tame lions, leopards and bears (declawed and toothless) into his guests' rooms.[8]

Trained animals appeared for commercial purposes as well as for political use. According to Isocrates, dancing bears were exhibited by showmen in Greece in the fourth century BC.[9] In Ptolemaic Egypt, baboons were taught to distinguish letters, to dance and to play the flute or lyre. Aelian had also seen a Barbary ape performing as a charioteer.[10] Performing monkeys are represented in Pompeian frescos, and one was to be seen in a Roman street armed with a helmet and shield, taught through fear of the whip to ride a goat and to throw javelins.[11] And nightingales were apparently trained to accompany musical instruments.[12]

Within the amphitheatre itself, the public exhibition of performances by trained animals assumed growing importance during the early Roman Empire. The animals were sometimes spared death, while their trainers increased in number and became ever more demanding.[13] Following capture by pit or net, iron-barred cages could be used for transport.[14] Pliny suggested that beating and starvation, mitigated by barley juice, had been the African way of taming captured elephants, and then for many animals, it was probably a matter of training by starvation and the whip, as mentioned by Philo.[15] Performing elephants were especially popular, being trained to throw weapons, fight duels, dance and even walk the tight rope.[16] Performing bulls fought and rolled over at the word of command, let themselves be caught and lifted by the horns, and even stood like charioteers in cars going at a gallop.[17] Martial's epigrams on animals in his *Book of Spectacles* may refer to games under Titus in AD 80 to mark the dedication of the Colosseum (Jennison 1937, pp. 72–78).

[7] Pausanias, *Description of Greece* 7.18.12, cited by Jennison (1937, p. 25).

[8] *Historia Augusta, The life of Elagabalus* 28.2, 25.1, 21.1, cited by Jennison (1937, p. 90).

[9] Isocrates, *Antidosis* 213, cited by Jennison (1937, p. 22).

[10] Aelian, *On the nature of animals* 5.26 and 6.10, cited by Jennison (1937, pp. 128–129).

[11] Juvenal, *The Satires* 5.154–5, cited by Jennison (1937, pp. 128–129).

[12] Pliny, *Natural History* 10.43, cited by Jennison (1937, p. 116).

[13] Plutarch, *On the Intelligence of Animals* 5, cited by Jennison (1937, p. 63).

[14] For example, Symmachus *Letters* 2.77, cited by Jennison (1937, p. 149), who notes: 'A cage of this sort when shuttered is kept warm by animal heat, and the close and foetid atmosphere does no harm, unless confinement in it is very prolonged. The change of scene on a journey keeps animals interested and alert, and when once they are reconciled to captivity, they retain their health and condition in their portable cages. In our modern travelling shows lions and tigers may live for years in a space of 8 × 4 ft; and they breed more rapidly and look more healthy than in the large cages and dreary monotony of a zoological garden'.

[15] Pliny, *Natural History* 8.7–9 and Philo *De Animalibus*, 90, cited by Jennison (1937, pp. 143 and 9).

[16] Pliny, *Natural History* 8.2–3 and Suetonius *Galba*, 6.1, cited by Jennison (1937, p. 65).

[17] Pliny, *Natural History* 8.70, cited by Jennison (1937, p. 71).

Trained bulls and horses appeared in a water show, the arena being suddenly flooded, and the animals swimming in the water and drawing boats shaped as cars. Martial reveals that under Domitian lions were trained to retrieve hares without harming them, having been made by their trainer to kill bulls in the arena beforehand.

According to Wiedemann (1992, pp. 86 and 186), displays by performing animals were termed *pyrrichae*, and in the eightieth animal fable of the poet Babrius, who probably wrote in the second century AD, a camel performs in one of them:

> A drinking master would his camel bring
> To dance to flutes, and brazen cymbal's ring.
> 'Would that I could on a plain road advance,
> Causing no laughter,' said she, 'much less dance'.

*Pyrrichae* could also help the less literate or less well educated understand episodes from Greek mythology that were re-enacted in the arena.

With the decline of the Roman Empire, performing animals left the large, fixed arena and became more itinerant, eventually reappearing in the context of the permanent theatre, and, ultimately, the circus. Animal acts in the meantime became regular components of fairs, one major example of which was London's annual Bartholomew Fair, which ran from 1133 to 1855 as 'a yearly concentration into one spot of entertainments that at other times were scattered over town and country' (Morley 1859, p. 316).

The medieval world, like classical antiquity, was unaware or negligent of the interests of its animals. The only exceptions were the marks of respect and affection shown by some nobles to particular hunting hounds or to highly trained horses used in tournaments or war. Minstrels, joculators, showmen and *ursinarii*, with dancing bears and other animals trained to imitate human behaviour, carried on into the Middle Ages the street traditions of the classical world and travelled large distances to offer their services in towns and villages and at the residences of rulers and nobility. Kiser (2011, pp. 103–126) concludes from extant records that 'medieval observers found the breaching of the traditional animal-human divide to be the most entertaining form of animal entertainment'.[18] As then also with animal fighting and baiting, this common form of entertainment was usually a service provided by individuals as an income-generating enterprise. Political investment and religious motives in the old, large-scale spectacles were replaced by commercial incentives based on small-scale visiting services to local communities, and, if fame earned it, to the courts of the nobility. However, as in antiquity, tamed, trained and exotic animals could still occasionally be used to emphasise the power and importance of an individual.[19]

---

[18] Kiser refers to a number of early sources describing the feats of performing animals such as dog-riding apes imitating tournament knights, and a bear learning the alphabet.

[19] For example, a twelfth-century monk recorded that after invading Normandy, Henry I of England attempted this in a triumphal parade including a lion and Ethiopian keeper, a leopard riding a horse and a panther drawing a chariot (Loisel 1912, pp. 166–168, cited by Kiser 2011, pp. 103–126).

From the tenth to fourteenth centuries, joculators built up high public regard and exhibited animals that tumbled, danced and performed a variety of human-like tricks 'contrary to their nature'.[20] A rope-dancing horse and horse-riding oxen holding trumpets to their mouths were described in the thirteenth century. In the fourteenth century, a horse was trained for mock combat with a joculator, with movements regulated by music, and another was trained to beat a drum with its front and back hooves. Birds were also traditionally trained, like the cock dancing on stilts to music in the fourteenth century, and itinerant trainers touring Europe with their dancing bears have been documented since the mid-thirteenth century (Figs. 2.1 and 2.2).[21] Boehrer (2007, pp. 2–3) believes that in the Middle Ages there was negligible acknowledgement of the emotional integrity of animals, their suffering or their interests; but from the fifteenth century, while animals were treated as commodities within an exploited natural world, there were incipient signs of concern for the welfare of some of them.

The sixteenth century provides us with what appears to be the first documented celebrity performing animal, the horse Marocco. Atkins (1934) has compiled a very extensive list of references to William Banks and his horse, demonstrating the great impression they made in their time.[22] Banks is recorded as presenting another horse in Shrewsbury in 1591, but following mention of a dancing horse by Shakespeare in *Love's Labour's Lost*, particular references to Marocco became frequent as a result of performances in London and around British towns and cities. Sir Walter Raleigh thought that 'if Banks had lived in older times, he would have shamed all the enchanters of the world, for whosoever was most famous of them could never master or instruct any beast as he did' (quoted in Manning-Sanders 1952, p. 23). Marocco would restore a glove to its owner 'after his master had whispered the man's name in his ear; would tell the just number of pence in any silver coin, newly shewn him by his master; and even obey presently his command, in discharging himself of his excrements, whensoever he had bade him'.[23] After the turn of the century, the pair travelled to the Continent. Atkins mentions Frankfurt, Orléans and Paris, and Schrickx (1982) describes a visit to Wolfenbüttel. On seeing Marocco perform in Paris in 1601, Isaac Casaubon reported that because performing animals were sometimes thought to be inhabited by evil spirits, Banks, in fear of local retribution against witchcraft, was persuaded to explain to him the secret method of

---

[20] See Strutt (1801), pp. 239–240, 243–245 (citing Bodleian MS 264), 249–250 (citing Bodleian MS 1344, Harleian Collection).

[21] See Ravelhofer (2002), who also cites, for example, Magnus (1555) and his unpleasant description of a 'modus domesticandi'.

[22] See also Griffith (2004) and Dando (1595).

[23] Sir Kenelm Digby, quoted in Toole Stott (1958), pp. 30–31.

**Fig. 2.1** A performing bear, by Jehan de Grise of Flanders, from the *Romance of Alexander* (1338–1344). © The Bodleian Libraries, The University of Oxford; MS Bodl. 264, fol. 76r (detail)

**Fig. 2.2** Dancing bears take contributions in 'De chorea ursorum Lithuanicorum' (Magnus 1555, p. 629)

his training (Pattison 1892, p. 445).[24] Banks was actually imprisoned and his horse impounded, before he had demonstrated how Marocco was trained with signs.[25]

Marocco even danced on the roof of the old St Paul's Cathedral in London, after a European tour. Bondeson confirms that Marocco remains the most famous and cited animal performer of all time. It appears that his training involved exclusive, patient, intensive and repeated instruction from Banks from young foalhood, together with the use of food incentives, so that he was trained to recognise increasingly discreet signals from his master, eventually ones consisting only of facial cues. Later trick horses such as Clever Hans were probably trained in the same way (Bondeson 2006, pp. 13–28). Those in Banks's time who might have wanted to train their own horses to perform like his were given instructions by Gervase Markham in 1607, using Banks's techniques, even though for a horse these are 'unnecessary and unnatural actions which more properly do belong to Dogges, Apes, Munkies, and Baboones'. The horse must come to love, fear and depend on its trainer, who must have exclusive dealings with it, undertaking one lesson at a time and correcting it only with two or three strokes of a 'sharpe rodde' and one verbal reprimand, and rewarding it with food together with only one word of encouragement and one means of petting. Such actions must be consistent to avoid confusion, and food must only be given when deserved: hunger will 'dubble his diligence'. As training develops, 'it is your eie and countenance, as well as your words, by which the horse is guided', and the horse must be made to look you in the face. Particular actions can be taught with a pricking in the buttocks or a pin in the lips, but the rod should only be used for discipline (Markham 1607, book 8, chapter 5, pp. 26–38).

In the late sixteenth century, it was understandable that the pamphleteer Philip Stubbes should have his attention drawn to the continuing and more extreme cruelty of baiting than to criticise any shortcomings in the training of performing animals, when he pronounced: 'for our paftimes and vain pleafures fake, wee are not in any wife to fpoyle or hurt them' (Furnivall 1879, p. 182). In the Induction of *Bartholomew Fair*, first performed in 1614, Ben Jonson describes the performance of a 'well-educated Ape'; and John Caius had written of a variety of dog tricks (1576, p. 35). According to Ravelhofer (2002), the collective imagination in the 1600s took flight when an animal's behaviour deviated from its instincts, and so alongside Marocco, other performing animals like these, especially bears, continued in popularity. There were monkeys dancing on ropes and walking on wires, dogs dancing minuets, 'learned pigs' forming words from letter cards, hares beating drums and birds firing cannons (Strutt 1801, p. xlix). Even a young walrus, a beast 'of strange docility, and apt to be taught, as by good experience we often proved',

[24] Atkins (1934) notes that Ben Jonson believed that Banks and Marocco had been 'burned for one witch' when in Rome, although Banks certainly did not suffer this fate and Griffith (2004) attributes the claim only to a playful allusion in Jonson's *Epigrams* (1616).

[25] Jean de Montlyard, in a note to a translation of *The Golden Ass* by Apuleius: *Les metamorphoses, ou, L'asne d'or*, 1602.

caused a stir in Jacobean London. And in September 1610, the Master of the Revels, Sir George Buc, issued a licence to Thomas Morris and William Grove to 'shew a strange lion, brought to strange things, as turning an ox to be roasted, &c'.[26]

Strunk (1917) provides a number of historic references of the early seventeenth century that he believes demonstrate that London's performing apes (often referred to as baboons), such as Jack of Paris Garden and Blind Gew, had ranked in favour with puppet shows and Banks's horse: 'A number of the ape's tricks—to play dead, to pretend to say his prayers, to skip at the crack of the trainer's whip, and to feign disobedience when commanded to perform for the Pope, the King of Spain, or the Great Turk—are recorded in the passages from Shakespeare, Jonson, and Barry'. Boehrer (2007, pp. 24–26) comments that in the Bankside 'liberties' of London, animal entertainments (typically baitings) served as a kind of lowbrow *doppelgänger* of the Elizabethan and Jacobean theatre, and both activities took place in similar, if not the same, premises. There were animal acts provided as entertainment in intervals or before or after a play, without any relevance to its plot. The dual use of Philip Henslowe's Hope Theatre for animal shows and regular plays is an indication of the importance of this type of spectacle (Wright 1927). Meanwhile, the 'carrier about' of apes did not confine himself to London, and a pamphlet of 1572 had protested against the practice of clergymen rushing through the Sunday service in order that the congregation might attend games or witness 'a beare or bull to be bated, or else Jack-an-Apes to ryde on horseback, or an interlude'.[27]

However, as in classical antiquity, animals began to feature once more in the plays themselves. A dog and cat are prominent among the entertaining features of Jonson's *Every Man out of His Humour* (1600). Wright (1927) suggests that Jonson created the role of the dog in this play to fit a trick dog in possession of the company. Between 1609 and 1611, even polar bear cubs 'acted' in plays (Grant 2001, 2007, pp. 95–96; Ravelhofer 2002). Henslowe and his partner Edward Alleyn exemplified the close connection between theatre and animal baiting as profitable public amusements based in dual-purpose premises such as the Hope Theatre, where different forms of entertainment—including animal performance—followed each other in the same programme. It is also quite possible that at this time trained or at least tamed wild animals such as bears appeared in the plays themselves, some of which, like *Mucedorus*, seem to have been re-edited to cater for extended animal parts. Bear cubs could be profitably trained before intractability set in and when they were not old enough to fight. Examples of their use in this way were set by itinerant bearwards; and, says Ravelhofer, 'the theatrical character of fights, and the peculiar architecture at the Bear Garden may have been factors encouraging the training of animals for tricks before an audience'. Ravelhofer also suggests that 'the intelligence and adaptability of bears and the public interest in "wild" animals with absurd skills provide conditions for Henslowe's establishment which could have led

[26] *Calendar of State Papers Domestic 1603–1610*, September 6, 1610, p. 631, cited in Warner 1881, p. 75.

[27] Murray, *Eng. Dram. Cos.*, ii, 402, cited by Graves (1920).

to diversified strategies of using animal capital beyond the straightforward way to the bear-pit'.

The mixing of baiting with performance is also suggested in the lines of John Taylor (1638, p. 55):

... At *Beare-Garden,* (a sweet Rotuntious Colledge)
Hee's taught the Rudiments of Art and knowledge.
There doth he learne to dance, and (gravely grumbling)
To fight & to be Active (bravely tumbling)
To practise wards, and postures, to and fro,
To guard himself, and to offend his foe.

Some years later, John Evelyn recorded in his diary for 13 September 1660:

> I saw in Southwark, at St Margaret's fair, monkeys and apes dance, and do other feats of activity on the high rope; they were gallantly clad *à la monde*, went upright, saluted the company, bowing and pulling off their hats; they saluted one another with as good a grace as if instructed by a dancing master; they turned heels over head with a basket having eggs in it, without breaking any; also, with lighted candles in their hands, and on their heads, without extinguishing them, and with vessels of water without spilling a drop. (Evelyn 1660)

On 1 September 1668, Samuel Pepys went to Bartholomew Fair and came across a successor of Marocco:

> So to the Fair, and there saw several sights; among others, the mare that tells money, and many things to admiration; and, among others, come to me, when she was bid to go to him of the company that most loved a pretty wench in a corner. And this did cost me 12d. to the horse, which I had flung him before, and did give me occasion to baiser a mighty belle fille that was in the house that was exceeding plain, but fort belle.

A few days later, he returned 'with Lord Brouncker (who was this day in an unusual manner merry, I believe with drink), J. Minnes and W. Pen to Bartholomew Fair; and there saw the Dancing Mare again (which today I find to act much worse than the other day, she forgetting many things, which her master beat her for, and was mightily vexed) ... ' (Morley 1859, pp. 250–251). A show bill in Queen Anne's reign advertised another of these acts, and Strutt (1801, p. 245) felt its claims were true because an 'equally scientific' horse could be seen in 1800 in Astley's Amphitheatre.[28]

Animal performances from the early eighteenth century continued to develop and ranged from the banal to the sophisticated, involving various degrees of cruelty. An advertisement in the *Postman* of September, 1701, announced: 'The Tiger in Bartholomew Fair, that yesterday gave such satisfaction to all persons of all qualities by pulling the feathers so nicely from live fowls, will, at the request of several persons, do the same thing this day: price 6d.' (Blake 1921). Crawley's act, 'The Ball of Little Dogs', was also presented in the early eighteenth century at

[28] Morley (1859, p. 315) notes of such show bills: 'They commonly profefs to defcribe things exhibited by his or her "Majesties' Authority"—like the quack medicines of our own day—becaufe a fee had been paid to Government by every fhowman for his licence'.

Southwark Fair, allegedly having performed before Queen Anne herself, and a 'wild hairy man' was exhibited at Charing Cross, dancing on ropes and walking on wires, tumbling, smoking and ale drinking (Strutt 1801, pp. 242–247). And there was an elephant that fired a gun, referred to in the *Medley* for 16 October 1710 (Morley 1859, pp. 391–392):

> So have I feen at Smithfield's Wondrous Fair,
> When all his brother Monfters flourish'd there,
> A lubber'd Elephant divert the town
> With making legs, and mooting off a gun.

However, alongside the exploitation and abuse of animals presented for entertainment in the eighteenth century, there appeared new signs of an interest in animals not just for what they could give but for what they were. Bondeson makes the interesting point that acquaintance from the eighteenth century with the achievements of animals such as trained pigs might have affected people's anthropomorphised prejudice against them and that animal performance might have encouraged philosophers to reconsider the status and characteristics of animals as compared with humans (Bondeson 2006, pp. 40–41).

The popularity of learned horses continued through the eighteenth century. In 1750, there was at 20 Hosier Lane, during the time of Bartholomew Fair, 'the Wonderful and Astonishing Arabian Poney', who could count the spots on cards and tell the time to a minute by the watch of any visitor. He could find out particular persons and do other clever things before an audience that had paid sixpence and threepence for admission (Morley 1859, p. 448). A variety of domesticated animals also now became specially promoted as 'learned'. Surprisingly, a lack of records suggests that dogs were for long little used for tricks, but they appeared more frequently during the eighteenth century, and by 1800, they were often used in dancing tricks in poor-quality street performances (Strutt 1801, pp. 246–247). Wiles (1967) cites an advertisement in *Harrop's Manchester Mercury* on 21 November 1752, aimed at 'Lovers and Admirers of the Docility and Sagacity of the Brute Creation' who could see the 'Amazing Learn'd English Dog', before it left for Preston, Warrington, Chester and other places:

> This entertaining and sagacious Animal, reads, writes, and casts Accompts, by means of Typographical Cards, in the same Manner that a printer composes; and, by the same method, answers various Questions in Ovid's Metamorphosis, Geography, and History; knows the Greek Alphabet, reckons the Number of People present, if not above 30 ... solves small Questions in the four Rules of Arithmetick, tells by looking on any common Watch of the Company, what is the Hour and Minute, ...

A team or, rather, an army of dogs appeared with monkeys in London in 1753 as *Mrs Midnight's Animal Comedians*. They are referred to in the anonymously written *Haney's Art of Training Animals* of 1869. A ladder-climbing sequence formed an impressive part of a celebrated siege scene, when a small stage was specially fitted up for their performances, furnished with scenery and decorations (Anon 1869, p. 92). Not long after, Scotsman Samuel Bisset taught kittens to 'strike their paws in such directions on the dulcimer as to produce several regular tunes, having music-books before them, and squalling at the same time'. At the Haymarket Theatre, his 'Cats Opera' made nearly a thousand pounds (Anon 1869, pp. 149–150).[29]

Of more significance as successful public attractions, still sometimes provoking charges of sorcery and witchcraft, were the feats of pigs, similar to those of the horses. As well as training horses, dogs, hares, monkeys and cats, Bisset, according to Bondeson, was the first to train a pig specifically for the stage. He bought a piglet in 1782 and prepared it for 2 years, and the poet Robert Southey was apparently told that it was not cruelly treated. It was later taken over by a man called Nicholson who had a theatrical background and who very successfully promoted and toured the animal to perform a range of tricks like those of the educated horse. It appeared to be relatively thin and was kept motivated by hunger. Acrobats who complained that it was degrading to perform with the learned pig in the circus at Sadler's Wells Theatre were sacked after presenting an ultimatum (Bondeson 2006, pp. 29–46).

Of a learned pig, 'well versed in all Languages, perfect Arethmatician & Composer of Musick', Robert Southey wrote that he was 'in his day a far greater object of admiration to the English nation than ever was Sir Isaac Newton' (Simmons 1951, p. 340). After triumphs at York, Scarborough and elsewhere in 1784, he made his London debut at 55 Charing Cross Road early in 1785. According to his publicity, reminiscent of that for the 'Amazing Learn'd English Dog' advertised in Manchester in 1752, 'he reads, writes, and casts accounts by means of typographical cards, in the same manner that a printer composes and by the same method ... sets down any capital or surname; solves questions in the four rules of Arithmetic' and so on (Lysons n.d.). He figured in contemporary prints, including Rowlandson's *The Wonderful Pig* of April 1785.

Later pigs included a number called Toby, one appearing in Bartholomew Fair in 1817 as the pupil of Nicholas Hoare. Between 1818 and 1823, a certain Pinchbeck exhibited Toby the Sapient Pig, who was alleged to have been tutored by Souchanguyee, the Chinese Philosopher, and shown in Spanish America. He could answer questions by pointing to cards, letters and persons in the audience (Lysons n.d.). In 1833, a pig of the same name was advertised as the

> Unrivalled Chinefe Swinifh Philofopher, Toby the Real Learned Pig. He will fpell, read, and caft accounts, tell the points of the fun's rifing and fetting, difcover the four grand divifions of the Earth, kneel at command, perform blindfold with 20 handkerchiefs over his eyes, tell the hour to a minute by a watch, tell a card, and the age of any party. He is in colour the moft beautiful of his race, in fymmetry the moft perfect, in temper the moft docile. And when afked a queftion, he will give an Immediate Anfwer. (Morley 1859, p. 480)

A rival 'Amazing Pig of Knowledge', owned by J. Fawkes and who could be seen for threepence, knew the value of money, could tell black from white,

[29] See also Bondeson (2006), pp. 7–11. The conjuror Cappelli would augment his performance with trained cats at Bartholomew Fair in 1832, and Bondeson refers to various performances of cats around the world to the present day.

distinguish colours, with a shrewd eye count his audience and even tell people their thoughts (Morley 1859, p. 481).

Alongside successor learned pigs in the early nineteenth century, there was also Castelli's dog Munito, who provided the same range of tricks as the learned pigs and some earlier dogs of the mid-eighteenth century. He was allegedly well treated, never struck and taken for walks. Such dogs would respond to discreet cues, like the horses and pigs, and Munito appeared all over Europe as at least three different dogs, being emulated ever since (Bondeson 2006, pp. 47–62). Charles Dickens wrote that Munito, as well as birds and pigs, selected cards and other objects because they had been marked with aniseed or 'sweet-wafer paste' (Dickens 1867).

Learned horses remained popular. One was exhibited at Exeter Change in 1760 and 1772,[30] and a conjuring horse also appeared at Philip Astley's premises by Westminster Bridge in the late eighteenth century (Dobson 1902, pp. 93–109). Indeed, fascination with the accomplishments of such horses eventually led to the phenomenon in Berlin of Clever Hans, who could apparently calculate, read and spell, and then to the scientific analysis to which his performance was subjected in the 1900s.[31] Exeter Change features again in Byron's journal of 14 November 1813 concerning a visit to its menagerie, where 'the elephant [probably Chunee] took and gave me my money again—took off my hat—opened a door—trunked a whip . . .'. Benjamin Silliman had encountered another elephant there in 1805 who did the same tricks and could count the number of men and women present (Marchand 1973–1982, pp. 206–207; Silliman 1810, pp. 159–162. Cited by Altick 1978, p. 309).

As the variety of animals trained or 'educated' for public performance increased during the eighteenth and nineteenth centuries, the sophistication of the acts themselves seemed also to develop. Strutt thought it 'astonishing what may be effected by constant exertion and continually tormenting even the most timid and intractable animals', such as the hare at Sadler's Wells trained to beat the tambourine—a long-established trick. Further impressive dog performances took place there and at Astley's: 'One of the dogs at Sadler's Wells acted the part of a lady, and was carried by two other dogs; some of them were seated at a table, and waited on by others; and the whole concluded with the attack and storming of a fort, entirely performed by dogs' (Strutt 1801, pp. 247–248) (Fig. 2.3). Between 1766 and 1773 Daniel Wildman's troupe of trained bees performed at several London rooms and tea gardens. Wildman was an equestrian as well as an apiarist. In one act, he rode about a ring with one foot on the saddle and the other on the horse's neck, 'with a curious mask of bees on his head and face'. So adorned, he would drink a glass of wine, and when he fired a pistol, the bees would break ranks, half of them marching

---

[30] Clippings, Enthoven Collection, Theatre Museum, London, Exeter Change file, cited by Altick (1978, p. 40).

[31] For an interesting summary of Oskar Pfungst's study of Herr von Osten's horse, Clever Hans, see Boakes (1984), pp. 78–81. Pfungst demonstrated that the horse responded not to auditory cues but to barely perceptible visual ones given unknowingly by his questioners.

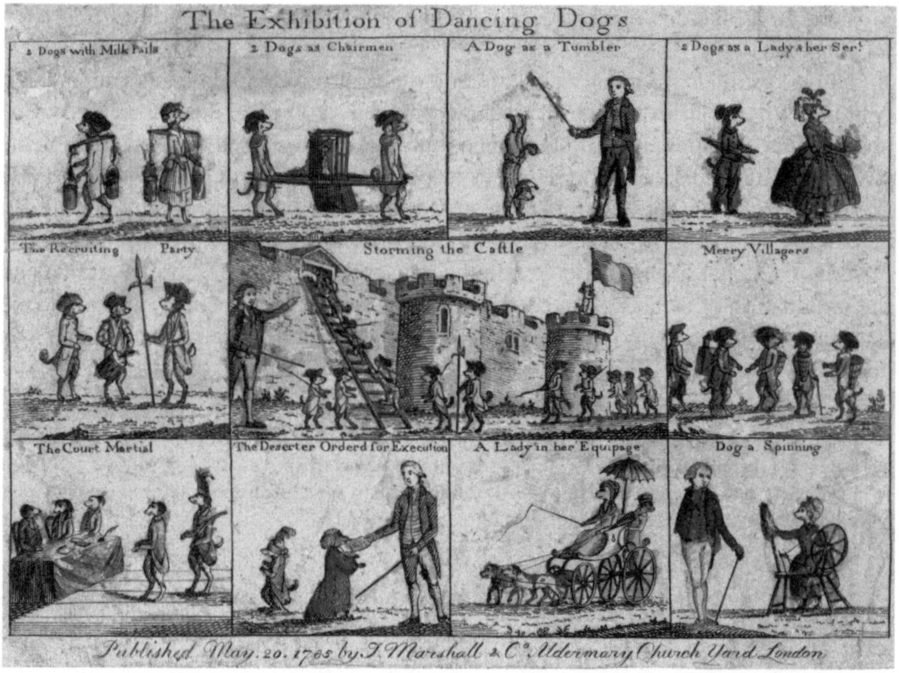


**Fig. 2.3** 'The Exhibition of Dancing Dogs' performed by Charles Benjamin Incledon, from a print published in 1785. © The Bodleian Libraries, The University of Oxford: John Johnson Collection; Animals on Show 1 (51)

over a table and the others returning to the hive (Lysons n.d., cited by Altick 1978, p. 40).

Strutt reports an even more surprising exhibition first shown in Pall Mall in 1789, at a charge of five shillings per person. This involved Breslaw's tabletop performance of 12–14 small birds in soldierly ranks, uniformed and capped and with miniature muskets under their left wings. They marched backwards and forwards in formation, when a deserter was produced and another grasped a lighted match to fire a cannon at him, whereupon he played dead (Strutt 1801, pp. 249–250). Dujon's canaries also appeared later at 23 Bond Street and performed a similar drama in which a uniformed deserter was court martialled, shot by another canary and then carted off in a wheelbarrow.[32] Prestidigitator Hoare presented Pinkey, a 'learned and scientific goose', from the late 1820s to early 1830s. And in 1838, Chinese birds could be seen playing cards and telling the time by clock or watch (*Mirror of Literature* 1838, cited by Altick 1978, p. 40). A troupe of 'Oiseaux Merveilleux' at Willis's Rooms in 1852 told the time, solved arithmetic problems, read minds and told fortunes (*Illustrated London News* 1852, cited by Altick 1978, p. 307). And the Oxford-educated 'scientific Java sparrows' at 23 Bond Street had met with the

[32] From a handbill referred to by Altick (1978, p. 307).

greatest approbation from the vice chancellor and were perfect in seven languages.[33]

Trained animals frequently appeared on the theatre stage in the early nineteenth century. When Martin's lioness appeared on stage at Drury Lane, it seemed to be either under the influence of sleep or opium and then wanted to escape from the 'well-known keeper, of whom she was considerably afraid'. Such use of live animals on stage in the legitimate theatre aroused the indignation of many human actors, and their objections would continue well into the twentieth century. *The Spectator* was not sympathetic: 'As to the nonsense about the legitimate drama and the profanation of a classic stage, the lions and elephants, by being real, violate the decencies of the drama no more than they would by being stuffed'(*Spectator* 1831); and

> The discomfited actors of the 'legitimate' order . . . are very indignant . . . They demand the restoration of the legitimate drama; but why don't they restore it themselves? How is it that they have suffered themselves to be hustled out of the way by the beasts? For, indeed, actors have proved no match for live animals. It is not worse to put a quadruped on the stage than a biped who as little resembles human nature. (*Spectator* 1847)

Towards the middle of the nineteenth century, the favoured locales for the kinds of entertainment involving easily portable exhibits of the sort that formerly occupied booths at the now declining or extinct fairs were taverns and assembly rooms in working-class neighbourhoods.[34] At 248 Regent Street in 1829 and in Bartholomew Fair at a house in Giltspur Street in the 1830s was Signor Cappelli of Tuscany with his learned cats, a *corps dramatique* consisting of mother, two sons, daughter and a 'jet-black and maternal-looking negress', who rang bells, beat a drum, worked a rice grinder, hammered on an anvil, drew a bucket up and down a well, roasted coffee, turned spits, played on an organ, sharpened knives and obeyed orders in French or Italian. There was also a 'Wonderful Dog' who would play any gentleman at dominoes. The admission charge was for gentlemen, 4d, and working people, 2d. The *Literary Gazette* asserted: '[Cats'] restoration to rational functions is the best proof we have yet met with of the progress of education and the march of mind' (*Literary Gazette* 1829; Morley 1859, p. 487). In 1843, the Cosmorama Rooms echoed with the melodies of a singing mouse, which was reported to warble 'incessantly' for a quarter of an hour: 'its notes are low but clear, and not unlike that of the nightingale'. Jenny, a seal originally captured off Africa in 1854, was exhibited at 191 Piccadilly in 1859. She could dance, ring a bell, stand on her tail, and say 'mama', 'papa' and 'John' (the name of her keeper, who may possibly have been a ventriloquist) (*Athenaeum* 1843; *Illustrated London News* 1859; Buckland 1868 vol. 2, p. 128, cited by Altick 1978, p. 306).

Strutt wrote that in England dancing bears were still common on the streets in 1800, 'but the miserable appearance of their masters plainly indicates the scantiness of the contributions they receive' (Strutt 1801, p. 241). Morley describes a bear

[33] From a handbill referred to by Altick (1978, p. 306).

[34] Handbills as referred to by Altick (1978, p. 307) demonstrate this.

dancing at Bartholomew Fair around 1808, accompanied by a street fight and an incident of bustle in the outer crowd. 'The bear now and then turned a good summersault and generally danced to the bagpipes. His companions were some little dogs dressed in red jackets and a monkey who usually rode on the bear's back' (Morley 1859, p. 478). However, in spite of the alleged decline in performing bears, Ravelhofer (2002) notes that the nineteenth century saw veritable bear schools across Europe, the bearward and his tame beast remaining a common feature to the present day. Such a school could be found in provincial France in the nineteenth century, when trainers took very young animals, and 'in the course of rigorous treatment they finally brought them to execute dance steps to a rhythmic tune, greet, applaud by clapping the paws, fight a mock battle with a stick, or play "dead"' (Ravelhofer 2002, citing Praneuf 1989, pp. 67–69). A traditional practice allegedly still applied in the Balkans doubtless once enjoyed greater currency throughout Europe:

> Bears had their claws and teeth filed down or even extracted. Sometimes their eyelids were burnt to close their eyes. A ring was driven through their nose. The trainer touched the animal with a hot iron on the backside or paws so that it executed the desired movement to avoid getting hurt. The bear was forced to sit on a hot stone slab, jumping up in pain and being pulled by reins into various movements and postures. During this procedure, somebody kept singing and playing the tambourine; thus, the rhythm and the tune conditioned the desired reactions of the animal. The sufferings were so great that after some weeks of this ordeal the terrorized bear would start dancing and crying spontaneously as soon as it heard the music and saw its master. (Ravelhofer 2002, citing Praneuf 1989, p. 73)

As Ravelhofer notes, any of these methods may have had predecessors or equivalents in earlier centuries in England, as well as on the European Continent.

The nineteenth century is readily associated with perhaps its most important and distinctive new form of animal entertainment, the exhibition of the lion with tamer. The public continued to be interested in exotic animals and retained age-old misgivings about the untamed wild; and the control of fierce wild animals fitted well with the nineteenth-century concern for religiously endorsed elevation of human status in the natural world, as well as for achieving civilising improvements through discipline and education. Capture and submission of fierce exotic beasts in the circus reflected an interest in the little known as well as politically correct ideas of imperial dominance over inferior, savage parts of the world in the nineteenth century.[35]

Isaac Van Amburgh allegedly invented the modern animal-tamer act in New York in 1833, and Joys studies his cultural significance (as well as, later, that of Frank Buck and Clyde Beatty) in making a commodity of the wild: 'because of their presumed abilities to freely and successfully cross back and forth between the dichotomous worlds of wilderness and civilization, [they] were able to demonstrate their heroics to the public largely through surrogates of the wilderness; captive wild animals, especially those deemed most dangerous' (Joys 2011,

---

[35] Since empire, circuses have tried to apply the commercially advantageous 'Noah's Ark' or conservation rationale. See Schwalm (2007, pp. 79–104).

p. iii). Van Amburgh was quoted as saying: 'They [his animals] believe that I have power to tear every one of them in pieces if they do not act as I say. I tell them so, and have frequently enforced it with a heavy crowbar' (*Times* 1838, cited by Joys 1983, p. 6). On one of her visits, Queen Victoria saw Van Amburgh 'beat severely' a leopard; the animals had not been fed since early the previous day and so were wilder than usual. She noted: 'He scarcely ever uses an iron bar to them, but only a stick made of rhinoceros hide, which he showed us . . ..' (Esher 1912, pp. 105–106, cited by Joys 2011, p. 36). Joys observes that 'since this was an era when discipline and corporal punishment of the unruly were very much part of the socialization and educational system, as the popular euphemism, "spare the rod, spoil the child," reflects, Van Amburgh would have been expected to willingly discipline his wild animals, if need be' (Joys 2011, p. 151). Starvation was another method of subjugation, which was also justified by the biblical explanation of man's dominion over animals (Durant 1957, p. 33, cited by Joys 1983, p. 7).

Some time after his death, Van Amburgh was deprecated by a former acquaintance who claimed he had declawed his animals in an 'excruciating process, the refined agony, and despicable cowardice' enabling him to become a 'Lion King'. The acquaintance was an artist staying at the same Paris hotel, who had been given permission by Van Amburgh to study his animals in quarters while he was recuperating after a bite, and who found them declawed. (He also alleges Van Amburgh's cowardice when about to be operated on for his own injury.)

> Here is beyond comparison the very handsomest and noblest collection of wild beasts ever seen together, tame, submissive, and tractable as domestic-bred animals, in most superb coat, fat as moles, and apparently as affectionate and grateful for kindness as would be the most intelligent and faithful of man's companions; the one great and accountable reason for this is that in themselves - their courage, their ferocity, and their savage natures - they are vanquished, annihilated, utterly undone and demoralised. Plundered of their weapons, offensive and defensive, their very heartstrings torn asunder, their quick, sensitive natures crushed out - cast off the rack, cowed, bleeding, benumbed and incapable, to obey the will of their torturer.

They were also trained to eat biscuits, lumps of sugar and only cooked meat, not being allowed the smell or taste of blood. 'Van Amburgh is now no more, but he died a natural death—not torn to pieces in revenge for unjustifiable brutality and vulgar daring' (*Gentleman's Magazine* 1873). But many of Van Amburgh's later imitators lost their lives, especially when their reactions became blunted by drink (Ex-Lion-King 1872; Joys 1983, pp. 13–14).

In the first half of the nineteenth century, travelling menageries of trained and exotic animals had also attempted to use religious feeling and imagery to promote themselves, as well as to defuse puritanical objections to them as unholy amusement (Joys 1983, p. 3). One of the most famous of these, George Wombwell's travelling wild animal menagerie and show, was established in the 1800s. His descendant, the trainer Frank Bostock, wrote that the animals were valuable and therefore as well cared for as limited knowledge of their 'ways and requirements' allowed (Bostock and Velvin 1903, pp. 24–29). Morley had earlier given a less favourable account of the management of the menagerie. He notes that Wombwell was sharply censured by Hone in 1825 for having exposed his fine lion, Nero, to be baited by dogs at Warwick, displaying a disgusting picture of the fight outside his show. He is described in the table book as 'undersized in mind, as well as in form, a weazen, sharp-faced man, with a skin reddened by more than natural spirits, and he speaks in a voice and language that accords with his feelings and propensities'. Of this man, who began life as a cobbler, Morley said he found only unfavourable record. 'He must, however, have had unusual ability and energy in his peculiar way. His collection was good.' He had a yellow card, with a tiger above his address: 'Wombwell, Wild Beast Merchant, Commercial Road, London. All sorts of Foreign Animals, Birds, &c, bought, sold, or exchanged, at the Repository, or the Travelling Menagerie' (Morley 1859, p. 486).

In 1847, Wombwell's menagerie was exhibited in front of the Royal Family and boys from Eton College. Mr Wombwell's 'exceedingly good' brass band performed while the 'Lion Queen' (Miss Chapman) performed in their cages with nine lions, tigers, panthers and leopards. These 'extraordinary performances' showed her 'astonishing command over the wild animals, highly to the gratification of her illustrious spectators'. She would sit on them, play with them and insert her head into the jaws of the noble lion 'Wallace', and perhaps with 'more self-possession than was exhibited either by Van Amburgh or Carter'. Apart from its containing about 500 animals costing £40 a day in upkeep, a merit of the menagerie was that 'the animals are kept in a condition of cleanliness, so that scarcely any unpleasant smell is perceivable' (*Illustrated London News* 1847).

Another major development of the nineteenth century in the production of exhibitions of performing animals was the introduction of new training techniques by Carl Hagenbeck. In 1887, Hagenbeck replaced the menagerie beast wagons wheeled into the circus arena with the large 'cage arena', allowing for the use of pedestals and ladders (Johnson 1990, p. 52). By the turn of the century, such wild animal performances would develop into much larger affairs (Bostock and Velvin 1903, p. 216). In 1891, Hagenbeck exhibited a group of trained wild animals at Crystal Palace, impressing an American so much that he offered $50,000 for them; however, they all soon died because of a lack of newly slaughtered meat (Shepstone 1902). From the 1890s, there was also a move from confrontational wild animal menagerie taming to circus training through conditioning, with claims of more humane methods from Hagenbeck and Bostock: perhaps influenced by Darwin's examination of the emotions (1872), they both began to identify human-like emotional traits in individual wild animals selected for training. Successful control of these emotions through training might then demonstrate human supremacy, represented by an ability to achieve evolutionary progress in the animals.[36] Tait observes that, paradoxically, 'an older creationist idea of human distinctiveness and of a God-given entitlement to control nature lingered' (Tait 2012, pp. 12–13, 16–17,

[36] *Haney's* had already implied this in 1869: a study of trained animals will show whether adaption to their natural environment has been developed into 'further ... or divergent advancement intellectually' (Anon 1869, pp. 182–183).

28). In exhibition and performance, this would result in the development of unnatural displays such as the mixing of incompatible animals or the lion-on-horseback act (first shown at the Paris Hippodrome in 1889). Although Hagenbeck stated that animals should be trained only to do what comes naturally, his own tricks included a tiger riding a horse, and when incompatible animals were mixed together, injury and death continued to be common (Stokes 2004).

Hagenbeck gave Raymond Blathwayt of *McClure's Magazine* a special performance when he was the only spectator, observing Hagenbeck's brother-in-law and his training method of 'kindness and coolness and firmness' (under which the animals responded smartly to the crack of his whip), with rewards of sugar. He saw a lion on horseback, an unwilling tiger riding a tricycle and balancing on a ball, a bear on a tight rope, a see-sawing lion and tiger with a bear in the middle, a mixed animal pyramid with a polar bear at the top and a lion dressed as a king in a chariot drawn by tigers. Blathwayt (1893) thought relations between staff and animals were affectionate.

The same privilege was afforded to another correspondent of *McClure's*, Cleveland Moffett, who went behind the scenes at Hagenbeck's show, but he describes bloody fights between lions kept in the same cage and separated with iron prongs by grooms, whose watch charms of lions' claws are illustrated in the article. Some of the old methods persisted. The trainer William Philadelphia would tease and prod the animals to provoke fierceness and then defend himself from attacks with cuts of the whip to the lion's head and with rod and club. Moffett was under the impression that lion tamers never carried revolvers, partly because they would be of little use and because of the danger to the spectators. The secret lay rather in making the animal so fearful of the trainer that it would not dare attack. Training the animal when very young would make it familiar with the trainer's presence. The trainer had to get gradually closer, repelling attacks with raps of a club on the head and flanks and giving meat before his departure a few hours later. Increasing familiarity over a course of months would lead to hand feeding, chaining, binding to the floor to tolerate stroking and detailed training. As a result, the animal was cowed by its awareness of the trainer's controlling power (Moffett 1893–1894). In a follow-up article, Moffett (1894) describes how a young tiger was taught to roll a ball. How much the whip was applied depended on the nature of the animal.

> The ball is placed upon an elevated wooden track, and one of the animals forced to mount it. Then, with Mr Mehrmann [Carl Hagenbeck's brother-in-law] in front and Willie Judge behind, the ball is sent on its course, while the animal on top scrambles and clutches to keep a balance. And in the instruction the trainer is none too tender. He cuffs the pupil under the jaw, and even whacks him resoundingly, while the assistant, holding the tiger's tail out straight, gives him sharp raps on the hind legs.

A tiger worth $500 would be worth $5,000 after 2 years' training. One trained wild animal group took 4 years to train and was a selection of 16 from 60 trained and assessed. However, Hagenbeck regarded animal training as an advertising medium, and his main business was supply to zoos, menageries and private parks (Shepstone 1902). And while he criticised the spectacle, Hagenbeck also supplied lions to fight with bulls in Spain (Hagenbeck 1912, p. 124).

## 2.2 Emerging Opposition

Although there was some feeling against animal suffering in the seventeenth century, until the eighteenth animals were considered soulless. The concept of 'cruelty' was not recognised, with indifference towards, if not deliberate infliction of, suffering. And, as a source of inconvenience if not fear, wilderness or wild nature was still often disliked in the eighteenth century, as before (Thomas 1983, pp. 148, 174, 257–258). This hostility was usually only tempered by curiosity, especially towards exotic animals under control in new and unnatural environments:

> This [the ape] is a native of Africa; I once faw a full-grown one fhewn as a curiofity in London: it was a male, and when it ftood erect, as it had been practifed to do with a ftick in it's hand, it had a very affecting, general refemblance of a decrepid old man. It was very quarrelfome, and would throw any thing that lay in it's way at thofe who came to feed it, if the eye of the fhowman were not continually upon it. (Hill 1752, p. 536)

Before the advent of the circus and music hall, animal performances in Britain could be seen in theatres, fairs and public houses, and such attractions readily lent themselves to the exaggerated advertisement of sensation and curiosity. Many such advertisements of the eighteenth century detailed graphically the promise of animal turns that no doubt later fell short of their claims and were probably a disappointment to any of the audience that had taken the posters and handbills literally. But the care with which such publicity was often drawn up reflected the commercial potential and profitability of animal training. For the same reasons, advertisements for animal acts remained equally uninhibited and confident until well into the twentieth century.

Scepticism (if not at that time unimpeded knowledge) about the training behind these acts started to take shape as well. During the eighteenth century, the human–animal relationship began to attract a small measure of attention within the spirit of Enlightenment thought. In *Millenium Hall*, a reformist novel or 'conduct book' of 1762 concerned with morality and humaneness, Sarah Scott, a woman whose design is to alter and regulate the conduct of men (Cruise 1995), uses her characters to expose the injustice of the confinement and exhibition of wild animals and human 'freaks':

> Lamont, not less curious, and more importunate, observed that the inclosure bore some resemblance to one of Lord Lamore's, where he kept lions, tigers, leopards, and such foreign animals, and he would be hanged, if the ladies had not made some such collection, intreating that he might be admitted to see them; for nothing gave him greater entertainment than to behold those beautiful wild beasts, brought out of their native woods, where they had reigned as kings, and here tamed and subjected by the superior art of man. It was a triumph of human reason, which could not fail to afford great pleasure.
>
> 'Not to us, I assure you, Sir,' replied Miss Mancel, 'when reason appears only in the exertion of cruelty and tyrannical oppression, it is surely not a gift to be boasted of. When a man forces the furious steed to endure the bit, or breaks oxen to the yoke, the great benefits he receive from, and communicates to the animals, excuse the forcible methods by which it is accomplished. But to see a man, from a vain desire to have in his possession the native of another climate and another country, reduce a fine and noble creature to misery, and confine

him within narrow inclosures whose happiness consisted in unbounded liberty, shocks my nature. There is I confess something so amiable in gentleness, that I could be pleased with seeing a tiger caress its keeper, if the cruel means by which the fiercest of beasts is taught all the servility of a fawning spaniel, did not recur every instant to my mind; and it is not much less abhorrent to my nature, to see a venerable lion jumping over a stick, than it would be to behold a hoary philosopher forced by some cruel tyrant to spend his days in whipping a top, or playing with a rattle. Every thing to me loses its charm when it is put out of the station wherein nature, or to speak more properly, the all-wise Creator has placed it. I imagine man has a right to use the animal race for his own preservation, perhaps for his convenience, but certainly not to treat them with wanton cruelty, and as it is not in his power to give them any thing so valuable as their liberty, it is, in my opinion, criminal to enslave them in order to procure ourselves a vain amusement, if we have so little feeling as to find any while others suffer.'

The 'inclosure' turned out to be an asylum for 'those poor [exhibited 'freak' human] creatures who are rendered miserable from some natural deficiency or redundancy' (Scott 1762, p. 10).

The consequences of the harsh training methods used to produce 'learned pigs' towards the end of the eighteenth century are referred to by Strutt (1801, p. 248). A pig

... filled his tormenter's pocket with money. One would not have thought that a hog had been an animal capable of learning: the fact, however, is another proof of what may be accomplished by assiduity; for the showman assured a friend of mine, that he had lost three very promising brutes in the course of training, and that the phenomenon then exhibited had often given him reason to despair of success.

Because a particular learned pig seen in Nottingham was said by Anna Seward to be as proficient in exhibition as dogs and horses, Samuel Johnson observed: 'Then, the pigs are a race unjustly calumniated. *Pig* has, it seems, not been wanting to *man*, but *man* to *pig*. We do not allow time for his education, we kill him at a year old.' Henry White told him that great torture must have been used before the 'indocility' of the animal could have been subdued. But Johnson replied: 'The pig has no cause to complain; he would have been killed the first year if he had not been *educated*, and protracted existence is a good recompence for very considerable degrees of torture' (Hill and Powell 1934–1950, vol. 4, pp. 373–374, cited by Altick 1978, p. 41).

In 1798, Thomas Young wrote:

I would sooner venture to affirm that no species whatever of animals is ugly, than to profess that I should have been much gratified with the performances of the learned pig. When I reflect upon the cruelty which must necessarily be used in order to produce these artificial monsters, I am still further confirmed in the above opinion. It is said in particular that bears are first taught to dance by placing them upon an heated floor and playing some tune to them in the meantime. (quoted by Swain 1933, p. 295)

However, it was not until the nineteenth century that concern for the welfare of performing animals gathered momentum, taking advantage of the expansion of the periodicals industry and the associated public sharing of critical opinion. Correspondents began freely to express views within a social context that had seen parliament debate the cruelty of animal baiting and when the Society for the Prevention of Cruelty to Animals had taken shape. One such periodical was *The*

*Spectator*, and it would remain critical of animal performance well into the twentieth century. As the nineteenth century wore on, the popular daily press would take on a significant role in the debate as well, responding to an increasingly literate and opinionated mass readership.

All kinds of creature now earned attention. In 1833, *The Spectator* carried a review of Bertolotto's 'industrious' trained fleas which were supposed to play in an orchestra and carry out a military siege: '"industrious" is a very pretty term, by the way, to apply to the struggles of a wretched little creature to escape from pain. The only process of "training" the flea to drag any load fastened to it, is that of putting it into a tiny tread-mill, where, if it indulges its natural agility, it knocks its head against a bit of wood, and so it learns to be content with a less lively movement' (*Spectator* 1833). At the other end of the scale, costumed performing monkeys on the Adelphi stage 'ape humanity with an edifying unconsciousness of the ludicrous figures they cut' (*Spectator* 1839).

Criticism of the conditions of captivity also began to make an appearance. A correspondent asked: 'Is a narrow and filthy cage worthy of an animal [the elephant] so large and so sagacious, good-tempered, and useful? Is it fair, that he should be confined for ever to such a space—in which he has scarcely the power to turn round?' (PD 1839). In this connection there was much public interest, encouraged by the press, in the fate of two elephants, Chunee and Jumbo, but these were not performing animals, rather captive animal exhibits (Bondeson 2006, pp. 63–140). Their owners made them into individual animal 'personalities', and their sad fates, each probably stemming largely from the effects of toothache, were accompanied by the encouragement of highly profitable public anthropomorphism. Although their cases had nothing to do with the rights and wrongs of training animals for performance, the related debate about captivity and confinement was given a new impetus, and this debate was revived during the hearings of the British parliamentary select committee in the following century.

But some cruelties were punished as well as criticised. Madame Poitevin was summoned to the Westminster Police Court by the Royal Society for the Prevention of Cruelty to Animals (RSPCA) and together with the manager of Cremorne Gardens was fined, she having ascended from Cremorne in the guise of Europa on the back of a bull attached to a balloon and suspended by girths. A pony had also been used. While theatres and the National Gallery were empty, said *The Spectator*, the places that collected delighted audiences were precisely those like pleasure gardens: 'the public mind is depraved . . . it is in the better training of the people that these idle cruelties are to be cured' (*Spectator* 1852). In a satirical pseudo-biography of Van Amburgh which has been described as 'erroneous fodder for future historians'(Joys 2011, p. 55),[37] his appearance at Astley's was to be followed by a balloon ascension from Vauxhall Gardens with a tiger as a passenger, which

[37] The work was presented as written by a certain Ephraim Watts, but it was actually a satire by Richard Horne, who set out to ridicule Van Amburgh because he regarded his popularity as evidence of a decline in public taste and as a threat to more culturally respectable entertainments.

would subsequently parachute down. Although local magistrates stopped the stunt, the trainer still hoped to descend in a diving bell with his favourite lion (Horne 1838, pp. 29–31).

Approval was also given where it seemed due. Of Leonard's dogs, performing at the Lowther Rooms, Strand, *The Spectator* said:

> Instead of the animals manifesting only a dread of the lash by their servile obedience and mechanical manner of going through their tricks, these dogs evince a lively docility and aptitude in the doing of their tasks, implying delight and eagerness to please their master, and precluding all idea of suffering and coercion. (*Spectator* 1841)

Charles Dickens wrote that when he was younger he used to go to exhibitions of learned animals to discover how they were taught. 'It is merely a work of time and patience to teach animals various feats of docility ... Although harsh treatment, beating, and half-starving, are too often resorted to, petting and kindness are more successful.' Even if punished, rewards were always necessary, and he had seen in the Cirque Napoleon in Paris the ape Jacko on horseback, trained with 'a mixture of petting and flogging' (Dickens 1867). But of the training of wild animals at that time, *The Times* complained:

> The subduing of wild beasts ... is merely the result of merciless thrashing when they are young. The application of the heavy cudgel, the iron bar or the red-hot ramrod on the tender limbs gives an impression which the threatening glance of him who wields these weapons keeps for ever fresh in the brute's memory. (*Times* 1869)

It was in the nineteenth century that the pattern of assertion and denial of cruelty in training became properly established, the confusion being exacerbated by a climate of secrecy and, therefore, suspicion. Conflicting accounts applied to and sometimes distinguished between individuals and organisations, and wild and domesticated animals. Commentators were usually quick to form an opinion about the general state of affairs, but there were those who admitted uncertainty about the truth. One of these acknowledged the suspicion that trained animals were taught their 'business' by cruel means and that much had been written in confirmation of this suspicion, but seemed convinced that the animals trained by the clown Mr Felix were happy at their circus in the grounds of Alexandra Palace. For his elephants, poodles and monkey, his motto was 'No knocking about', and he believed that if an animal were not intelligent enough to be taught by kindness, it was not worth teaching at all (*Spectator* 1877). But for others, there were philosophical as well as practical considerations. Edward Nicholson, soon to be Bodley's Librarian at Oxford University, argued 'It does not seem warrantable to shut up animals caught in places where man does not dwell' and 'Whether those taken about in wandering shows are treated with a like care and kindness [as in zoos], and whether such shows serve any good end, is to my mind very far from sure' (Nicholson 1879, pp. 48–49). Some also began to compare the positions of performing children and performing animals. There was an old circus rhyme that described the traditional misery of performing children and animals: 'Risley kids [child acrobats] and slanging buffers [performing dogs], Lord only knows how they suffers' (Westacott 1962, p. 10). *The Illustrated London News* reported that allegations had been denied that child circus performers were trained with the whip:

> Performing animals, like children, have to be taught by kindness, and it is not likely that people who have passed their lives training unwieldy elephants to ride tricycles, or coaxing horses into walking on tightropes, would have recourse to cruel measures with their own species. (*Illustrated London News* 1885)

By this time, in the closing decades of the nineteenth century and after the assimilation of Darwinian thought, a further and growing moral complication for some was the newly proposed kinship between humans and other animals. An alternative but related complication is suggested by Tait, who proposes that some acts were opposed because they compromised human status and dignity in the natural order, as when the head of a trainer was put between the jaws of a lion: such a stunt in New Zealand in 1894 was opposed 'because it undermined civilised society, rather than such opposition indicating recognition of animal rights' (Tait 2009). The performing animals controversy then soon took its place in a succession of public animal welfare issues of the late nineteenth and early twentieth centuries that had so far centred on vivisection and then on the extravagant use of feathers in fashionable women's clothing (Wilson 2008). These issues benefited from a symbiosis between press coverage and public interest, resulting not just in political involvement but also in the better organisation of those representing opposing viewpoints. Furthermore, the performing animals controversy in Britain came under the influence of developments taking place abroad, influences that would make it all the more intense.

In the United States, as in Britain, trained animals in the nineteenth century offered cheap entertainment that was popular and provided an experience of the exotic and the wild and of politically correct human control of nature. There was usually no cruelty evident to its large general public, who were consequently disinclined to criticise enterprises which presented themselves as the victims of organised objection. This was the case in the confrontation between Henry Bergh, president of the American Society for the Prevention of Cruelty to Animals (ASPCA), and the showman P. T. Barnum from 1867 and in 1880, involving charges of unjustified capture from the wild, unacceptable accommodation and humiliating performances. Barnum claimed his accommodation was comfortable and safer than the wild, shortly before a second fire destroyed its trapped occupants. However, his business continued to prosper, and Bergh lost the support of the press as a result of some unsubstantiated allegations which Barnum successfully exploited, in the meantime discrediting the anti-circus campaign (Beers 2006, pp. 79–81). It appears that Barnum knew Bergh had reason to be suspicious. In a letter of 11 April 1883 to Professor Spencer F. Baird, who had become secretary of the Smithsonian in 1878, about the dead animals he was donating to him, Barnum added at the end: '*Mum*. The elephant was cut up soon as possible to prevent Bergh's men from looking for wounds' (Saxon 1983, p. 233). Later, in what may have been a sincere gesture, Barnum willed $1,000 for a statue in Bridgeport honouring Bergh (Wallace 1959, p. 259).

The national press in England continued to be sympathetic with critics, but the trade press now also began to represent its readers' interests. In 1896, after Bensusan had called attention to 'barbarities practised' in animal training for the music-hall stage, *The Daily Chronicle* attempted a systematic investigation of allegations of cruelty, and *The English Illustrated Magazine* published Bensusan's article, 'The torture of trained animals' (Bensusan 1913b, c). In response to this attention, 'Professor' Chard wrote as an experienced trainer of all types of animal to *The Era* from Poole's Myriorama, Hengler's Circus, Hull, in May 1896, criticising the ignorance of those who claimed that training was cruel and the mischief they might cause: 'An animal that requires his tricks to be beaten into him is no performer at all, or such a one as no master would care to exhibit.' Careful selection and maintenance of a suitable animal, followed by patience and gentleness combined with strictness, was the most effective means of training dogs, and the whip was used only for signalling: 'If I throw my whip away any of my dogs will run and bring it to me'. Chard offered to demonstrate his methods with any young dog before a critical committee, and claimed that the dogs of performers such as himself were far better cared for than those of the majority of dog owners. In the same issue of *The Era*, 'Professor' Devereaux (the son of 'Professor' Peterson, for 50 years a dog trainer) of Reynold's Exhibition in Liverpool said that animal trainers had been shamefully slandered and confirmed that successful training was only possible through kindness and patience: 'I was presented with a medal by a member of the Humane Society bearing the following inscription: "To Devereaux, for his great kindness to dogs, Plymouth, 1890"' (*Era* 1896a).

*The Era* devoted an editorial to the subject the following week, referring to its 'sensational' treatment in *The English Illustrated Magazine* and in *The Daily Chronicle* and emphasising the futility, and therefore general absence, of cruelty in training. At an interview, the well-known theatre manager Charles Morton had repeated to *The Era* his opposition to exhibitions of animals on the stage, using general statements rather than ones that *The Era* thought could result in libel. Meanwhile, *The Era* said it had received no letters endorsing the statements in *The English Illustrated Magazine* and criticised the lack of evidence and the use of hearsay. Any incident of cruelty would result from 'wanton brutality' rather than a successful training process, and there were 'black sheep in every flock'. The editorial concluded:

> We can only express our regret and indignation that such serious and disgraceful charges should have been made against a class of caterers for the public amusement without the exact evidence which alone could have excused the publication of such a ghastly story. (*Era* 1896b)

But on 22 March 1899 *The Sketch* published claims by a first-hand witness of severe beating during training by Permane to his performing bears: Joseph Simpson, consulting engineer to the Empire Theatre Company at Blackpool, was told by Permane that there were at that time no English laws protecting wild animals, so he could do as he liked. Then, *The Observer* on 23 June 1912 included a letter from 'Pom', who had been on the staff of one of the largest variety theatres for over 7 years and referred to poor transport, lack of exercise, short-tempered and brutal foreigners and cruel training devices (Wilson 2008).

The press in the United States had neglected the cause of the critics since Bergh's counterproductive spats with Barnum, but news reporting and the letter writers remained active. One complained that at Hagenbeck's show an eye of a polar bear was closed after 15 min of chastisement to get him in a docile mood: 'I cannot understand what refined parents are thinking of who are willing their children should attend a performance where they are liable to witness such revolting cruelty' (AJS 1902). The proprietor of New York's Starrett Circus and one of his clowns were arrested for cruelty after the judge had been invited to attend the performance on the intervention of the ASPCA. Four of the show's horses were found to be unfit for duty and confiscated. Starrett was fined $100 for cruelty to animals and sent to jail for 20 days (*Variety* 1910a, b).

Ryder notes that there had been a boom in animal training and performance in the nineteenth century because of growing public interest in the 'sagacity' of animals, but that the resulting concern about treatment was seen by ordinary working people (typical of the music-hall audience) as a preoccupation of the sentimental rich; also that in the late Victorian and Edwardian periods, animal protection had become associated in the public mind with the campaigns of women rather than men (Ryder 2000, pp. 132 and 141). The advertisements of many animal protection societies emphasised the involvement of prominent members of society, perhaps to give added legitimacy and dignity to their contested aims.[38] But the interests of animals were also winning attention from radical-intellectual journals such as *The Westminster Review*. In an article largely devoted to the cruelty behind fur, skin and feather in women's adornment, Elizabeth Saville remarked that seals' feats of skill cannot be performed without long drawn-out torture and distress as well as unnatural surroundings. She complained of:

> the travelling caravans of pent-up beasts . . . demoralizing spectacles . . . hoarded together in unwholesome surroundings, groaning in pain and filth, caged within a few yards of space, carrying disease and contagion wherever they go. Some of the more intelligent are trained under most inhuman and painful conditions to perform unnatural feats, and go through manoeuvres causing utmost fright and suffering. (Saville 1910)

Literary figures in particular played a part in opposing cruelty of various kinds. Thomas Hardy protested to *The Times* in 1913 that it seemed 'marvellous that the twentieth century, with all its rhetoric on morality, should tolerate such useless inflictions as making animals do what is unnatural to them' (Hardy 1913). He elaborated on these comments in a letter to Florence Henniker soon afterwards:

> The words 'performing animals' do not clearly indicate the matter: what I object to most are performances *with* animals – in which they are passive – e.g. bringing live canaries, rabbits, pigeons, &c. out of the sleeve or handkerchief. Every spectator can see that the wretched

[38] For example, the London and Provincial Anti-Vivisection Society in 1913 listed 46 vice-presidents, of whom 32 were titled, and 40 medical vice-presidents. In recent times, the aristocrat has been replaced by the celebrity.

**Fig. 2.4** John Galsworthy, by James Russell & Sons, c. 1916. © National Portrait Gallery, London. Ref. Ax39222. Reproduced with permission

> creature is in the greatest misery, & I believe that a great many are 'used' in these tricks – that is, tortured to death. (Purdy and Millgate 1984, pp. 330–331)

On 15 December 1913, John Galsworthy (whose cousin would marry Florence) had given a speech at a crowded meeting in Kensington Town Hall on the 'Treatment of Performing Animals'. He objected to the immorality of unnecessary suffering: 'All suffering that is inflicted merely for our pleasure, distraction, and even for our convenience, as distinct from our preservation, is unnecessary and an abomination' and 'it is hardly ever to the interest of anyone to draw attention to cruelty, certainly not to fabricate such a charge'; meanwhile, being sentimental meant being moved by pity and anger. Galsworthy considered 'this sometimes distressing and always unnatural training of performing animals' as the cause of unnecessary suffering and as the use of animals not for human self-preservation but for unnecessary 'pleasure, convenience and distraction'; and he pointed out that the recent 1911 Protection of Animals Act made unnecessary suffering (against which the nation had ranged itself) illegal (Galsworthy 1913) (Fig. 2.4). He claimed his speech was characterised by a trainer present as 'very fair' (Marrot 1935, p. 385). The next day *The Times* reported that he had encouraged the public to hiss as a sacred duty because cruelty was a blasphemy against conscience and human society. A report from Boston claimed that many vaudeville animal acts were already being hissed by the audience (Bensusan 1913a, pp. 124–127). The RSPCA's *The Animal World* later quoted Galsworthy:

> We who are getting on in years were brought up in very unimaginative times. For instance, we were never taught to think what a wretched thing it is to train animals to perform difficult tricks and drag them from country to country, town to town, platform to platform, under conditions which are always unnatural, and may be horrifying. (Fairholme 1919)

At this time some of the humane societies issued publications criticising the state of affairs and public ignorance of it.[39] Ernest Bell, a founding member of the Humanitarian League and a friend of Henry Salt, complained that performing

---

[39] For example: Bensusan (1913a), Robinson (1913) and Johns (1914). Bensusan called for registration and inspection (Bensusan 1913b, p. 113).

**Fig. 2.5** Ernest Bell, by Elliott & Fry. Published in the *Vegetarian Messenger and Health Review*, October 1933. Courtesy of Henry Salt Archive

animals suffered debasement at the hands of trainers whose 'triumph it is ... to control them to do silly things' (Bell 1918, p. 99) (Fig. 2.5).[40] British public concern about cruelty to performing animals (in capture, accommodation, transport, training and performance) was, however, only given exclusive representation for the first time in 1914 with the creation of the Performing Animals' Defence Committee by Ernest Bell, John Galsworthy, Miss Jessey Wade, Mrs Albert Bradshaw, 'who purposely became a Variety Artist in order to find out for herself how performing animals lived and were "trained"' and others (MacMichael Papers). On 31 July 1914, it brought its first prosecution at Tower Bridge Police Court in London of Les Marcellis for terrifying a dog, when the National Canine Defence League (NCDL) gave evidence. The case involved a dog hand balancing and climbing an oscillating ladder. Arthur Case, a veterinary surgeon, confirmed that the dog's appearance indicated abject terror, no doubt resulting from gross cruelty in training. A small fine resulted, and confiscation of the dog to prevent its intended use in France. This was one of the few cases in which advantage had so far been taken of the provision in the Act of 1911 by which the magistrate could deprive the owner of an animal which he had ill-treated (Bell 1918, *New York Times* 1914). But Bensusan criticised the lack of convictions secured by the RSPCA and their 'pathetic announcement that they pay for prosecutions and do not get the fines' (Bensusan 1913b, p. 113). One recent conviction it had obtained was against

> ... a band of Serbians who have been travelling in the country with a so-called dancing bear, who is hauled about, sometimes behind a cart by a ring through his nose, which, as the evidence showed, was tender and swollen. The man's best excuse was that it had been

[40] The Humanitarian League was founded in 1891 'to enforce the principle that it is iniquitous to inflict avoidable suffering on any sentient being' (Bell 1907). Founding members included Bell, Salt and Galsworthy, and the League was supported by, among others, George Bernard Shaw, Thomas Hardy and Christabel Pankhurst. One of its aims was 'A more vigorous application of the existing laws for the prevention of cruelty to animals, and an extension of these laws for the protection of wild animals as well as domestic'. The League easily associated its concern about the captivity of animals and birds with its campaign for prison reform. See Weinbren (1994).

> treated in the same way for some years. Yet the bear's tormentor was still allowed to keep possession of it. (Bell 1918)

The first stirrings of parliamentary interest were also soon evident. On 5 March 1917, Colonel Lockwood drew attention in the House of Commons to the imminent arrival of ships containing hundreds of wild animals captured for show purposes, including very many apes and eight pandas ('very rare'), calling for the prohibition of such imports and asking which official was responsible for allowing the current importations. He was advised that no restrictions existed but that these would be considered (Wilson 2008).[41] It is ironic that such imports were taking place at this time, as during March 1917 Britain was approaching the highest wartime monthly losses to U-boats of merchant ships involved in the import–export trade. The trade in essentials (materials and food) was seen as decisive in the successful prosecution of the war, and the nation as a whole was gravely concerned about the success of the German campaign of unrestricted submarine warfare.[42] That valuable accommodation in merchant ships under extreme threat should be allocated to animals for exhibition rather than to those dwindling essentials vital to the war effort is perhaps a reflection on the continuing strength of the business in wild animals.

Following Bergh's loss of influence on the press and public opinion, the American Anti-Vivisection Society began to publicise circus abuses, but its call for boycotts was unsuccessful. The situation then changed, as described by Beers (2006, p. 105),

> when a self-described 'hard head, inured to hardship, cruelty, and brutality' stepped forward to fight for circus animals' rights. That hard head was none other than Jack London, and the famous writer's outrage over trained animal acts sparked a reaction that reverberated all the way to the tents of the Ringling–Barnum and Bailey Circus.

He was horrified by the cruel methods used to remove the natural spirit of wild animals in order to make them perform as short-lived mockeries of their true selves. Apart from the burnt, dancing bears and the beaten, subdued lions, docile animals like horses were tormented with spikes under their saddles and elephants shackled for life. London's love of the wilderness and its inhabitants was matched by his hatred of this unnatural human interference, making animals degraded spectacles and perverting their characters: 'taming—or breaking—the spirit of a wild creature was like conquering the wilderness' (Mighetto 1991, p. 69). His level of personal knowledge of what happened 'behind the scenes' at the circus would be questioned, but based on what he claimed to know, the message in his novel *Michael, Brother of Jerry* had a sudden and long-lasting effect on both sides of the Atlantic. He stated in its foreword: 'Never have I been so appalled and shocked by the world's cruelty as I have been ... in the midst of happy, laughing, and applauding audiences when trained-animal turns were being performed on the stage' and 'cold-blooded, conscious, deliberate cruelty and torment' were behind 99 % of the turns.

---

[41] *Parliamentary Debates (Commons)* 1917–1918, vol. 91, column 28.

[42] An account of the U-boat threat itself, and of an experimental attempt to thwart it through the use of performing sea lions, is given in Wilson (2001).

**Fig. 2.6** Jack London. By courtesy of Special Collections & Archives, Merrill-Cazier Library, Utah State University

He continued: 'Practically all of us will weep red tears and sweat bloody sweats as we come to knowledge of the unavoidable cruelty and brutality on which the trained-animal world rests', but very few people usually did anything, so he called for temporary walkouts from programmes when animal acts appeared (London 1917, pp. vi–viii) (Fig. 2.6).

In *Michael, Brother of Jerry*, a dog loses his master and is abducted into the world of the circus, when he experiences and witnesses a variety of cruelties associated with training for public entertainment in the United States. London's very graphic and intense style enabled a work of fiction to act as a successful medium in drawing attention to the horrific cruelties he believed still to exist.[43] His understanding of training methods was dismissed as ill-informed and out of date by most animal trainers, but the work was a source of inspiration to their opponents.[44] Copies of the book were carried and held high by protestors and distributed as propaganda, much as was Peter Singer's *Animal Liberation* in the late 1970s, or Mao's *Little Red Book* during the Cultural Revolution. Contemporary British reviews of the book tended to concentrate on its subject and purpose rather than on its literary qualities. The *Times Literary Supplement* claimed that its story belonged to a 'class of fiction popular in America—the class which denounces some specific institution or practice' (*Times Literary Supplement* 1918). The staff reviewer in *Punch* was more openly enthusiastic about Jack London's 'sound purpose', and his 'sincerity and skill' in this 'intensely moving' work, admitting 'Never again can I watch a performance of trained animals' (*Punch* 1918). More recently, Joys has questioned London's knowledge and motive. She doubts that London saw the kind of brutal training he described and suggests he drew from an article by Maurice Brown Kirby of 1908, making *Michael, Brother of Jerry*

[43] Trevelyan Thomson would ask his select committee witness, R. H. Gillespie: 'If the suggested cruelty is not a fact, how are we to explain that such a well-known writer as Jack London should say, "Cruelty as a fine art has attained a perfect flower in the training of performing animals"?' (*1945*). (See note 54 for explanation of numerical references to select committee reports.)

[44] In the United States, the American Humane Education Society distributed free copies of London's books to all members (Beers 2006, p. 107).

'avowedly and blatantly propagandistic', written when the author was ill, depressed, under the influence of narcotics and labouring in agony to finish it before he died (Joys 1983, pp. 54–55; Kirby 1908; O'Connor 1964, pp. 368–387).[45]

The significant practical effect of the novel was the creation of the international Jack London Clubs, the first of which was devised in 1918 by Francis Rowley of the Massachusetts SPCA. This first club served as an example for many that soon followed. It had no subscription or officials and depended on a pledge to oppose trained animal acts, distribute literature, write in protest as necessary to show managers and the press, recruit sympathisers and hiss or leave when a trained animal act appeared. Beers believes the Jack London Club mirrored a shift in societal attitudes about nature, the wild and the ethical consideration of nature:

> London stirred up that nascent concern with his plea for captive wild animals, and through the circus campaign, Americans tried to give form to what was at the time a nebulous desire to strike a balance between exploitive consumption activities and a blossoming appreciation of the natural world. (Beers 2006, pp. 106–108)

During the 1920s, the journal *Our Dumb Animals*, published by George T. Angell of Boston, president of Massachusetts SPCA, would usually feature a lead article on the Jack London Club, which in less than a decade attracted 300,000 members, many of whom distributed *Michael, Brother of Jerry* in support of their cause (Mighetto 1991, p. 68).[46]

In the United Kingdom, pressure on performing animal interests also increased on the founding of the Jack London Club in 1918, following publication there in the previous year of *Michael, Brother of Jerry*.[47] A public controversy about performing animals then started to emerge as the First World War ended, and on the development of the Performing Animals' Defence League (PADL) out of the previous Committee.[48] Other contemporary pressure groups with related interests or with some shared memberships also began to comment on this growing controversy, and the RSPCA began to formulate an approach to the issue, but it was later to suffer serious criticism from the PADL, whose director became bitterly opposed to the measures of the RSPCA's chairman, Sir Robert Gower. The creation of the Jack London Club and the dramatic tactics of its members secured greater press

---

[45] London actually died before his work was published. In *The Wild Animal Trainer in America* (Joys 1983), a well-sourced and illustrated study of practice mainly in the United States, Joys states that her favourite aspect of the circus has always been the wild animal act and, although she claims to support most conservation and humane legislation, she hopes it will continue (Joys 1983, pp. ix–x, 280). The prologue and epilogue are both supplied by animal trainers.

[46] Estimates of the level of membership vary. By 1924, the Jack London Clubs claimed a membership of 400,000 (London 1939, p. 363). Beers (2006, p. 107) writes that by the mid-1920s, the number approached a remarkable 750,000.

[47] During a later debate on the second reading of the Performing Animals (Regulation) Bill, a member cited the RSPCA's annual report for 1919 indicating 1,494 members of the Club and its steady growth and influence (*Parliamentary Debates (Commons)* 1923, vol. 161, column 2978).

[48] The PADL later also entitled itself the Performing and Captive Animals' Defence League, but both titles continued to be used, and the abbreviation PADL is given throughout the present work.

attention so that the issue gained momentum. Both the RSPCA and the PADL were quick to associate themselves with the club, although its simple form of 'direct action' was probably more in keeping with the League's methods. Throughout 1919, the RSPCA's *The Animal World* promoted the Jack London Club, its concerns and its aims, with articles and advertisements, but was at first pessimistic about reforming legislation: 'Frankly, we can see no immediate prospect of successful legislation [for prohibition] . . . The rights of animals even in these days are only slowly emerging from obscurity' (*Animal World* 1919b). The council of the RSPCA had decided to start a Jack London Club, referring to the pledge but assuring readers that it 'entails no unpleasant publicity or aggressive propaganda', and a handsome enamelled badge of membership was available for 1s 3d (Fairholme 1919; *Animal World* 1919a). The PADL also promoted the club regularly in its own published material and sought to cooperate on a practical basis with its members, who were no doubt often shared as a result of the closely aligned aims and methods of the two organisations.[49]

Professional interests were keen to discredit the basis of London's views. A letter from J. B. Cuthbert in *The Performer* criticised references to Jack London in the press as an authority on animal training and claimed that he was not because he only came across crude imitators (*Performer* 1921). Bob Anderson, owner of Polo, the educated pony, claimed never to have seen what Jack London claimed as 'colleges of pain' (*Era* 1921a). Joseph Woodward, the sea lion trainer, saw the agitation coming to Britain from America as reducing managers' contracting of animal acts (*Era* 1921b). His father, James, complained that Jack London exaggerated but his views were treated as 'Gospel' (*World's Fair* 1921a).[50] And in *The Aberdeen Free Press*, Mrs Frank Bostock, wife of the famous trainer and menagerie owner, criticised members of the Jack London Club, who would not sit through a performance but who would nevertheless condemn what they had not seen (*World's Fair* 1921b).

Kean has noted that during this period the Humanitarian League, which concerned itself with human and animal interests alike, did not survive beyond

[49] For example, early in 1925 members of the League and the Club combined at a meeting in Central Hall, Westminster, and resolved to urge the government to make progress with the agreed bill affecting the training and exhibition of performing animals. With Roy Horniman presiding, concern was also expressed about the growing use of cruelty to animals in films and that the government should forbid their exhibition (*World's Fair* 1925). George Bernard Shaw spoke at one of the League's meetings on 10 February 1921, when rodeos and bullfights were included as objects of criticism, and the League prepared a petition against rodeos (*World's Fair* 1924). The association between the Club and the League would last, so that by 1962 all members of the League were *ipso facto* members of the Club, too, and the League distributed the reprinted *Michael, Brother of Jerry* (Westacott 1962, pp. 8 and 9).

[50] Joseph would later write to *The Spectator* that London knew very little about the subject and that he 'copied from a 4d. booklet, entitled The Confessions of an Animal Trainer (a story paid for to a down-and-out circus hanger-on) producing, in regular sequence, from its pages a well-padded book called Jerry, the Brother of Michael, which is still the fictitious basis upon which most anti-performing-animal propaganda exists' (Woodward 1930).

1919 because of the differences of opinion of its members on the recent war (Kean 1998, pp. 179 and 191). Instead, some of its membership began to concentrate on the development of new campaigns, and a case in point was Ernest Bell, who as a joint founder of the Humanitarian League transferred some of his energies to the new PADL. Meanwhile, Methodists and Nonconformists continued to oppose animal cruelty inflicted for what they regarded as frivolous purposes and took part in the criticism of animal performance in the 1920s (Wilson 2008).

The First World War itself encouraged sober attention to the condition of humanity and its capacity for destructiveness, when service animals were recognised as victims, too, but a specific result of the war was a tendency to blame cruelty to performing animals on foreigners, especially Germans, as part-justification for a trade policy to boycott 'alien acts'.[51] After the war, there was more open discussion of emotive ethical issues such as this, and others, such as women's suffrage, sometimes shared support.[52] As had been the case in earlier animal welfare issues, opponents of animal performance were characterised as middle class or above, well educated, especially in the arts, and often religious, and they were now supported by much of the national press as well as by means of campaigning groups, direct action and graphic propaganda (Wilson 2009).[53] During the course of the controversy, the opposition of the groups on either side of it was sometimes also explained as reflecting a division of class as well as of perceptions of humane behaviour or of legitimate economic interests, when it was proposed that privileged and hypocritical members of society forgot their interest in field sports in order to interfere with legitimate and harmless working-class entertainment. However, the daily press was now generally critical of animal performance, and the alleged class factor seems not to have affected their editorial treatment: for example, the socialist *Daily Herald* was especially critical of animal acts. *The Era*, as a trade journal, complained:

> The daily press ranged themselves on the side of the objectors to animal performances, and on the eve of the debate in Parliament inflammatory articles with huge scare headlines were published in order to carry the day. (*Era* 1923; Wilson 2008)

[51] In this post-war setting, particular attempts were made to blame any cruel practices on Germans as the 'alien enemy', Germany having been the pre-war source of many trained animal acts.

[52] In the Press, this especially applied to *The Daily Herald*, which took a strong line against animal performance, just as it had—uniquely—given full support for women campaigning for the vote.

[53] Bertram Mills, as proprietor of the International Circus and Fun Fair at London Olympia, in proposing a toast to the Corporation of the City of London, said that some of 'the first people in the country' had received letters from opponents, asking them not to visit circuses and opposing the Olympia event, and that Lord Lonsdale had told him the Royal Family said it had also been approached (*World's Fair* 1923). Toole Stott (1971, p. 63) suggests that Mills himself was eventually denied a knighthood because 'during his last years the anti-animal people were very vocal, and it was this probably which induced the body politic to take the diplomatic line'. (A similar assumption was recently made concerning the failure of Colin Blakemore to achieve in 2003 the knighthood usually attached to the post of chief executive of the Medical Research Council, because of his advocacy of animal experimentation.)

**Fig. 2.7** 'The supreme showmen. M. Briand and Mr Lloyd George (together). "Cruelty to animals! Why, it's all done by tact and kindness."' (1921) Reproduced with permission of Punch Ltd, www.punch.co.uk

The professional interests and their political agents therefore set out to defend from 'cranks' what they advocated as decent and economically important entertainment for working people. They acted through their trade organisations, such as the Variety Artistes' Federation (VAF), and expressed their views in their periodicals. Both sides were represented in parliament and in the membership of the forthcoming select committee.

The controversy was developed and sustained because of political and press interest. Two key political figures, Joseph Kenworthy and Sir John Butcher, promoted legislation, and the controversy soon provided metaphors for cartoonists illustrating current affairs (Fig. 2.7). The trade organisations and their periodicals mounted a strong defence, and the controversy was soon used by them to highlight the contemporary economic and social condition of the nation, with the charge that to ban animal performance would cause destitution and debilitate much of the entertainments industry; it would also be a victory for those 'faddists' who some thought were infecting society at the time.

Following attempts by Kenworthy to bring this controversy to a head with the introduction of a prohibiting bill in 1921, the Select Committee on Performing Animals was set up in the same year and given clear terms of reference for its investigation.[54] The minutes and reports enable a full analysis of the controversy. Members of the committee and its witnesses were almost equally divided in outlook, and the full range of available evidence and prejudices was revealed. In

[54] The select committee's brief was 'to inquire into the conditions under which performing animals are trained and exhibited, and to consider whether legislation is desirable to prohibit or regulate such training and exhibition, and, if so, what lines such legislation should follow' (*Parliamentary Debates (Commons)* 1921, vol. 144, column 1243). Reports were made after two hearings. It met seven times between 19 July and 11 August, 1921, but did not manage to conclude its investigation before the end of the Session. It was therefore reappointed early in the following parliamentary Session and met 12 times between 2 March and 9 May 1922. The findings of the select committee were the basis of the Performing Animals (Regulation) Act, passed in 1925. In the present work, numerical references indicate the numbered paragraphs of the minutes in the reports, those in italics referring to that of 1921 and those in roman to that of 1922 (United Kingdom Parliament 1921, 1922).

the light of the findings of the committee, it is possible to examine the use of wild and domesticated animals; training methods and equipment; training to timetable; professional tensions within the entertainment industry, individual trainers, owners and managers; and arguments, strategies and tactics.

Just as there had been some protest against performing animal acts well before the First World War, such protests continued on a larger scale in the 1930s and later, after the legislation in 1925 that many thought inadequate. Because sympathies among members of the select committee were almost evenly split, and because the music-hall and circus professions mounted a campaign of defence of their interests equal to the agitation of the PADL and Jack London Club, the idea of legislative prohibition of performing animal acts had soon been reduced to one of licensing, and the licensing provision of the resulting Act of 1925 has so far remained unaltered in the absence of any new legislation.[55] The incentive for further legislation was perhaps reduced because of the decline of the music hall with the rise of cinema and radio entertainment in the 1920s and 1930s and because both the general public and local authorities began gradually to turn against animal performances: restriction of these has come to depend on policies adopted at local government level. Animal acts also came to take up a smaller and less significant part in the British circus, and these acts started to become associated more with foreign circuses and trainers on tour in Britain. The Performing Animals (Regulation) Act of 1925 was therefore the first and last act of its kind, in spite of subsequent attempts to amend or replace it.[56] The Act itself resulted from the weakening of a series of bills and was not thought by critics of animal performance to reflect the findings of the select committee. For those defending the performing animals industry, it was accepted as innocuous and therefore overtly supported (Wilson 2008).[57]

[55] As part of the Animal Welfare Act (2006), new regulations on the use of animals in circuses would soon be considered, although only 35 wild animals were left in circuses in the United Kingdom in 2009 (RSPCA 2009).

[56] For example, in 1996, the Born Free Foundation was invited by the Associate Parliamentary Group for Animal Welfare to join the Circus Working Group which published a report on the UK animal circus industry in 1998 (Circus Working Group and RSPCA 1998). In 2006, the RSPCA joined forces with the Foundation to present a report to MPs outlining the reasons why the welfare of wild animals in the circus cannot be satisfactorily guaranteed due to the nature of circus life (Born Free Foundation and RSPCA 2006).

[57] The development of performing animals legislation was reminiscent of that for vivisection in the 1870s. At that time, because of the last-minute organised opposition from the medical establishment, the 1876 bill was weakened and became in the same year the Cruelty to Animals Act that not only protected vivisection through licensing but could also remove the requirement for anaesthesia through exempting certificates (Ryder 1983, p. 135; Turner 1992, pp. 209 and 210).

# References

## *(Articles in periodicals are anonymous or untitled unless otherwise shown)*


AJS (1902) Cruelty in training animals. Letter The New York Times 19 October, p 30

Altick RD (1978) The shows of London. Belknap, Cambridge, MA

The Animal World (1919a) Trained animal performances, June, pp 69–70

The Animal World (1919b) September, pp 99–100

Anon (1869) Haney's art of training animals. A practical guide for amateur or professional trainers. Giving full instructions for breaking, taming and teaching all kinds of animals including an improved method of horse breaking, management of farm animals, training of sporting dogs; serpent charming, care and tuition of talking, singing and performing birds; and detailed instructions for teaching all circus tricks, and many other wonderful feats. Jesse Haney, New York, NY

The Athenaeum (1843) 16 September, p 845

Atkins SH (1934) Mr Banks and his horse. Notes Queries 167:39–44

Beers DL (2006) For the prevention of cruelty. The history and legacy of animal rights activism in the United States. Swallow Press, Athens

Bell E (1907) Christmas cruelties. Reprinted from the 'Humane Review'. Humanitarian League, London

Bell E (1918) The performing animal. The Contemporary Review 113:93–100

Bensusan SL (1913a) The torture of trained animals. In: Trist S (ed) The under dog. A series of papers by various authors on the wrongs suffered by animals at the hand of man. The Animals' Guardian Office, London, pp 109–130

Bensusan SL (1913b) The torture of trained animals (reprinted from *The English Illustrated Magazine*). In: Trist S (ed) The under dog. A series of papers by various authors on the wrongs suffered by animals at the hand of man. The Animals' Guardian Office, London, pp 109–117

Bensusan SL (1913c) The case of performing animals (reprinted from *Black and White*, 1899). In: Trist S (ed) The under dog. A series of papers by various authors on the wrongs suffered by animals at the hand of man. The Animals' Guardian Office, London, pp 118–121

Blake IA (1921) Animal shows of bygone days. The Spectator 9 September, pp 326–327

Blathwayt R (1893) Wild beasts. How they are transported and trained. McClure's Magazine 1:126–135

Boakes RA (1984) From Darwin to behaviourism. Psychology and the minds of animals. Cambridge University Press, Cambridge

Boehrer B (2007) Introduction: the animal renaissance. In: Boehrer B (ed) A cultural history of animals in the Renaissance. Berg, Oxford

Bondeson J (2006) The cat orchestra and the elephant butler. The strange history of amazing animals. Tempus, Stroud

Born Free Foundation, RSPCA (2006) It's time Parliament changed its Act. An examination of the state of UK circuses with wild animals. Born Free Foundation and RSPCA, Horsham

Bostock FC, Velvin E (ed) (1903) The training of wild animals. Watchmaker, Seaside, OR (reprinted 2009)

Buckland FT (1868) Curiosities of natural history, vol 3, 2nd edn. Richard Bentley, London, 2 vols

Caius J (1576) Of Englishe dogges (trans. Fleming A (1969)). Da Capo, Amsterdam

Circus Working Group, RSPCA (1998) A report into the welfare of circus animals in England and Wales. Circus Working Group and RSPCA, Horsham

Cruise J (1995) A house divided: Sarah Scott's Millenium Hall. Stud Eng Lit 1500–1900 35 (3):555–573

Dando J, Runt H (pseuds) (1595) Maroccus exstaticus, or Bankes bay horse in a trance. Scarlet, London
Darwin C (1872) The expression of the emotions in man and animals. John Murray, London
Dickens C (1867) All the year round, 26 January, pp 105–106
Dobson A (1902) Side-walk studies. Chatto and Windus, London, pp 93–109
Durant J and A (1957) Pictorial history of the American circus. A S Barnes and Co., New York, NY
The Era (1896a) Are trained animals tortured? 2 May, p 16
The Era (1896b) The alleged animal torturing 9 May, p 17
The Era (1921a) 2 March, p 16
The Era (1921b) 9 March, p 15
The Era (1923) Editorial, Looking back, 1 March, p 13
Esher V (ed) (1912) The girlhood of Queen Victoria—a selection from Her Majesty's diaries between the years 1832 and 1840. Longmans, Green, New York, NY
Evelyn J (1660) The diary of John Evelyn. http://way-mark.co.uk/bygonedays/evelyn/diary/166/16600913.htm
Ex-Lion-King (1872) Lions and lion-taming. Every Saturday 17 February, pp 173–176
Fairholme EG (1919) Trained animal performances. An insult to animals and human beings alike. The Animal World, March, pp 28–29
Furnivall FJ (ed) (1879) Philip Stubbes's anatomy of the abuses in England in Shakespeare's youth, A.D. 1583. Trübner, London
Futrell A (2006) The Roman games. Blackwell, Malden, MA
Galsworthy J (1913) Treatment of animals. Being a speech delivered at the Kensington Town Hall on December 15, 1913, at a meeting called to protest against cruelties to performing animals. In: Trist S (ed) AF Pamphlet series. Animals' Friend Society, London
The Gentleman's Magazine (1873) Three months with a Lion King, March, pp 254–261
Grant T (2001) White bears in Mucedorus, The winter's tale, and Oberon, the Fairy Prince. Notes Queries 48:311–313
Grant T (2007) Entertaining animals. In: Boehrer B (ed) A cultural history of animals in the Renaissance. Berg, Oxford
Graves TS (1920) The Elizabethan trained ape. Mod Lang Notes 35(4):248–249
Griffith E (2004) Banks, William. In: Matthew HCG, Harrison B (eds) Oxford dictionary of national biography. Oxford University Press, Oxford
Hagenbeck C (1912) Beasts and men: being Carl Hagenbeck's experiences for half a century among wild animals, trans. Elliot HSR, Thacker AG. Longmans Green, London
Hardy T (1913) Letter The Times 19 December, p 9c
Hill GB, Powell LF (eds) (1934–1950) Boswell's life of Johnson. Clarendon, Oxford
Hill J (1752) An history of animals. Thomas Osborne, London
Horne RH (1838) The life of Van Amburgh: the brute tamer, with anecdotes of his extraordinary pupils. By Ephraim Watts, a Citizen of New York .... Robert Tyas, London
The Illustrated London News (1847) Wombwell's Menagerie at Windsor Castle, 6 November, p 297
The Illustrated London News (1852) 24 July, pp 22, 62
The Illustrated London News (1859) 7 May, p 443
The Illustrated London News (1885) Our Note Book, 25 April, p 420
Jennison G (1937) Animals for show and pleasure in ancient Rome. Manchester University Press, Manchester, Reprinted 2005 University of Pennsylvania Press, Philadelphia
Johns CR (1914) By an ex-trainer, the pitiful story of the performing animal. Animals' Friend Society, London
Johnson W (1990) The rose-tinted menagerie. Heretic Books, London
Jonson B (1600) Every man out of his humour. Wiliam Holme, London
Jonson B (1616) Epigrams. In: The workes of Benjamin Jonson. William Stansby, London
Joys JC (1983) The wild animal trainer in America. Pruett, Boulder, CO

Joys JC (2011) The wild things. Dissertation, Bowling Green State University
Kean H (1998) Animal rights: political and social change in Britain since 1800. Reaktion, London
Kirby MB (1908) The gentle art of training wild beasts. Everybody's Magazine 19(4):435–446
Kiser LJ (2011) Animals in medieval sports, entertainments, and menageries. In: Resl B (ed) A cultural history of animals in the medieval age. Berg, Oxford
Kyle DG (1998) Spectacles of death in ancient Rome. Routledge, London
The Literary Gazette (1829) 21 February, p 132
Loisel G (1912) Histoire des menageries de l'antiquité a nos jours, vol 1. Octave Doin et Fils, Paris
London J (1917) Michael, brother of Jerry. Mills and Boon, London
London J (1939) Jack London and his times. University of Washington Press, Seattle, WA
Lysons D (n.d.) Collectanea: or, A collection of advertisements and paragraphs from the newspapers, relating to various subjects. Public exhibitions and places of amusement. Kirkgate, London
MacMichael Papers (uncatalogued and unpublished material relating to the director of the Performing Animals' Defence League between 1929 and 1968, kindly lent to the present author by Edmund MacMichael's daughter, Mrs Esther Denham)
Magnus O (1555) Historia de gentibus septentrionalibus earumque diversis statibus, conditionibus, moribus, ritibus, superstitionibus, disciplinis ... et rebus mirabilibus nec non universis pene animalibus in septentrione degentibus eorumque natura. Book 18 'De animalibus sylvestribus' Chapter 32 'De chorea ursorum Lithuanicorum'. Giovanni M. Viotto, Rome. pp 629–630
Manning-Sanders R (1952) The English circus. Laurie, London
Marchand LA (ed) (1973–1982) Byron's letters and journals, vol. 3. Belknap, Cambridge, MA
Markham G (1607) Cavelarice, or the English horseman. Edward White, London
Marrot HV (1935) The life and letters of John Galsworthy. Heinemann, London
Mighetto L (1991) Wild animals and American environmental ethics. University of Arizona Press, Tucson, AZ
The Mirror of Literature, Amusement, and Instruction (1838) 32, p 441
Moffett C (1893–1894) Wild beasts and their keepers. How the animals in a menagerie are tamed, trained, and cared for. McClure's Magazine 2:544–561
Moffett C (1894) Wild beasts in captivity. How they are watered and fed; the special dangers in handling them. McClure's Magazine 3:71–88
Morley H (1859) Memoirs of Bartholomew Fair. Chapman and Hall, London
The New York Times (1914) To end exporting of worn-out horses ... [etc.], 2 August, p S6
Nicholson EWB (1879) The rights of an animal: a new essay in ethics. Kegan Paul, London
Nicolson H (1950) Marginal comment. The Spectator 17 March, p 336
O'Connor R (1964) Jack London, a biography. Little, Brown and Co., Boston, MA
Pattison M (1892) Isaac Casaubon 1559–1614. Clarendon, Oxford
PD (1839) The elephant and his keepers. The Spectator 28 December, pp 1229–1230
The Performer (1921) 19 May, p 12
Praneuf M (1989) L'ours et les hommes dans les traditions européennes. Imago, Paris
Punch (1918) 13 February, p 112
Purdy RL, Millgate M (eds) (1984) The collected letters of Thomas Hardy, vol 4. Clarendon, Oxford, pp 1909–1913
Ravelhofer B (2002) "Beasts of Recreation": Henslowe's white bears. Engl Lit Renaiss 32 (2):287–323
Robinson, Chief Inspector (Northumberland and Durham Society for the Protection of Animals) (1913) How a foreign trainer was brought to book. In: Trist S (ed) The under dog. A series of papers by various authors on the wrongs suffered by animals at the hand of man. The Animals' Guardian Office, London. pp 133–140
RSPCA (2009) Circuses—help animals leave the circus. www.rspca.org.uk. Accessed 19 August, 2009
Ryder RD (1983) Victims of science. The use of animals in research. National Anti-Vivisection Society, London

Ryder RD (2000) Animal revolution. Changing attitudes towards speciesism. Berg, Oxford
Saville E (1910) The dew of mercy. The Westminster Review 173(1):86–95
Saxon AH (ed) (1983) Selected letters of P T Barnum. Columbia University Press, New York, NY
Schrickx W (1982) Richard Banks and his horse in Wolfenbüttel in 1605. Notes Queries 277:137–138
Schwalm T (2007) "No circus without animals?" Animal acts and ideology in the virtual circus. In: Simmons L, Armstrong P (eds) Knowing animals. Brill, Boston, MA
Scott S (1762) A Gentleman on his travels. A description of Millenium Hall and the country adjacent together with the characters of the inhabitants and such historical anecdotes and reflections as may excite in the reader proper sentiments of humanity, and lead the mind to the love of virtue. J. Newbury, London. Project Gutenberg online version
Shepstone HJ (1902) The scientific training of wild animals. Scientific American, October, pp 260–261
Silliman B (1810) A journal of travels in England, Holland, and Scotland . . ., vol 1. D and G Bruce, New York, NY
Simmons J (ed) (1951) Robert Southey. Letters from England. Cresset Press, London
The Spectator (1831) The Lions of Drury Lane, 22 October, pp 1025–1026
The Spectator (1833) The fleas, 16 March, p 236
The Spectator (1839) 9 March, p 225
The Spectator (1841) M. Leonard's intelligent dogs, 3 July, p 638
The Spectator (1847) Arrival of foreign artistes for Drury Lane, 3 April, pp 324–325
The Spectator (1852) Europa at Cremorne, 4 September, p 848
The Spectator (1877) Four-footed pupils, 29 September, pp 1203–1205
Stokes J (2004) Lion griefs. New Theat Q 20(2):138–154
Strunk W (1917) The Elizabethan showman's ape. Mod Lang Notes 32(4):215–221
Strutt J (1801) The sports and pastimes of the people of England. In: Hone W (ed.) (1833) The sports and pastimes of the people of England . . . A new edition . . . by William Hone. T T and J Tegg, London
Swain J (1933) Brutes and beasts. Noel Douglas, London
Tait P (2009) Controversy about a human–animal big cat stunt in Fillis's Circus. Early Popul Vis Cult 7(2):199–211
Tait P (2012) Wild and dangerous performances: animals, emotions, circus. Palgrave Macmillan, Basingstoke
Taylor J (1638) Bull, bear, and horse, &c. Henry Gosson, London
Thomas K (1983) Man and the natural world: changing attitudes in England 1500–1800. Allen Lane, London
The Times (1838) Van Amburgh, the lion tamer, 19 September, p 6
The Times (1869) 24 August
The Times Literary Supplement (1918) Review of Michael, brother of Jerry, 17 January, p 32
Toole Stott R (1958) Circus and allied arts: a world bibliography, 1500–1957. Harpur, Derby
Toole Stott R (1971) Circus and allied arts: a world bibliography, 1500–1970. Harpur, Derby
Turner ES (1992) All heaven in a rage. Centaur Press, Fontwell
United Kingdom Parliament (various dates) Parliamentary Debates (Commons). HMSO, London
United Kingdom Parliament (1921) Report from the Select Committee on Performing Animals, together with the Proceedings of the Committee and Minutes of Evidence. HMSO, London
United Kingdom Parliament (1922) Report from the Select Committee on Performing Animals, together with the Proceedings of the Committee and Minutes of Evidence. HMSO, London
Variety (1910a) 17 September, p 13
Variety (1910b) 24 September, p 13
Wallace I (1959) The fabulous showman: the life and times of P T Barnum. Knopf, New York, NY
Warner GF (1881) Catalogue of the manuscripts and muniments of Alleyn's College of God's Gift at Dulwich. Longmans, Green, London

Weinbren D (1994) Against *all* cruelty: the Humanitarian League, 1891–1919. Hist Work J 38:86–105
Westacott E (1962) Spotlights on performing animals. Being extracts of evidence given before the Select Committee of 1921 and 1922. Selected by E Westacott, with reference to existing conditions. C W Daniel, Rochford
Wiedemann T (1992) Emperors and gladiators. Routledge, London
Wiles RM (1967) Crowd-pleasing spectacles in eighteenth-century England. J Popul Cult 1 (2):90–105
Wilson DAH (2001) Sea lions, greasepaint and the U-boat threat: admiralty scientists turn to the music hall in 1916. Notes Record Roy Soc 55(3):425–455
Wilson DAH (2008) Politics, press and the performing animals controversy in early twentieth-century Britain. Anthrozoös 21(4):317–337
Wilson DAH (2009) 'Crank legislators', 'faddists' and professionals' defence of animal performance in 1920s Britain. Early Popular Visual Culture 7(1):83–101, http://www.tandfonline.com/doi/full/10.1080/17460650902775393
Joseph Woodward (1930) Letter The Spectator 27 December, p 1014
The World's Fair (1921a) 12 February, p 8
The World's Fair (1921b) 10 September, p 7
The World's Fair (1923) 29 December, front page
The World's Fair (1924) 3 May, front page
The World's Fair (1925) 7 March, p 25
Wright LB (1927) Animal actors on the English stage before 1642. PMLA 42(3):656–669

# Chapter 3
# Open Debate in the British Parliament and Beyond

**Abstract** In 1921 Joseph Kenworthy presented his Performing Animals (Prohibition) Bill. This was weakened by standing committee amendments and therefore withdrawn, but a select committee was then appointed in 1921 and 1922 to examine conditions of training and exhibition and possible legislation. Subsequent bills were based on regulation of animal performance through prohibition of categories of animal and through inspection and licensing, leading to the Performing Animals (Regulation) Act of 1925. The reports of the investigations of the select committee's members and minutes of the information given to them by witnesses (mainly representing the circus and music-hall industries and critics) reveal the full range of concerns applying to the use of trained animals in public entertainment. The problems of confinement and transport also began to receive greater attention. As critics and trade interests developed their campaigning strategies, other arguments were introduced, such as the role of foreign training or the effect on children's perception of animals. This chapter discusses the nature of the evidence presented to the select committee and the difficulty experienced in reaching a satisfactory conclusion to the controversy, one which continued to attract much attention from the press and public.

## 3.1 Parliamentary Enquiries, Bills, and Legislation[1]

The Performing Animals (Regulation) Act came into effect on 1 January 1926, following protracted investigation and debate, both in Parliament and in the press. Earlier related legislation consisted of the Wild Animals in Captivity Protection Act, 1900, and the Protection of Animals Act, 1911, drafted by the RSPCA and introduced by George Greenwood MP, one of its council members. Under the 1900 Act, it became illegal to cause any unnecessary suffering or to cruelly abuse, infuriate, tease, or terrify; and the 1911 Act made illegal, again by cause or permission, similar mistreatments of any animal, including those of a more

[1] Elements of this section are taken from Wilson (2008).

 
D.A.H. Wilson, *The Welfare of Performing Animals*, Animal Welfare 15,
DOI 10.1007/978-3-662-45834-1_3

everyday nature, such as beating, overloading, or stressful transportation, and also staged animal fighting and baiting, poisoning, and cruel operations.

Joseph Kenworthy[2] had originally raised the subject of performing animals in Parliament on 21 February 1921, asking the Prime Minister if he was aware of the:

> ... many proved cases of cruelty involved in the breaking and training of wild and domestic animals, other than horses, for performances on the stage and in travelling circuses, and that an increasing number of people object to these performances.

He asked if the Prime Minister would consider introducing legislation on this subject. The response was that the 1911 Act offered sufficient protection.[3] Kenworthy thereupon promptly presented the Performing Animals (Prohibition) Bill, supported by, among others, Trevelyan Thomson,[4] who would also later serve on the select committee. The bill sought to prohibit the exhibition of performing animals and birds in places of public entertainment. A standing committee amendment allowed for the exhibition of horses and of animals ordinarily used in military tournaments and the exhibition of animals 'by any person licensed to train and exhibit animals by any justice of the peace in such form as may be prescribed by a Secretary of State'.[5] Although Kenworthy withdrew his weakened bill, as a result of debate in Parliament and special encouragement from Sir John Butcher, it was decided to appoint a select committee and nominate members. Meanwhile, Charles Stanton was growing impatient with protesters inside and outside Parliament:

> Hon. Members should not be so ready to lend an ear to the hysterical screechings of people, many of whom show far more sympathy for animals than for human beings who risk their lives in Ireland and elsewhere. ... I would be as ready as anyone to stop a show which is cruel, or which is calculated to brutalise the minds of the people, but I am not going to listen to the sentiments and fads of all sorts of Tom Noddies.[6]

The select committee was appointed on 12 July 1921 and met seven times between 19 July and 11 August. The RSPCA still hoped prohibition might result. On the failure of Kenworthy's amended bill, it commented: 'The experience of the Society, given on its behalf by Mr C. Wilson before a Committee at the House of Commons, is that licensing would be valueless. It is therefore a matter for little regret that the amended Bill failed to pass, and we may hope that a new Bill soon to be introduced will render trained animal turns illegal' (*Animal World* 1921b). At the conclusion of the select committee's work that Session, a report with minutes of evidence had been brought up under 'oral answers', read and ordered to be printed (United Kingdom Parliament 1921a); but it was pointed out that the inquiry was not

---

[2] MP for Hull Central in 1919, serving as a Liberal until 1926 and then as a Labour MP until 1931.

[3] *Parliamentary Debates* (Commons) 1921, vol. 138, column 545.

[4] Independent Liberal member for Middlesbrough West between 1918 and 1928.

[5] Sir John Butcher, Conservative MP for York, had opposed the decision on horses, as he believed they were abused in presentations such as those involving a revolving stage turntable (United Kingdom Parliament 1921b).

[6] *Parliamentary Debates* (Commons) 1921, vol. 142, columns 1425 and 1432; vol. 144, column 1243. Stanton was Coalition National Democratic and Labour Party MP for Aberdare.

**Fig. 3.1** Sir Richard Beale Colvin, by Walter Stoneman, 1934. © National Portrait Gallery, London. Ref. x166703. Reproduced with permission

completed.[7] Soon after, Brigadier-General Colvin (Fig. 3.1)[8] asked the Prime Minister whether the select committee he had chaired, and which had taken a large amount of important evidence at the end of the last Session, would be reappointed forthwith so that the evidence might be completed and the committee enabled to issue its final report.[9] The Secretary of State for the Home Department replied in the affirmative, and the minutes of evidence of the select committee of the first Session of 1921 were later referred to the reappointed select committee.[10]

As was usual, the select committee was empowered to send for 'persons, papers and records', through an informal invitation by means of press notices to individuals to give personal evidence or to interested organizations and individuals to send memoranda on the subject of the enquiry. The select committee, being an extension of the House, possessed substantial powers, greater than in a court of law, to require answers to questions and the production of documents. The members of the select committee approached witnesses according to their own prejudices or preconceptions. A witness was required to answer questions and to be truthful. Evidence was not given on oath; therefore, punishment for contempt rather than perjury would apply to giving false evidence. Witnesses were protected from the consequences of giving evidence, such as slander. According to convention, members of the select committee co-operated and did not unreasonably interrupt during the investigation, although a question put to a witness could be objected to by a member, much like the interaction of prosecution and defence in a court (Limon and McKay 1997, pp. 646–654).

[7] *Parliamentary Debates* (Commons) 1921, vol. 146, column 657.

[8] Brigadier-General Sir Richard Beale Colvin, Unionist member for Epping between 1917 and 1923.

[9] *Parliamentary Debates* (Commons) 1922, vol. 150, column 605.

[10] *Parliamentary Debates* (Commons) 1922, vol. 150, column 2065.

With a quorum of five, the select committee comprised 15 members.[11] Twenty witnesses were examined over 6 days. As recommended, the select committee was reappointed on 22 February 1922 to continue its work. It heard evidence from a further 23 witnesses over 6 days and met 12 times between 2 March and 9 May 1922, when a draft report was produced.

Amendments to the draft were extensively argued over and so further meetings were held on 11, 16, and 18 May before confirmation of the report (United Kingdom Parliament 1922). In each division over these amendments, the committee members were nearly always almost evenly split according to their differing views on the desirability of new legislation. Broadly, the views reflected on the one hand those, some representing the humane societies, who believed that cruelty was an existing problem, and on the other those, some representing professional interests, who believed it was not. Those eager for legislation were in the slight majority, and amendments were therefore made that resulted in a final report that was a little more powerful than the draft. These amendments served to emphasize that ill-treatment was still common, removed reference to benefits to children from any animal performances, suggested that cruelty had to be obvious to spark public disapproval, and enabled inspection without previous notice given, and the prohibition both of performances of animals thought to be cruelly trained abroad and of conjurors' devices involving cruelty. It was now also recommended that training and performance of anthropoid apes be prohibited. An amendment from Sir John Butcher to include larger carnivora had been narrowly defeated, but it was agreed that they should have the special attention of the proposed Committee of Supervision. Dissatisfaction with the sum of the amendments in the final report on the part of those opposed to legislation led them to vote against its submission to the House of Commons in amended form, but they were defeated by five votes to seven.

The report includes an account of the nature of evidence given and the committee's reaction to it, a summary including a general expression of opinion, and recommendations. The committee reported that cruelty existed, that the public was opposed to it, and that the profession was willing to co-operate against

[11] Captain George Bowyer, Conservative member for Buckingham between 1918 and 1937, when he was created 1st Baron Denham of Weston Underwood; Sir John Butcher; Brigadier-General Sir Richard Beale Colvin (chairman); Sir Walter de Frece, Unionist member for Ashton-under-Lyne between 1920 and 1924 and for Blackpool between 1924 and 1931; Lieutenant-Colonel Sir Raymond Greene, Conservative member for North Hackney between 1910 and 1923; Mr Charles Jesson, Coalition National Democratic and Labour Party member for Walthamstow West between 1918 and 1922 and late organizer of the Musicians' Union; Lieutenant-Commander Joseph Kenworthy; Captain James O'Grady, Labour MP for Leeds South-East (1918–1924) and formerly secretary of the National Federation of General Workers; Mr Alfred Raper, Coalition Unionist member for East Islington between 1918 and 1922; Mr Frederick Roberts, Labour member for West Bromwich between 1918 and 1931, and 1935 and 1941; Mr James Seddon, Coalition Labour member for Hanley between 1918 and 1922 and chairman of the Trades Union Council in 1914; Mr Charles Stanton (the only nominee who subsequently failed to take part); Mr John Swan, Labour member for Barnard Castle between 1918 and 1922; Mr Trevelyan Thomson; and Lieutenant-Colonel Claud Willoughby, Unionist member for Rutland and Stamford between 1918 and 1922.

it. Evidence covered the previous 25 years, but difficulty was experienced in obtaining evidence of cruelty from persons who considered that their chances of employment might be endangered. The committee received a large number of letters as well as sworn declarations, but this evidence was of less value because it could not be made subject to cross-examination. The committee remarked that its enquiry had drawn press attention, and audiences of animal acts had become very critical as a result. A critical public was thought to be the animals' only existing source of protection, since there was no legal right of inspection of practices. It was noted: 'It is not contended that admission [for inspection] would be denied . . ., but a warning might be given that would render such a visit useless'. There was also a need for closer inspection of transport and accommodation arrangements, and it was felt, too, that animals trained for use in film productions should benefit under the proposed measures. Two categories of animals were identified, 'savage and untrustworthy' and 'more domesticated and docile': cruelty to the first could be involved especially in subjugation and to the second in training for tricks. The effect on animals of performance to a timetable was thought to vary according to their temperament. Finally, the committee stated that in 1914, about 20 % of trainers had been British, and they were considered more humane.

The committee considered that wholesale prohibition was not necessary, but that supervision was needed for trainers, training establishments and animal performances, and that existing penalties were inadequate (Fig. 3.2). It recommended that the proposed Committee of Supervision should have powers to prohibit, restrict, suspend, and modify performance or training, including prohibition of acts resulting from suspect training methods abroad. Trainers and training establishments should be registered, and local authority, RSPCA, and police officers should have right of access at all times, without previous notice given, to performances and training establishments. The use of cruel appliances by conjurors and the training and performance of anthropoid apes should be prohibited. Finally, stiffer penalties should be applied to those convicted of cruelty. The report of the select committee therefore marked the point at which the possibility of early general prohibition was replaced by an intention to regulate. A leading article in *The Times* of 4 December 1922 commented:

**Fig. 3.2** The dogs' employment bureau: 'What we may expect if the prohibition of animal performances bill is passed'. (1922) Reproduced with permission of Punch Ltd, www.punch.co.uk

> The Report contains more than sufficient grounds for immediate legislation ... we accept, although with reluctance, the decision that the time has not yet come for total prohibition of this particular form of exploitation of animals for the pleasure of men.

It also disagreed with the distinction made by the select committee between wild and domestic animals (giving special attention to the circumstances of the larger carnivora), because domesticated or docile ones were at least as susceptible to cruelty.

Following the work of the select committee, whose final report was published in May 1922, a new bill to regulate the exhibition and training of performing animals was presented in 1923 by Colvin and supported by Butcher, Thomson, Greene, Bowyer, Roberts, and Kenworthy. It called for the registration of trainers and exhibitors of performing animals; power for the Secretary of State by regulations to prohibit or make conditions on the use of apes, (and now also) large carnivora, and appliances for conjuring tricks; power of a court order to prohibit or restrict an activity; power of a police constable or representative of an approved humane society to inspect premises ‘at all reasonable times’ (the select committee had recommended such inspections ‘without previous notice given’); possible deregistration following conviction; and an advisory committee for the Secretary of State concerning regulations, consisting of a chairman with two humane society members and two from the profession (instead of the staffed Committee of Supervision proposed by the select committee, which would have been costly, requiring offices).

During the debate of 3½ h on the second reading on 23 March 1923,[12] Colvin said that legislation was necessary because evidence had been presented that there was cruelty, that job losses would only affect the cruel, and that the public needed reassurance about such entertainment, which was becoming less attractive. Referring to the findings of the select committee, he noted that some animals such as seals were suitable for training without cruelty or intimidation, while others, such as large carnivora and apes, were not. There was also the problem of inept use of conjurers’ apparatus, such as the small collapsible cage in the disappearing-bird trick. At the same time, he was against the ‘frivolous’ interruption of performances by protestors, and would support a special amendment to suppress them.

O’Grady blocked the bill on behalf of the VAF and moved for its rejection, claiming that the select committee had been given no proof of cruelty in evidence, resulting in an almost even split of members when asked to sign approval of the report (*Performer* 1923b). He said the bill would effectively result in an Act of prohibition unless exemptions were made by the Home Secretary, who would feel uncomfortable because of the behaviour of vindictive societies, as in the case of Lord Lonsdale.[13] There had been few recent convictions, and he claimed that

[12] *Parliamentary Debates* (Commons) 1923, vol. 161, columns 2961–3020.

[13] Lord Lonsdale had been criticized by opponents of animal performance because he was at the same time a vice-president of the RSPCA and president of Bertram Mills International Circus at London Olympia. (He was also president of the Shikar Club. See McKenzie 2000.) He declared that he had accepted the circus presidency because he knew how every trick could be taught, and by accepting it he could ensure that no trick was allowed where cruelty was entailed or necessary. ‘I made it a condition that all what we call “acrobatic tricks” were eliminated from the show, and,

cruelty before the war had usually involved Germans (often blamed and referred to during this controversy as the 'alien enemy') (Wilson 2009a). Representing the VAF and being one of its originators, he confirmed that the organization would expel a member for cruelty, that the bill would destroy livelihoods, and that the RSPCA already had full access for inspection. He attempted to pull apart the evidence of some select committee witnesses, saying that the bill was promoted by irrational prejudice (Wilson 2009b).

Major Thomas Paget, Conservative MP for Bosworth, questioned the scope of performing animals as understood in the bill, and the interference of inspections. As a warning, Major Leonard Molloy, Conservative MP for Blackpool, cited the RSPCA's president at the annual meeting in 1921 who considered the bill of that time as a step towards the total abolition desired by some, and Molloy also criticized undesirable direct-action tactics.[14] He regarded the current bill as an assault on the working class and its entertainment and felt that general sympathy with the movement towards humane treatment of animals could be lessened as a result. An extension of the law was not necessary, and the select committee recognized the opposition of the profession to cruelty. The bill was 'the attempt of a noisy, but socially influential, class to interfere with the legitimate amusements of the people', resulting from the 'fulminations of a lot of hysterical men and women' and the 'senile arguments of a lot of old women, of both sexes'. But George Buchanan, Labour MP for Glasgow Gorbals, believed the bill should go further and include other amusements such as steeple-chasing, fox-hunting, and human performances. Employment and the cinema industry should not have to depend on cruelty, he said, and supervision was necessary because professionals placed popularity and profit above animal welfare (Wilson 2009b).

Sir Walter de Frece, speaking for the Society of West End Managers, the Theatrical Managers' Association, the VAF, circus interests and the entertainment world, asserted that it was only a Private Member's Bill because the government did not believe in the report of the divided select committee [which nevertheless acknowledged as a whole that cruelty existed], and it reminded him of the Puritans

having seen the turns and things that were done, I am quite convinced that all the tricks performed are done without any cruelty to animals at all' (*World's Fair* 1923a). He was later hissed and booed at an annual meeting of Our Dumb Friends' League (ODFL) because of his attitude to the rodeo, and Lady Lumb unsuccessfully opposed his re-election as its president on the grounds that he saw no cruelty in it (*World's Fair* 1924c). Soon after, he stated to the press that he was instrumental in inducing the promoters both to cease the public displays of steer roping at the rodeo, and to allow the RSPCA full access (*World's Fair* 1924d).

[14] The following month, Margaret Bradish complained to *The Performer* that she had been quoted by Molloy in his parliamentary speech from a private letter addressed to her by Miss Jessey Wade, secretary of the PADL: 'May I ask the trainers and their friends: by what right they read and showed to others my private letter? How they obtained possession of it? Its publication has thrown a curious light on the trainers' methods of obtaining information!' A trainer, Harry Rochez, replied that the VAF said the letter was mistakenly put in with another 'usual' one from the PADL to a theatrical manager. He referred to the 'lady who is good at shouting' (Mrs Speedwell Massingham) as the paid organizer of the PADL, and Miss Bradish as formerly its secretary (*Performer* 1923d, e).

who were not concerned by animal cruelty but by human pleasure. Unemployment would result because exhibitors would not want the bother of inspection. Although he did not object to registration and honest investigation and inspection, he objected to inquisition by people who would 'have to make a case to keep their jobs, and who are not honest critics of what they see', leading not just to sentimental interference but also to widespread state control through the autocratic regulatory powers given to the Home Secretary via the bill's proposed advisory committee (Wilson 2009b).

Butcher said that the select committee did hear evidence of cruelty, and all had agreed to the reference to it in the report, including Sir Walter de Frece. He commented that 'under the existing law, the facilities for inspection, obtaining evidence, prosecution, and so on, were insufficient to stop this cruelty' and that training took place in secret, making cruelty easier and inspection necessary. What animals to include (e.g. as mentioned in the debate, Grand National horses or snakes) could be decided on at Committee Stage, and the idea of wholesale prohibition was a bogey of no consequence. He also noted that 80 % of trainers were foreign, many being Germans whose cruelty produced what they called good results, and he claimed:

> You do not get rid of the strong determination in this country to suppress every form of cruelty towards animals—a determination which exists both in this House and very largely throughout the country, and not least amongst the newly enfranchised voters, namely, the women—. . .by calling it sentimentalism.

Bowyer admitted that it was very difficult to expose evidence of cruelty, which was often caused by animals being forced to perform on demand and to a timetable, as explained by Peter Chalmers Mitchell, whom he thought the most important witness examined by the select committee (Figs. 3.3 and 3.4).[15] Benjamin Tillett, Labour MP for Salford North and an ex-circus boy,[16] thought training was only possible by kindness, and cruel people were ostracized by the circus, where chivalry was shown to animals. He asked why that day's Grand National, or hunting, or slums, were not the object of concern: this was an attempt at class legislation. Sir Walter Greaves-Lord, Conservative MP for Norwood, asked what exactly would be the grounds for objection leading to a conviction and was worried by the subjectivity of magistrates, an appeal against whose decision would mean interim loss of earnings, and expense. Thomas Groves, Labour MP for Stratford West Ham, believed that cruelty was involved in unnatural performances and that legislation in a civilized society necessarily interfered with individual liberty.

William Bridgeman, Conservative MP for Oswestry and Secretary of State for the Home Department, said that the government would not use its parliamentary whips in the vote, but that the Home Office welcomed the bill, and opposition to it

[15] Mitchell was an anatomist who was appointed secretary of the Zoological Society of London in 1903 until 1935. He was elected a Fellow of the Royal Society in 1906, and later served between 1923 and 1927 as president of the Society for the Preservation of the Fauna of the Empire. He worked also for *The Times* between 1919 and 1932, writing leaders and scientific articles.

[16] He was also a founder of the Labour Party and helped found *The Daily Herald*.

**Fig. 3.3** George Edward Wentworth Bowyer, 1st Baron Denham, by Bassano Ltd, 1920. © National Portrait Gallery, London. Ref. x120110. Reproduced with permission

**Fig. 3.4** Peter Chalmers Mitchell. Unattributed

that day had been weak. There was strong public feeling against cruelty to animals, and those in societies opposed to cruelty to animals were also in societies opposed to cruelty to children. Stephen Walsh, Labour MP for Ince, saw no evidence that the 1911 Act had not had effect, but approved of the large attendance in the House as a reflection of British society's wish to reduce unnecessary animal suffering. James Gould, Conservative MP for Cardiff Central, said that everyone had been badgered by societies, but that only an increase in penalties was necessary. The bill then passed its second reading and was put to standing committee. Subsequent amendments by standing committee removed the proposal for the Secretary of State's regulatory powers and the associated advisory committee and allowed for inspection only by a police constable or a local authority official, neither of whom could go on or behind the stage during a performance.[17]

In the next Session of Parliament, Kenworthy asked the Under-Secretary of State for the Home Department whether the government would introduce this bill, as

[17] The RSPCA thought these amendments too drastic because its inspection provision was removed, and its officer could not now even make a complaint. The maximum penalty remained at £50, but there was now no mention of the £5 per day extra if an offence continued (*Animal World* 1923).

**Fig. 3.5** Sir George Kynaston Cockerill, by Bassano Ltd, 1922. © National Portrait Gallery, London. Ref. x121362. Reproduced with permission

"WILL IT HATCH?"

[The Performing Animals Bill was introduced by Brigadier-General COCKERILL, supported by Sir SYDNEY HENN.]

**Fig. 3.6** 'Will it hatch?' (1924) Reproduced with permission of Punch Ltd, www.punch.co.uk

amended [as Bill no. 208] to meet the views of all as a compromise, which would have passed through the last Parliament as an agreed measure had not time run out because of Dissolution. Rhys Davies replied that the government was now too busy and had no time, but that if agreement had been reached in the last Session, he hoped a Private Member's Bill might be successful.[18] The bill (no. 81) was duly presented in 1924 by Brigadier-General Sir George Cockerill, Conservative MP for Reigate, and supported by, among others, former select committee members Bowyer and Kenworthy and by Sir Sydney Henn (Figs. 3.5 and 3.6). It failed to compete successfully with other business before Dissolution, but it had been debated and had passed its second reading in the Lords on 3 April 1924. (In that debate Lord Raglan disapproved of police powers of inspection without warrant.)

In the Session of 1924–1925, the Performing Animals (No. 2) Bill was presented as a Private Member's Bill by Cockerill and supported by select committee members Bowyer, de Frece, Kenworthy, and Thomson. It was substantially the same as the latest form of the previous bill, except that registration was now made specific to the exhibition of performing animals at any entertainment to which the

[18] *Parliamentary Debates* (Commons) 1923, vol. 169, column 1187.

**Fig. 3.7** FitzRoy Richard Somerset, 4th Baron Raglan, by Bassano Ltd, 1923. © National Portrait Gallery, London. Ref. x122446. Reproduced with permission

public were admitted, whether on payment of money or otherwise, and to training for such exhibition; and the Act would not apply to the 'training of animals for *bona fide* military, police, agricultural or sporting purposes, or for the exhibition of any animals so trained'. The bill then still remained substantially the same after amendment by standing committee and Lords amendments,[19] although only public, rather than RSPCA or private, potentially 'vexatious', prosecutions would be allowed. Invertebrates were excluded, thereby denying benefit to the performing flea—*de minimis non curat lex*—and the possible reasons for the decision on fleas were offered by Cockerill in Consideration of Lords Amendments.[20] The bill was now passed as the Performing Animals (Regulation) Act.[21] During the Lord's Debate on its second reading, Lord Raglan (Fig. 3.7) had again taken the opportunity of expressing his disapproval with it and describing its origins:

> Some four or five years ago a number of hysterical women suffragettes, being in search of an emotional appeal and having no excuse for burning down buildings or molesting the noble Earl [perhaps Lonsdale] opposite, formed themselves into bands and went round to circuses, where they howled and generally made themselves objectionable. Eventually they succeeded in getting the sympathy of the noble Lord [probably Danesfort, the recently ennobled Sir John Butcher], and other members of the House of Commons and a

[19] *Parliamentary Debates* (Lords) 1925, vol. 61, columns 254ff, 473ff and 528ff.

[20] *Parliamentary Debates* (Commons) 1925, vol. 185, columns 967 and 968.

[21] Soon after, the sea lion trainer Joseph Woodward drafted and sent a letter to Cockerill, signed by himself (on behalf of the VAF), Frank Glenister (Entertainment Protection Association), R.H. Gillespie (Moss Empires Ltd), Sir Oswald Stoll (Stoll's Circuit), Walter Payne (Syndicate Halls), Percy B. Broadhead (Provincial Entertainments Proprietors' and Managers' Association), John Swallow and Frank Ginnett (Metropolitan Circus Industry), James Sanger and George Harrop (Provincial Circus Industry), Bertram W. Mills (Olympia Circus, London), E. H. Bostock and G. Tyrwhitt-Drake (Provincial Zoological and Menagerie Industry), W. Savage (Showmen's Guild of Great Britain), and Albert Voyce (VAF). The letter congratulated Cockerill and expressed support for the legislation. In reply, Cockerill said he recognized the spirit of cooperation shown after his taking charge of the unsettled 1924 bill, and he understood the resentment at indiscriminate attacks that had led to earlier dissension over the bill. He thought the bill good for the industry, and there was no excuse now for interrupting performing animal acts because safeguards were in place (*World's Fair* 1925a).

> Committee was set up . . . It would seem that this Bill is required, not so much to protect the animal from the cruelty of the trainer as to protect the trainers from the cruelty of Lord Danesfort and his friends. . ..[22]

## 3.2 Select Committee Witnesses and Reliability of Evidence

Forty-three witnesses were examined during the life of the select committee.[23] Witnesses critical of animal training, performances, accommodation, or transport arrangements were: Charles William Rodgers, unofficial assistant to the secretary, and cashier and investigator of the NCDL; Mrs Albert Bradshaw, a founder of the PADL who had given up keeping pets, was opposed to field sports, and who was engaged in independent private investigations and campaigns—'I wrote a sketch and went on the halls for the purpose of investigating' (*214ff*); Charles Wilson, registrar at head office of the RSPCA, formerly an inspector for nearly 20 years; Charles Butler, music-hall singer; Charles E. Haverly, ex actor-manager; Samuel James, former artiste, theatre manager, and lessee of 15 theatres; Sydney Arthur, booking manager; Athole Stewart, actor on the legitimate stage; Basil Tozer, former theatrical publicity agent; Mrs Lily Grahame, ring girl/lady groom, pianist, and chorus mistress; Mrs Mollie Lawrence, former wife of an artiste; Mrs Margaret Martin, former music-hall artiste; John Oscar Mansley, journalist, who like others had been encouraged to offer himself as a witness by the PADL; John Edward Bird, journalist; Albert Hill, theatre electrician; Sidney Davis, stage carpenter; Jack Thompson, composer and musical performer; Arthur Valentine, amateur conjuror; Miss Jessie Hales, music-hall and theatre performer; Dr Peter Chalmers Mitchell, secretary of the Zoological Society of London; George Frederick Venables, civil servant; and Henry Trengrouse, magistrate.

Those witnesses defending the status quo against intrusive legislation and tending to deny the existence of cruelty were: James Sanger of Sanger's Circus[24]; George Henry Harrop, general manager of the Blackpool Tower Company; George Lockhart, elephant trainer and equestrian director of the Tower Circus, Blackpool; the American Carl Hertz, conjuror and member of the executive committee of the VAF; Richard Henry Gillespie, managing director of Moss Empires and other individual theatres; Jack Russell Vokes, owner of 'Don the Drunken Dog'; Arthur Aitchley, horse trainer; Frederick Trussell, former manager of the London Hippodrome; Edward Henry Bostock, owner of menageries, music halls, and theatres and member for several years of Glasgow corporation, which appointed him senior adviser to Glasgow Zoo (*3162*); L.O. Liesching, amateur conjuror; Garrard

[22] *Parliamentary Debates* (Lords) 1925, vol. 60, column 932ff.

[23] Their selection was in the hands of the chairman (864).

[24] His father, 'Lord' George Sanger, was described as the greatest advocate for and user of all sorts of animals on the stage, and one who set the fashion (Tyrwhitt-Drake 1946, p. 163).

Tyrwhitt-Drake, owner of Garrard's Circus and secretary of Ringlands Zoo Company; Reuben Castang, professional animal trainer; James Burton Cuthbert, trainer; Fred Ginnett, circus proprietor; Harold Laurence Hilliard, manager-handler acting for the owner of Consul the chimpanzee; Harry Rochez, trainer; Henri Rey, owner of a training establishment at Brixton; Everard R. Calthrop, amateur trainer; Bertram W. Mills, circus proprietor; Alfred Roelgin, bird trainer; and Albert Voyce, former artiste, journalist (leader writer of *The Performer*), and salaried chairman of the VAF.

In addition, the select committee received documents in the form of written declarations and letters, firstly from some foreigners claiming to have been witnesses of cruelty to performing animals in France and Holland, sworn at the British consulates (when no British trainers were accused, but charges were made against members of the Hagenbeck family, suppliers of ready-trained animals to British exhibitors), and secondly from members of the Hagenbeck family itself who denied to Bertram Mills charges of cruelty after he had sent them details of such charges presented at the select committee in 1922 by 'old women'. The family also sent Mills a copy of a letter of support from Professor Dr G. Heck, manager of the Zoological Garden at Berlin, who applauded the family's pioneering of kind 'taming' methods and who suspected anti-German prejudice in the British press.

The minutes of the select committee show that witnesses more usually provided opinion rather than the first-hand evidence sought after, and this reflected the secret nature of much of the preparation of animals for performance. A parallel might be drawn with the suspicion in which vivisection has been held by its critics: even authorized inspection has been limited and difficult, and the general public has been disinclined to trust the accounts of those with vested interests in practices they wished to protect from outside interference. The only witness able to offer an objective scientific interpretation of the issues addressed by the committee was Peter Chalmers Mitchell, and he was accepted by many on the committee as their most valuable witness. The emerging discipline of animal psychology made no contribution to the investigation, although during the course of it attention began to encompass the possible effects of stressful environmental conditions on performing animals as well as allegations of physical cruelty. Because of a lack of hard evidence, the findings of the committee were the subject of disagreement between the two 'camps' within it, and prospects for prohibitive legislation began to evaporate.

Mitchell later wrote in 1929:

> Opinions differ as to the amount of pleasure to be derived from seeing animals doing actions that are unnatural to them, and, although there has been much pressure on us both from Fellows of the [Zoological] Society [in this case mainly consisting of trainers, some of whom appeared before the select committee] and from outside, to associate the Society with the defence of circus and music-hall performances in which trained animals are concerned, I do not think that there is any probability of the Society ever doing so. For my own part, I am strongly opposed to 'performing animals', both as a zoologist and as a humanitarian. (Mitchell 1929, pp. 244 and 245)

There is no mention in his autobiography of the select committee or the resulting legislation, but he noted that Hagenbeck was awarded the silver medal of the Society, referring to him as a great showman and trainer of animals, the lucrative part of his business being the formation of troupes and circuses which he sent to every part of the world; but, he added:

> On the training of animals for shows I was never able to agree with him. All my life I have tried to avoid these emotional judgments which interpret the behaviour of animals in terms of human feelings . . . For many years I seldom lost an opportunity of seeing exhibitions of performing animals, but I came to the conclusion that, with the possible exception of sea lions, terror of the trainer lay behind the tricks, a conviction that was not disturbed by the rewards of food that were given.[25]

At the select committee itself, he stated that he would prefer to be cross-examined because he had refused to represent either side despite their pressing requests (2216) and that he would not express a view on total abolition unless he had heard and considered all the evidence (2252), but, with much experience in the audience and behind the scenes, he very much disliked animal performances and believed in a ban in the use of dangerous animals because of the cruelty needed to subdue them (2219, 2302, 2305).

Fellowship of the Zoological Society was open to those who kept and studied animals in the broadest sense, and an example of one Fellow who as a witness presented to the select committee a commercial viewpoint different from that of Mitchell was Garrard Tyrwhitt-Drake:

> My 'first love', and the one which originally brought me into touch with shows, was that of collecting wild animals, which I started on a fairly large scale on the death of my father in 1908 . . . From the menagerie to the side-show (freak animals and suchlike) is a very short step, and thence to the circus was but one step further. Supplying wild animals, horses, ponies, camels, etc., to theatres, pageants, films and spectacles of all sorts was another step that soon followed. (Tyrwhitt-Drake 1936, p. 12)

In 1914 he supplied the menagerie for the World's Fair at the Agricultural Hall in London, when he had negligible experience either of travelling menageries or the show business, but he held the contract successfully until 1928 (Tyrwhitt-Drake 1946, p. 176). When before the select committee, he explained that his Garrard's Circus had begun as an amateur hobby which had grown in size and which he had then to make pay its way, although it was not his livelihood (222). Although he disliked certain tricks, such as the use of horses on revolving tables and the jumps of dogs from great heights, he was concerned less about possible cruelty than by the risk and the temperamental suitability of the individual animal chosen. He thought that the judgement of the public should be relied on rather than a prohibition of certain tricks (421–422).

In his evidence to the second sitting of the select committee, Fred Ginnett explained that he had trained almost every animal, especially the horse, including 750 'mad' military horses for the War Office. He denied the existence of cruelty,

[25] Mitchell 1937, pp. 123 and 124, as originally stated in his introduction to Hagenbeck 1912.

claimed to be unaware of actual convictions within his specialized circle of trainers, and believed that convictions for cruelty had resulted from 'crank' behaviour on the part of magistrates: it was cheaper not to appeal (695, 806ff). Later Albert Voyce prefaced his evidence with an account of the moral principles of the VAF, its association with the London Council for the Promotion of Morality, and its work to prevent the exploitation of young girls in the entertainment world (2427). On being reminded to keep to the point, he asked to be cross-examined. He implied that among others, Sir Charles Hawtrey, an opponent of animal performances, was neurotic (2491) and considered Mitchell fastidious, prejudiced, and ignorant of the music hall's practical business, having himself ten times more working knowledge of performing animals (2447–2449).[26] But Voyce also claimed that some sensitive VAF members had refused to appear on bills that included animal turns involving apparent cruelty, thereby demonstrating that the profession could ensure humaneness (2458ff).

The proceedings of the select committee revealed the possibility of opposition to animal performances resulting from grudges. Bertram Mills was later able to say that he had not suffered in this way. At the Showmen's Guild of Great Britain (SGGB) annual meeting at Caxton Hall in 1937, as president he defended showmen from charges of cruelty and claimed that there had been no comeback concerning cruelty from any who had been discharged by him over many years and who might have held a grievance (*Times* 1937). Nevertheless, the members of the select committee attempted to establish the possibility of spiteful secret informing, especially because such accusers were usually unwilling to stand by their charges for the purposes of bringing an open action; but the reason for this behaviour was given as concern for job security (*78ff*). When this concern no longer applied, witnesses were keen to deny a grievance motive (e.g. *2660*). In the case of Mollie Lawrence, however, some doubt arose because she gave evidence against her estranged husband's turn (*2828*, *2837*).

Another possible motive for objection to animal performances was rivalry between, on the one hand, actors, singers, and so on and, on the other, those presenting animal turns. Frederick Trussell believed that singers and dancers only objected in a jocular way (*2992*). But Helen Trevelyan, herself an actress (and founder of the Fellowship of Faithful Friends), said that performing animals put human artistes out of work (Trevelyan 1936, p. 91). This belief had been examined by the select committee (*619*). James Sanger and George Lockhart referred to the resentment of theatrical people who felt themselves superior when they arrived on the music-hall stage and were jealous if billed less prominently than animal turns or if interrupted by the noise of the animals during a performance (*794*, *1485*). Matilda de Frece (Vesta Tilley) later explained, 'There was a time when the actors and actresses of the legitimate theatre looked down upon the Music Hall, and scorned the idea of appearing there', and also noted, 'There was a big invasion of the

[26] The actor-manager Hawtrey and others signed a letter from the PADL to *The Spectator*, appealing for money to help with the parliamentary campaign (Birmingham et al. 1922).

Variety theatre by legitimate stars, much to the indignation of the Music Hall artistes, who complained that the long sketches they presented kept many Music Hall artistes out of work' (de Frece 1934, pp. 93 and 96). E.H. Bostock attributed the recent 'harsh' prosecution and conviction of his son (who worked in his menagerie but had since died) over the case of a bear to professional jealousy on the part of a cinema proprietor and his operator (*3207ff*). Athole Stewart OBE as an actor associated himself with the legitimate stage and claimed that fellow members of the Garrick Club and actors and actresses of his kind were generally opposed to animal performances, largely for reasons of the cruelty of the unnatural environment in which the animals lived and not because of jealousy (*2296*, *2318*, *2340*, *2352*); however, he was at a loss to explain why these colleagues did not come forward in greater numbers (*2317*). Basil Tozer indicated that it was because theatrical people did not feel entitled to interfere with another part of the entertainment profession (*2478*).

## 3.3 Press and Political Interest[27]

The performing animals controversy was particularly appropriate material for the national press, with popular newspapers, especially providing for readers the excitement of intense altercation between rival groups, pronouncements on a familiar form of entertainment about which ordinary people could be expected to take a moral view, the emotional stimulation of alleged cruelty to animals, and news of the possibility of government intervention. The trade journals occasionally also presented the points of view of those opposing animal performances. 'Nil Desperandum' wrote to *The Performer* explaining his opposition: animals disliked performing and were not created to do abnormal tricks that gave them no benefit (*Performer* 1921j). *The Era* published a letter from E.J. Holland, editor of *The Anti-Vivisection Journal*, in which he criticized the increasing complexity of tricks, performing to timetable, and the stress of transport. He also supplied a campaign leaflet although the editor of *The Era* thought it contained no real evidence, but, rather, reference only to a band of gypsies and a dog left in a box at a railway station and to common ailments (*Era* 1921a). On 11 June 1921, *The World's Fair*, in one of its acknowledgements of the views of the opposition, reproduced a *Times* leading article on performing animals that asked why they amused, suggesting a shortcoming in human nature. Two years later, its columnist 'Cosmopolite' wrote in *The Era* welcoming the idea of RSPCA inspection and licensing and to express sympathy with those who left their seats at unnatural animal performances of dogs, monkeys, or cats, because in Europe (as opposed to Britain) he had seen cruelty, remembering the 'hidden hand' of the whip and the use of techniques such as water deprivation (*Era* 1923a).

[27] Elements of this section are taken from Wilson (2008).

**Fig. 3.8** 'A popular turn.' (1921) Reproduced with permission of Punch Ltd, www.punch.co.uk

A POPULAR TURN.
LIEUTENANT-COMMANDER KENWORTHY HAS PRESENTED A BILL FOR THE ABOLITION OF ANIMAL PERFORMANCES.

The two figures who played the most important part in pursuing new legislation on performing animals were Joseph Kenworthy and Sir John Butcher, both of whom served on the select committee. Kenworthy (later 10th Baron Strabolgi) had retired from the Royal Navy in 1920. He made the first move in Parliament on this issue and had become a favourite victim of satire in the 'Essence of Parliament' columns of *Punch*, where it was suggested that he asked too many questions as Supplementaries and broke procedural rules and that he was a long-winded bore—over-confident, indiscreet, and unrestrained—and had 'bees in his bonnet' (*Punch* 1921b, c, 1923, 1924).[28] His introduction of the Performing Animals (Prohibition) Bill provided a good opportunity for sarcasm about his popularity as a performer in parliament (*Punch* 1921a) (Fig. 3.8).

Like many of those who joined his cause and served on his side in the select committee, he saw no conflict between this interest and his enthusiasm for hunting and shooting. The suggestion of such conflicts was repeatedly made by those defending animal performance, and perhaps they helped to bring to the foreground an inconsistency which today would smack of hypocrisy. At the time, as Kenworthy and many others might have seen it, hunting was regarded as a natural, traditional, socially respectable, healthy, manly, unsentimental, and character-building activity, sometimes involving the invigorating challenge of danger. It should require knowledge and understanding of wildlife in its own context, and it was meant to be sporting, being based on codes, fair play, something approaching equal contest, and with little cruelty because of the quick despatch of the animal.[29] It could also fulfil a diplomatic function abroad, when government representatives relaxed in the neutral but bonding environment of the field.

[28] An example of Kenworthy's clumsiness was his question to the select committee witness, George Lockhart: 'Are you the gentleman who was said to have been killed by an elephant?—No. That was my father' (*1308*).

[29] See McKenzie (2000). In defence of his father, David Kenworthy, 11th Baron Strabolgi, rightly points out that in the early 1930s, big game shooting was still considered an acceptable activity by most people: 'Autres temps, autres moeurs'. Personal communication, 16 February 2008.

F.G. Aflalo, the writer on fishing and shooting, had compared animal performance with field sports at the turn of the century and believed a change in attitudes to duties towards animals resulted from a decline in old-church anthropocentric conceit. And if dominion had been seen as a right, little use had been made of it because of such limited domestication of species, in spite of many failed attempts; therefore, 'If the beasts and birds are not here merely for the purpose of supplying man's wants and necessaries, how much less were they developed to amuse his leisure!' Training performing wild animals to show human dominion had no practical or psychological value, while breaking a horse was acceptable, because useful.

The laws made inhumane distinctions between the treatment of the few domestic and the many wild animals. Aflalo would extend these laws to apply to performing animals (as judges seem to have wished in the case of tortured bears), but not to game because of 'the complications arising from the just claims of sport' based on the need for sustenance and safety, even though the best sportsmen agreed it was cruel. But a clean shot was less cruel than much of the slow natural death in the wild, which was comparable to the cruelty to harshly treated or culpably neglected captive wild animals. In training, as in other fields, the cruel quacks must outnumber the (more civilized) masters. The hot irons seen in performances were indicative of horrors behind the scenes, involving degrading and criminal human deaths and accidents, alongside animal suffering. We should not encourage ghoulishness and voyeurism in spectators nor display animals that obey through fear of the severe punishment they have experienced, making criticism of foreign pastimes (like bullfighting) hypocritical. Meanwhile, he continued, professional secrecy meant there was little of educational value regarding animal psychology. Aflalo only objects to the exploitation of wild and dangerous performing animals (especially in the mixed 'happy families' displays, which he thought a brutal outrage on the instincts of the victims); he approved of trained dogs and the diving horses at Crystal Palace (unlike the Press) because he believed one could not train a dog or horse successfully with cruelty, and, not being wild, they were protected under the law at the time (Aflalo 1900).

In contrast to the healthy sport of hunting, urban animal performances for working-class spectators were therefore criticized for being unnatural, cruel, and undignified (for animal and human): a cowardly advantage was taken of a captive animal beneath a show of fake anthropomorphism and sentimentality, and children at the circus were liable to be deceived and corrupted by such spectacles. Meanwhile, those who depended on and defended animal performance would attempt to organize themselves to protect their livelihoods and way of life from external threats.

There are frequent and enthusiastic references to hunting expeditions abroad in Kenworthy's autobiography, in which he deals mainly with the navy and British foreign policy (Kenworthy 1933) (Fig. 3.9). His political concern for the welfare of performing animals (not itself mentioned at all in the autobiography) appears today as a contradiction that would not have been as well recognized at the time. Butcher, Bowyer, Colvin, and others who took Kenworthy's side over performing animals also shared his enthusiasm for hunting and shooting. However, those who wished to

PANTHER SHOT IN NAWANAGAR STATE

**Fig. 3.9** Joseph Montague Kenworthy, hunting in India in 1926. From Kenworthy 1933. Reproduced with permission of David Kenworthy, 11th Baron Strabolgi

protect their profession from legislative interference lost no time in highlighting the contradiction and perhaps in so doing inadvertently extended future attention to a wider range of problems in human treatment of non-human animals. Kenworthy described hunting on elephant-back in India, in the foothills of the Himalayas in 1926:

> We shot anything which came along . . . the bag comprised wild boar, wolf, four varieties of deer, a porcupine (which I wanted as a trophy), jungle cat, three varieties of partridges and quail . . . I got a good tiger in Alwar State. He charged my elephant and then swerved off and I got in two .420 express bullets. [The Maharaja of] Alwar gave him two more .470s, and I changed rifles and hit him once more through the backbone with a .350 bullet. Each one of those five shots should have killed him. The first four were behind the shoulder, and my second bullet just missed the heart. Yet he ran back thirty yards, and then lay in a nullah. We walked him up with the elephants and found him stone dead, but he had charged back thirty yards . . . His skin measures nine feet nine inches in length (Kenworthy 1933, pp. 244 and 247).

There is no doubt that Kenworthy expended much energy in drawing attention to the cruelties that he seemed genuinely to believe animals suffered in the course of training and performance, if not in hunting. However, given his extrovert character and possible tendency to self-advertisement, a cynic might wonder if he had decided to exploit the controversy for opportunistic political purposes. The press soon began to play its part in the debate, and Kenworthy's name appeared frequently. He told *The Sunday Chronicle* of the use of electricity to make performing animals dance and jump and of heated irons to make wild animals snarl (cited in World's Fair 1921d). Writing in *The Daily News*, he described the performances as degrading to the animal and the audience, and that long before the introduction of his bill, he had heard much from the entertainments profession about the subject (cited in *World's Fair* 1921p). Complacency among his supporters, including the Prime Minister, meant that too few had turned up to vote for the bill, and its opponents then narrowly defeated it by introducing a licensing amendment that could have given immunity to licensed trainers from inspection by the RSPCA and the police. Kenworthy therefore withdrew the bill, but because of the interest generated, the select committee was formed late in the parliamentary Session.

Shortly before the first meeting of the select committee, *The Daily Herald* devoted an editorial to the subject:

> It is high time that the cruelty depended on the comedy of its clowns [involving fake cruelty], not on the tragedy of its dogs and horses lashed, goaded, or starved into cutting dismal capers [involving real cruelty]... No decent person can gain pleasure from seeing a wild beast tamed or a tame beast taught senseless tricks. Such things are an affront to Nature and an insult to the audience ... We hold them in trust from Nature, and to exploit their intelligence and suffering for our profit is a contemptible breach of that trust. We are opposed to all exploitation, whether of human or non-human life. Let us clear the stage and clear our conscience too. (*Daily Herald* 1921b)

The following day's editorial decried the change in the bill from prohibition to 'ludicrous' licensing and advocated the 'easy and effective form of direct action' as practised by the Jack London Club. The editorial was particularly critical of the 'nauseous buffoonery' of Charles Stanton during the Commons debate on the third reading, when he opposed the bill. (Formerly a militant miners' leader, Stanton had abandoned the Independent Labour Party during the war in order to adopt a jingoistic pro-war stance and denounce many of his former colleagues for their pacifism.) On the same day *The Daily Express* published a critical article by Basil Tozer, the former theatrical publicity agent who later served as a select committee witness (*2515f*), entitled 'How performing dogs are taught'. *The World's Fair* later reported that *The Nation* (a weekly) regretted that this original bill had been weakened (*World's Fair* 1923i).[30]

Kenworthy's interest in this subject attracted as much hostility as support. *The World's Fair* published a letter from 'Captain' Lloyd, an animal trainer for 30 years, who called him a crank, claiming there was great love between animals and trainers and referring to the case of the late 'Lord' George Sanger. Sanger had told Queen Victoria of a lion that had died 2 days before his visit to Windsor, when he utterly broke down and the Queen, putting her hand on his shoulder, said, 'The love between animal and trainer must be great indeed'. Heated irons were used 'to gull such asses as the Kenworthy click [probably meaning 'clique'] to make them

[30] The radical-Liberal *The Nation* merged with *The New Statesman* in 1931.

**Fig. 3.10** John George Butcher, Baron Danesfort, by Lafayette, 1926. © National Portrait Gallery, London. Ref. x48840. Reproduced with permission

think that the performance was a most dangerous one'. Finally, to eliminate suspicion of cruel training methods, Lloyd challenged the press to witness the training of specially bought new animals (World's Fair 1921f). Other such challenges were also made directly to Kenworthy.[31] An American trainer, 'Captain' W.K. Bernard joined in the criticism and suggested Kenworthy would be better engaged 'if he looked into some of the affairs of poor children whose fathers gave up their lives possibly under his command' (World's Fair 1921j). Nevertheless, in spite of this antagonism, Kenworthy claimed later that in Parliament he served the interests of showmen, as when he obtained some fiscal concessions on taxes on their road transport; and he records great help in the election of 1929 from the SGGB and from the son of Pat Collins, the showman who had entered parliament as Liberal MP for Walsall in 1922 and resisted new performing animals legislation (Kenworthy 1933, p. 279).

Kenworthy's main ally in Parliament in opposing animal performance was Sir John Butcher, created Baron Danesfort in 1924 (Fig. 3.10). He was less colourful than Kenworthy but shared his attitude to animals. *The Times* recorded:

> He was fond of hunting and shooting, and, like many another sportsman, where animals were concerned he was always their warmest champion; he was vice-chairman of the RSPCA, and an ardent opponent of vivisection. ... All his life he had been a sportsman. He shot and rode well into old age.

He was described as a diehard Tory who earnestly maintained the most extreme Ulsterman's attitude and was no opportunist, being unswerving, of passionate sincerity, outspoken, and having honest conviction (*Times* 1935).

Of those in Parliament who defended the professional interests of the circus and music hall and who attempted to limit any new legislation, leading parts were played by Sir Walter de Frece, Captain James O'Grady, and Charles Jesson.

[31] For example, from Leslie Walter and his horse and goat (*Era* 1921f); and from George Dalmere (*World's Fair* 1921k).

**Fig. 3.11** Sir James O'Grady when Governor of Tasmania, 1924–1930. State Library of Victoria. Reproduced with permission

O'Grady (Fig. 3.11) had addressed a letter to every member of the House of Commons on the performing animals matter. He had helped found the VAF, and in an article 'Triumph of the Animal Trainers', he was described as a great friend of the music-hall profession (*Encore* 1921c).[32] De Frece wrote a letter to *The Daily Telegraph* against the bill,[33] and on the commencement of Standing Committee 'D', he gave all the current reasons for opposing it (World's Fair 1921h). A letter in *The Daily Herald* from Ernest Bell, chairman of the PADL, referred to Jesson's challenge to witnesses in sympathy with the PADL to substantiate their claims about cruelty under cross-examination before music-hall managers and trainers at the House of Commons on 31 May 1921 in Committee Room 6. This challenge was accepted, but the witnesses were shown into Committee Room 12, which was empty, and they were refused entry into Room 6 itself, as if there had been a change of heart over the feasibility of disproving cruelty (*Daily Herald* 1921a). The explanation given was that an MP objected to Bell's entrance, perhaps because of lack of time (Era 1921j). *The Performer* reported this meeting of MPs, organized by O'Grady at the request of the VAF, about the possible government Committee of Inquiry into the training of animals, the result, it said, of 'hypersensitive and morbidly sentimental humanitarians' (*Performer* 1921m). It attributed the subsequent failure of the bill, against which O'Grady had acted on behalf of the VAF, to flimsy evidence for cruelty (*Performer* 1921o). But it now warned of the forthcoming select committee and the legislation that might result. The VAF arranged once more for O'Grady to watch its interests (*Performer* 1921p).

[32] On 30 March 1922, the Executive Committee of the VAF donated ten guineas from its political fund to Jesson's election expenses. On 26 October, the balance of the political fund, £28.0s.1d, was donated to O'Grady for the same purpose, and a letter of thanks for his great service to the VAF in the last Parliament was approved on 9 November (VAF Archives).

[33] See (*Performer* 1921i).

## 3.4 Professionals' Perceptions of the RSPCA, PADL, and Public Opinion

The RSPCA was usually referred to with respect by performing animal interests, in contrast to the resentment felt for other societies, perhaps because these interests were happy with a comparatively lenient and voluntary system of inspection as the status quo. Those with animal training interests were keen both to secure its approval and to criticize any support it gave to new legislation. J.B. Cuthbert wrote to its secretary, Captain E.G. Fairholme, to assure him of the excellent lifestyle of his horse, Beauty (*Era* 1921b); then soon after *The Era* complained that in order to bolster the current bill it had brought a 'desperate' and unsuccessful case against Victor Niblo for cruelty to his parrot (*Era* 1921h). But the ineffectual nature of the RSPCA's ability to ensure humane standards in the treatment of performing animals was revealed in Charles Wilson's evidence before the select committee. It also maintained a rather low profile in comparison with the direct action of the Jack London Club and the PADL. Furthermore, the role of the RSPCA at this time was affected by the fact that it had a vice president in Lord Lonsdale, who was also president of Bertram Mills Circus and who supported the staging of the rodeo in London. But the RSPCA itself did get as far as threatening prosecution if C.B. Cochran's rodeo, managed by Tex Austin, arrived at the British Empire Exhibition at Wembley (*World's Fair* 1924a).[34]

The performing animals controversy revealed some lack of unanimity in the animal welfare organizations, beyond the embarrassing problem of Lord Lonsdale. For example, the Reverend Dr Rosslyn Bruce, a trustee of the RSPCA, vice-president of Our Dumb Friends' League (ODFL) and chairman of the Birmingham Zoological Committee of the Botanical Gardens, in an article in *The People*, and claiming knowledge of animal performance, attacked the 'old women' who opposed it: he confirmed that firmness in all animal training was necessary and that inspection was already adequate. He thought that performing animals were happier than those in the wild, but 'to them [the opponents] aviaries are prisons and dog-leads chains of slavery, and so as a result their homes are wholly destitute of the companionship of birds or animals' (*World's Fair* 1921q).

The PADL engaged in more aggressive lobbying and tactics. A select committee witness, Harry Rochez, owner of a monkey orchestra, complained that organized 'claques'[*sic*] were opposed to performing animal acts even when there was no cruelty and in spite of public enjoyment, and they were not open to reasoned discussion. The trouble started, he said, when the Society for the Prevention of Cruelty to Performing Animals [*sic*, presumably the League] was founded and then pressed for the introduction of the bill (1454/7/8). Albert Voyce said that, while the RSPCA had been greatly respected, two or three extreme societies had affected the attitude of the VAF (2425). Bertram Mills reported that he had received many anonymous letters in 1922, including 33 standard postcards in 2 days. When visiting

[34] In *The World's Fair* of 5 July (1924b), p. 29, it was reported that E. H. Bostock had failed in his effort to get Cochran's rodeo to Glasgow.

Ginnett's Circus at Crystal Palace, he had seen two ladies hissing and booing animal acts and applauding all non-animal ones (2150/1). Such tactics persuaded Harry Rochez to confine his acts to America (1430/3). Reference was made in *The Performer* to a letter from the chairman of the PADL, Ernest Bell, to Joseph Woodward in which he said that in many cases the innocent have to suffer with the guilty (*Performer* 1921k). In the same journal the following week, Woodward noted: '...there are fifteen small societies, competing with the RSPCA, all trying to get as much cheap advertisement out of any cases that can be brought against animal owners just about this time...' (*Performer* 1921l). But organized opposition came from other quarters, too. *The Times* in 1923 reported that a resolution urging the government to support the Performing Animals (Regulation) Bill, the second reading stage of which was then being reached in the House of Commons, had been signed by the following societies: National Union of Teachers, National Union of Trained Nurses, Women's Auxiliary Service, Women's International League, Conservative Women's Reform Association, Union of Jewish Women, British Women's Temperance Association, National Council of Women, National Organization of Girls' Clubs, Young Women's Christian Association, Professional Union of Trained Nurses, Social Institutes for Women and Girls, Women's Total Abstinence Union, Midwives' Institute, Association of Assistant Mistresses, Federation of Women Civil Servants, Women's Freedom League, Women's Village Councils Federation, and RSPCA (*Times* 1923).

*The Performer* objected to the PADL's organizer, Mrs Speedwell Massingham, since she asked sympathisers to write to MPs in support of the bill but without mentioning the League itself, because this would result in protests that would seem to be, but were not, voluntary and spontaneous (*Performer* 1923f). The League had also campaigned with posters in the London Underground, but these were later removed or obscured in anticipation of libel actions (*Era* 1921g). Such posters had encouraged Lily Grahame and George Frederick Venables to offer themselves as witnesses (*2655*, 2327).[35] Other witnesses claimed that their objections had not been stimulated by society propaganda (951). An editorial in *The Era* reported that as a result of the submission of the report of the select committee, Kenworthy said that the League would have to accept keen disappointment and compromise on total abolition in order to enable progress. *The Era* disagreed with him that this keen disappointment would extend to the RSPCA and all animal lovers and called for the removal of red-hot bar and spiked-collar posters, hoping that public reservations were now allayed because of the select committee findings and that managers could make bookings of animal acts without fear of audience disapproval (*Era* 1922c). However, the PADL was only at an early stage in its life and remained uncompromising.

The professional press played a leading part in articulating defensive arguments and propaganda against those who threatened to interfere through criticism, direct action, or new legislation. It also served to rally support, promote unity,

[35] In a general promotional pamphlet dated 12 December 1958, MacMichael claimed that the PADL initiated the original select committee enquiry, when 19 of the 22 witnesses to cruelty were introduced by the League's parliamentary organizer, Mrs Massingham.

communicate tactics, and maintain morale in the profession. Regular attention was paid to the issue between 1921 and 1923, especially in *The Performer* and *The World's Fair*. Those who were blamed for creating it were described as cranks, killjoys, faddists, puritans, teetotalers, mercenary subscription chasers, sentimental, neurotic, pussyfoots, old women, and politically ambitious. When faced with any powerful evidence or argument, the tendency was to blame foreigners (especially Germans). Albert Voyce wrote in *The Performer* of 'sob-stuffed cranks ... comprised mainly of bureaucratic men and neurotic women' who diverted the government from dealing with industrial and trade problems. He noted that there were 15 societies for animals and one for children (*Performer* 1922b). Sufferers of 'virulent crankitis' were 'brothers to the anti-milk drinker, the anti-fruit eater, the anti-laughter, and the anti-what not—a class of men and women who are kill-joys in excelsie' (*Performer* 1922c). *The Performer* criticized the lack of corroboration of allegations of cruelty which had been rejected by the select committee but which were nevertheless reported in the daily press. The agitators consisted of, firstly, those who deserved support against cruelty; secondly, those financially interested in agitation; and, thirdly, 'those who are using the Entertainment profession to obtain public notoriety and political kudos by means of the grossest form of sensationalism'. There was hypocrisy because, animals being servants of man, agitators eat them, and their use for transport was accepted as legitimate. Why was it wrong to train them to amuse as well as use them to feed us? The prohibitionists were a majority on the select committee and their recommendations might ridicule them (*Performer* 1922e).

*The World's Fair* described the reformers as 'a class of people who have neither love for animals, nor even for human beings', noting that the SGGB helped opposition to them 'although not interested in a big way' (*World's Fair* 1921m). B.A. Mason wrote that animals were trained for all manner of reasons and the bill should not discriminate against performing animals, as they suffered no more or less cruelty: the promoters were actuated not by kindness but by prejudice and supported 'class legislation' (*World's Fair* 1923b). Was it cruelty to the cow to eat cheese and not to eat butter? And why were activities like the Grand National not opposed? No trainer would treat animals like that (*World's Fair* 1923d, e). *The World's Fair* reported that Sanger's elephants had been fatally poisoned with arsenic and that he suspected a fanatic wanted to injure him by killing his animals because of his select committee evidence in support of animal performances (*World's Fair* 1923l). The same was suspected by F. Ginnett when his horse suffered a similar fate (*World's Fair* 1923m). *The Encore* described their attacks as a livelihood for the attackers, who failed to provide specific names, dates, and evidence (*Encore* 1921a). It proclaimed: 'We believe that our animal trainers and users have been sadly maligned by these subscription-chasers. We further believe that as a result of the recent big ventilation of the question ideas have been clarified and a great deal of public education accomplished' (*Encore* 1921b). Although *The Era* believed that the proposed restriction of potentially cruel mechanical devices would result in the vexatious revelation of secret magic methods, licensing was no problem because there was no cruelty: the RSPCA had only really found cruelty in itinerant showmen with bears, organ-grinding monkeys, and so on. A licensing regime should reduce the problem of protests, posters, and hissing by 'plants' that had resulted in nervous managers and a reduction in the appearance of animal acts (*Era* 1923b).

The motives and behaviour of the 'cranks' was similarly criticized by those witnesses to the select committee in support of animal performances. It was claimed that people were paid to report cruelty (*2087*) or that they were ignorant or malicious (*2898*). Women would comment 'Oh, how cruel' for something to say (*3037*). People of a nervous disposition imagined things, and the very tender-hearted could deceive themselves over non-existent cruelty (*3135*, 123). Convictions had been based on prejudices and neurotic persons' statements and imaginings (2446). One witness believed there was much nonsense talked about cruelty, but felt he could comment about it objectively because he had his own business and was not an employee (*2419*). Jack Russell Vokes received many anonymous letters about the alleged cruel training of his 'Don, the Drunken Dog'. He believed these came from teetotal people because he had once called his dog 'Anti-Pussyfoot' (*2103*, *2106*).

Soon after the Act was passed in 1925, E.H. Bostock, a former select committee witness, reflected on the controversy from the point of view of his profession, denying that cruelty was necessary in training, that he had ever practised it, and that he had ever come across it in the form described by the 'cranks': 'It is the extremist in every instance that one has to fear, and who looks for a case, and is disappointed if he does not find it, and tries to conjure one up'. He and his brother, F.C. Bostock, 'both loved and valued our animals, or we should not have stuck to the zoological business as he did until his death and I have done for nearly 60 years . . . This, in itself, answers many of the stupid questions that one hears on this subject' (Bostock 1927, pp. 303 and 304).

The performing animals controversy was the only high-profile public animal welfare issue in the 1920s Britain. Ryder has commented on the effect of the First World War in changing lightweight public sentimentality towards animals:

> The mud and blood of Flanders could not have been further removed from precious tales such as these [anthropomorphic ones of animals in Edwardian times]. The stark contrast of reality at its most hellish with the cosy world of Edwardian nursery stories was, sadly, to leave the latter looking discredited. Many serious people could no longer take the subject seriously. (Ryder 2000, p. 126)

He proposed that the animal welfare movement in Britain was 'almost paralysed by its fear of appearing sentimental or extreme' and he saw 'a transition in Northern Europe from strong sentiment and idealism to excessive caution, practicality and a decline in serious interest' (Ryder 2000, pp. 145 and 163). Nevertheless, the PADL would buck any such post-war trend. Ryder further comments that after the war the British working class had not yet taken animal protection to its bosom and did not see nonhumans as a kind of exploited proletariat but rather as the sentimental concern of wealthier members of society. This situation was only to change in Britain in the 1970s. After the First World War (as after the Second), humans reflected more on their own suffering. Between the mid-1920s and the 1960s, there was therefore a lull in animal welfare progress and animal interests were associated with moderate middle-class women, who were often regarded as rather disoriented (Ryder 2000, pp. 141, 142 and 146). For example, in 1933 a report of the Blackwell Colliery Company described the good conditions for the 150 ponies at 'A' Winning Colliery and criticized the boycotting by some buyers of coal of collieries using ponies. This was the result of an appeal by 'distinguished men and women' of the Pit Ponies' Protection Society. The report commented that they were predominantly novelists, musicians, actresses, and clergymen, perhaps suggesting some middle-class effeminacy (Baker 1995).

Supposed public opinion and audience reaction concerning animal performances were cited by both sides to support their case. This resulted in a grey area of counter-claims and contradictions that remained unresolved in an age before objective public opinion polls had become commonplace. The profession was clearly concerned to maintain the popularity of animal acts and became worried by the effects of the critical opposition. 'Captain' Permane, the bear trainer, wrote to *The Performer* and pointed out that *The Daily Mail* claimed nine out of ten did not want to see animal turns (*Performer* 1921r). *The Performer* soon after published a half-page notice from 'Captain' Woodward, advising that:

> Proprietors of Performing Animal Acts should now take the public into their confidence and instruct them, on common sense grounds, regarding the libellous accusations that have been levelled against ALL animal trainers. Patrons should be invited daily to inspect the animals themselves, questions invited, and short demonstrations given to show the simplicity of training by reward and the entire needlessness for anything approaching cruelty. (*Performer* 1921n)

*The Encore*, under the heading 'Performing Animals Cranks Routed', claimed that opposition societies were interested in financial gain through prosecutions leading to more public support, subscriptions, and contributions (*Encore* 1921b). An editorial in *The Era* called for the suppression of hostile posters damaging to business and admitted that hissing made managers timid about booking animal turns: this was the cause of the decline in animal acts, not public antipathy (*Era* 1922a).[36]

Interviewed in *The Era*, 'Captain' Woodward said that the public was the best judge, that animal performances were better than some dull non-animal acts, and that the hissing 'plan' was only carried out by six or seven people (*Era* 1922b). The select committee was told by some witnesses that audiences had not been critical (*210–211*, *3055*), that for up to 30 years they had received no complaints from a very large public (*1141*), or that there had been no 'walkouts' (*1271*). (One reported that there were, on the other hand, complaints about cruelty to the audience from acts not involving animals (*1906f*).) Perhaps, as other witnesses believed, this was because audiences were deceived and ignorant of the cruelty that lay behind what

[36] But Charles Wilson, giving evidence to the select committee (*390*), believed that rapidly growing public antipathy was the real cause.

entertained them (*258*, *667–668*, *708*). It was claimed that an audience could not see wounds covered by an animal's hair (863), or that, in the case of Consul the chimpanzee, it was reassured that he had pyjamas and a comfortable bed (912). In her criticism of the legislation, Trevelyan later stressed the continuing need to change public opinion and quoted John Galsworthy as wanting to effect a 'gradual educational infection' against performing animals (Trevelyan 1936, p. 115). He had found it ironical that animal lovers should tolerate or enjoy this product of misery (*Literary Digest* 1920, cited by Joys 1983, p. 58).

James Sanger, as a witness, thought it is better to legislate only against what the public considered as cruel acts rather than against all animal acts, and he agreed to provide a list of tricks he thought should be eliminated as subject to public doubt (*926ff*, *1135f*). Indeed, some witnesses defended animal acts by asserting that audiences were in reality critical and perceptive and that they thereby ensured that cruelty was avoided. It was alleged that an audience could detect fear in performing animals (*992*), as well as signs of cruelty behind the scenes, and would boycott transgressors (*1277*, *1463ff*). Audiences were said to be suspicious of certain tricks (*1406ff*), and reputable professionals would not want to offend them (*1422*), if only because of the effect on earnings (*1715*). Consequently, although whips were supposed only to be used for ornament (*2966*), performers were sometimes asked not even to use a switch (*2963*). Moreover, music-hall audiences were always changing and would therefore not become complacent about dubious performing animal acts (*1933*). T.H. Gillespie later proposed that circus and music-hall turns encouraged audience sympathy with animals and that the audience was quick to object to anything questionable (Gillespie 1934, pp. 146 and 147). Some witnesses confirmed negative reaction from the general audience, involving more than just agitators: in the case of a whipped pony, it protested for 5 or 10 min (919), and there were spontaneous objections to what was considered an unreasonable trick for an elephant that did not result from simple prejudice against animal acts (2348–2350).

It was of course claimed by many select committee witnesses that audiences were supportive of and enthusiastic about animal acts and that there was a demand for them. The acts' actual popularity was said to be crucial to the profession (*837*) and that the public would resent a ban (*3169*), seals, dogs, and birds being especially favoured at the time (*1807*/*1895–1896*) and the circus being more popular than 10 years previously (416); and there had been counter-protests from the audience against protesters (1449)—neither had propaganda affected the popularity of the menagerie (*3171*). The public wanted comedy rather than complexity, for example, in the 'drunken dog' act (345–346), and preferred 'homely' to sensational animal acts (2196).

However, some public enthusiasm might have resulted from a degree of callousness or voyeurism in an audience. Wild animal acts that pandered to the voyeurism of spectators had already been criticized at the turn of the century on both sides of the Atlantic and would later be (Aflalo 1900; Peck 1906, pp. 76–77; cited by Joys 1983, p. 37; 2011, pp. 365–366). For example, all-time attendance records were broken at the Public Hall in Cleveland, United States, when the big cat trainer Clyde Beatty staged a 'committee' of four brother lions to punish Sammy, a renegade lion who had killed a lioness (*Time* 1934). Tyrwhitt-Drake said that the public also liked some unnatural acts he did not (295), and another witness to the select committee said there was no audience protest at a cruel dog act at the London Palladium (2382/2405). Trevelyan thought that public protests by individuals were potentially counterproductive, as the audience might resent having its pleasure spoilt, regarding the protesters as deranged and the act correspondingly reasonable (Trevelyan 1936, p. 28). Stanley W. Wathon, managing director of the Stanley W. Wathon Universal Variety and Circus Agency, wrote to Edmund MacMichael of the PADL on 31 May 1932 assuring him that it was not he who wanted to book wild animal acts. He would avoid them if possible, but for the 'blood-thirsty' public as at the Tower Circus, Blackpool, where attendances had dropped without them (MacMichael Papers).

Voyeuristic public enthusiasm for circus confrontations with wild animals in the United States had been frustrated for a few years by a decision taken by Ringling Bros. and Barnum & Bailey Circus (RBBBC). It suspended wild animal acts between 1925 and 1929, as a response to pressure from the Jack London Club membership and press objections, explaining: 'There has been enough criticism by the public of wild-animal acts to warrant us in withdrawing them, as a quite common impression is prevalent that tigers, lions, etc., are taught by very rough methods, and that it is cruel to force them through their stunts' and 'The public seems to prefer animal acts in which the animals themselves seem to take an interested and playful part, as do dogs, seals, horses, elephants, etc.' (Mighetto 1991, p. 69; Chindahl 1959, p. 209; *Billboard* 1924). Other circuses were encouraged by the press to take the same action, but the opposite happened. Because in the 1920s American fairgrounds were concentrating on mechanical rides, their many once-popular wild animal trainers joined other circuses than RBBBC and contributed to a heyday for them: 'Fierce competition among the circuses and trainers caused many of these men and women to push the envelope in recklessness and daring, and led to the wild animal act's extreme popularity and dominance over other traditional circus acts' (Joys 2011, p. 292). RBBBC therefore ended its suspension and hired Clyde Beatty. Beers argues that circus protests had in the meantime become weakened by a reaction from owners and employees of RBBBC who had formed in 1926 a public relations group, the Circus Fans Association of America. As in the UK, the circus was defended as healthy and inexpensive entertainment. The effectiveness of the Jack London Club was also later reduced by the public distraction of the Great Depression and the need for cheap entertainment in hard times, and it receded into obscurity (Beers 2006, pp. 107–108).

In the 1920s, Clyde Beatty had developed his fighting act by unnaturally mixing lions and tigers on a very large scale, further increasing the conflict by mixing the sexes, believing that wild animals could be trained but never tamed (Joys 1983, pp. 107–108). A writer to *The New York Times* in 1931 objected to these confrontational acts as exceptionally cruel, with the cats being loudly clubbed, causing children to cry. Ringling and Beatty thereupon announced that the act would be changed and objectionable parts removed (*New York Times* 1931a, b, cited by Joys

1983, p. 116), although Joys records that other newspapers as well as the curator of mammals at the Bronx Zoo saw nothing objectionable in the performance, and 'contrary to the beliefs of some sentimentalists, cruelty is not an efficient circus method' (Ditmars 1933, cited by Joys 1983, p. 117). However, when a new member of the Ringling family took over the circus's direction in 1938, he introduced a lasting policy of presenting milder European-style acts instead of the brutality he and many others saw in the dramatic provocation of the fighting act, which thereafter declined in the United States (Joys 1983, pp. 136–140, 147). Meanwhile the heavily scarred Terrell Jacobs defended his 'whip and pistol' act against the long, jabbing forks attributed to training for the foreign acts (Joys 1983, pp. 150 and 153). As if acknowledging past methods, Ringling now advertised Alfred Court's mixed acts as being produced by 'understanding, kindness and patience . . . without use of gunfire, crowbars, electric persuaders, pyrotechnics . . .' (*New York Times* 1940, cited by Joys 1983, p. 159).

## 3.5 Organized Defence of Animal Performance[37]

As a result of the public influence of the Jack London Club, widespread criticism of training animals for popular entertainment had become well established by the 1920s, both in Britain and the United States. A predictable reaction was an attempt by those on the receiving end to deflect such criticism, as in the creation of the Circus Fans Association of America. This reaction varied in quality and motive, but included some well-considered responses, as from William Hornaday, conservationist and director of the New York Zoological Park. Hornaday's opinions can be gauged from his *Minds and Manners of Wild Animals* of 1922. He regarded the followers of Jack London as limited in their effect, sentimental, and misguided. He believed it was right that animals, like men and women, should work for a living, in captivity as in the wild: 'We hold that it is no more "cruelty" for an ape or dog to work on the stage than it is for men, women, and young people to work as acrobats, or actors, or to engage in honest toil 8 h per day. Who gave to any warm-blooded animal that consumes food and requires shelter the right to live without work? No one!' But he went on to propose a wild animals' bill of rights, including a legal requirement for good standards of accommodation and maintenance in captivity; and 'an animal in captivity has a right to do all the damage to its surroundings that it can do, and it is not to be punished therefore'. He claimed: 'It is the opinion of the vast majority of civilized people that there is no higher use to which a wild bird or mammal can be devoted than to place it in perfectly comfortable captivity to be seen by millions of persons who desire to make its acquaintance'. Nearly all captive wild animals were either born in captivity or captured in their infancy and had no

[37] Elements of this section are taken from Wilson (2009b).

idea of freedom to pine for. Some of the articles in his bill of rights referred specifically to performing animals:

> Art. 19. While the use of trained animals in stage performances is not necessarily cruel, and while training operations are based chiefly upon kindness and reward, it is necessary that vigilance should be exercised to insure that the cages and stage quarters of such animals shall be adequate in size, properly lighted and acceptably ventilated, and that cruel punishments shall not be inflicted upon the animals themselves.
>
> Art. 20. The training of wild animals may, or may not, involve cruelties, according to the intelligence and the moral status of the trainer. This is equally true of the training of children, and the treatment of wives and husbands. A reasonable blow with a whip to a mean and refractory animal in captivity is not necessarily an act of cruelty. Every such act must be judged according to the evidence.
>
> Art. 21. It is unjust to proclaim that "all wild animal performances are cruel" and therefore should be prohibited by law. The claim is untrue, and no lawmaker should pay heed to it. Wild animal performances are no more cruel or unjust than men-and-women performances of acrobatics. Practically all trained animals are fed and tended, they welcome their performances, and go through them with lively interest. Such performances, when good, have a high educational value, but not to closed minds.
>
> Art. 24. The use of monkeys by organ-grinders is cruel, it is degrading to the monkeys, and should in all states be prohibited by law. (Hornaday 1922, pp. 52–53, 204–205)

In Britain at the time, witnesses summoned before the select committee, as well as its political members, broadly reflected two sides with little common ground for agreement. Those who resisted new legislation represented the commercial interests and claimed to defend innocent working-class entertainment from the 'killjoys' that they believed threatened a variety of freedoms without any evidence or justification. It was inevitable that allegations concerning the presence or absence of cruelty in training, accommodation, transportation, and performance for the circus and music hall should involve an organized professional response. As with vivisection, professional secrecies in the work and methods of animal trainers excited suspicion and fuelled the activities of the campaigning societies. It was therefore necessary for the professional interest to justify its work and to oppose the threat of new, restrictive legislation. This defence was mounted by the VAF,[38] the SGGB,[39] their parliamentary representatives, music-hall owners and managers,

[38] The VAF campaigned for better contracts and working conditions in the music-hall and variety profession and led the fight against the all-powerful theatre managers, as in the 'Music Hall Strike' of 1907. By early 1923 it had 4,000 members, and during 1925 it added under its title: 'incorporating actors, concert and circus artistes'. It amalgamated with the British Actors Equity Association in 1957. Its official organ, the weekly *Performer*, ran from 1906 to 1957, its editor in 1921 being John Warr.

[39] The Showmen's Guild of Great Britain was originally known as The United Kingdom Van Dwellers' Protection Association, and from the early 1900s it was usually referred to simply as the Showmen's Guild. It was founded in 1888 or 1889 as the trade association of travelling funfairs, became a registered trade union in 1917, and between 1918 and 1927 its part-time general secretary was William Savage. Frank Mellor, its treasurer, also served as editor of *The World's Fair* (founded in 1904) between 1909 and 1930. Pat Collins, Liberal MP for Walsall between 1922 and 1924, was a co-founder of the Guild and succeeded 'Lord' George Sanger as president, serving between 1909 and 1927. Collins was an outdoor amusement caterer and theatre and cinema

circus and fairground proprietors, and individual trainers. Those who defended animal performance were clearly concerned for livelihoods and objected in principle to the ignorance, prejudice, and spirit of interference they saw in their opponents, claiming their work to be legitimate and educational entertainment of importance to employment, the economy, and the very future of the music hall and circus.[40]

Although the energetic campaigning efforts of a few individuals within the VAF and SGGB had to overcome lethargy in their general memberships, with the support of representatives in Parliament and full coverage of the professions' side of the issue in their journals, their response ultimately achieved a weakening of successive bills, leading to the Performing Animals (Regulation) Act of 1925, based on licensing rather than prohibition. On 2 June 1921, the Executive Committee of the VAF had resolved that the actions of its permanent officials in supporting the opposition to the prohibiting bill be endorsed; and on 21 July it instructed Monte Bayly, the VAF's national organizer:

> To watch the interests of artistes on the Select Committee appointed to enquire into the Cruelty alleged against the Trainers of Performing Animals and to use his endeavours to bring about a co-operation between managers and the VAF both financial and otherwise, such co-operation to be if possible, without the aid of legal representatives (VAF Archives).

The SGGB relied mainly on propaganda through *The World's Fair* and *The Era*, while the VAF, speaking through *The Performer*, organized a more active opposition, but one which did not appear to stimulate much energy among its professional readership.[41] Performing animal acts were a specialist area and were resented or disapproved of by some in the profession because there was competition for popularity and employment with actors, singers, dancers, and others on the music-hall stage; and so the VAF and its specialist interests had to call on the

---

proprietor. He was concerned in the early development of steam roundabouts, and helped to introduce and popularize the cinema industry in Britain.

[40] An effect of the war had been to reduce the availability of animals for the music hall, which, together with growing competition from the cinema and home radio, accelerated its decline and made its professional representatives especially sensitive about the added impact of any bad public perception caused by the controversy.

[41] *The Era* served as a trade paper of the theatrical press and travelling fairs, but was superseded by *The World's Fair*, a trade paper for travelling fairs and circuses, and by *The Performer*. In correspondence of 4 August 1924 from W.H. McCarthy, secretary of The Performer Ltd, to F. Herbert, secretary of the VAF, it was confirmed that the VAF had significant financial and executive influence on the publication, that 'complete control of all political matter in *The Performer* is vested in the VAF officials', that around two pages each week were earmarked for free VAF propaganda, that there was a strict policy to support the VAF in every possible way and that 'articles or items of news in criticism of the Federation are not to be printed in the paper'. However on 15 August 1927, F. Russell, managing director of The Performer Ltd, wrote to the Executive Committee of the VAF that being the VAF's official organ was no longer a commercial asset because advertisements and subscriptions from VAF members had fallen off, and because 'many people who advertise freely in the other trade papers refuse to advertise in *The Performer* because they do not agree with the VAF or have some grievance with the Organisation' (VAF Archives).

principle of united defence against external interference. In the ‘Variorum’ column of *The Performer* by ‘Artifedera’, a reply was given to a member of the VAF who had questioned its representation of performing animal interests:

> It is the duty of the Variety Artistes’ Federation to protect the interests of every section of those earning their living as programme items. One section of our membership has been attacked, and as an Organisation we have a legal and moral obligation to protect them. To us it matters not whether the ‘cranks’ are ‘stunting’ against ‘red-nose comics’, ‘tango-dancers’, ‘acrobats’, or ‘sketches’, when the occasion arises, we will be there to do all in our power to see that the class affected gets a square deal. The VAF represents everybody who matters in British vaudeville; it represents all shades of opinion, likes, dislikes and prejudices. (*Performer* 1922c)

There was also concern that other professional associations should not compromise the campaign of defence: *The Performer* reported in 1922 that the VAF was protesting to the Actors’ Association because it allowed a letter in *The Actor* from the chairman of the PADL (*Performer* 1922a). In 1923, the Magicians’ Club joined in the opposition to legislation after campaigners Joseph Woodward, Albert Voyce, and James Sanger attended their meeting to represent animal performance interests (*Performer* 1923c).

The VAF attempted to show that it shared general concerns about the humane treatment of animals: ‘The VAF is against cruelty to all animals—more so against cruelty to performing animals’ (*Performer* 1922f); and *The Magnet* complained of the specially harsh measures threatening the profession: ‘It is very much like taking all the horses off the roads or out of the fields because one man happens to be charged with cruelty’ (‘Recorder’ 1921). It was frequently claimed by professional interests throughout the controversy that there were other forms of animal suffering that deserved attention instead. In a letter to *The Performer*, J.B. Cuthbert (a VAF member and owner of the horse ‘Beauty’) suggested that animals suffered more in other ways, and he reassured Lord Leigh (who had addressed the ODFL in Birmingham) that there was no cruelty in animal performance (*Performer* 1921c). Robert Bernard criticized cruelty in racing and hunting and defied anyone to prove there had been any sign of cruelty or neglect in his dog act of over 20 years (*Performer* 1921g). In an editorial in *The Era* complaining of the unfounded charges made before the select committee against magician Carl Hertz’s methods and against the cruelty of a ‘monkey orchestra’ and criticising the prevalence of propaganda posters during the select committee inquiry, a comparison was made with the cruelty of hurdle racing and fishing; and performing animals were said to enjoy ideal conditions off duty (Era 1921k). A later editorial argued that to avoid prejudice, and if animal performances were to be interfered with, regulation should also apply to military tournaments, the Grand National, educational films, and dog shows (*Era* 1923d). The VAF was itself accused of diverting attention to other forms of cruelty by ‘Londoner’ of *The Daily Graphic*, who quoted someone claiming to work for the VAF as a press agent in order to confuse the issue for public relations purposes. This was denied (*Performer* 1922f).

Professional journals such as *The World’s Fair* would also report parliamentary debates on the controversy and the proposed legislation associated with it (e.g. *World's Fair* 1923f, g, h). In its 'What we think' column, *The World's Fair* had alerted its readers to the introduction of the first Performing Animals (Prohibition) Bill, described as being handled by the Canine Defence League, and warned of the need to organize against it.

> For many years we have tried to rouse circus readers, animal performers and others to combine to protect the interests of their profession and to combat the numerous attacks that have, from time to time, been made upon them. (*World's Fair* 1921a)

This was followed soon after in *The Performer* by 'Animal Trainers, Look-Out!' by Albert Voyce, in which he reacted to an editorial in *The Anti-Vivisection Journal* on large animal circuses in London during the school holidays. He characterized this editorial, along with a letter he had seen from Captain E.G. Fairholme, chief secretary of the RSPCA, as another vague charge of cruelty without names, dates, or places. He also referred to illegal PADL posters, subsequently withdrawn, alongside those of a circus in Birmingham, and complained that:

> The Pussyfoot-minded persons behind this movement and others of a similar kind, have little or no knowledge of the profession they denounce ... Certain intensely sincere but parrot-brained persons [and] fanatical busybodies [had as their main object] to condemn and not to consider.

He asked why they were not against cattle transport instead and warned that unity was needed against the agitators and 'crank legislators' and their imminent prohibiting bill (*Performer* 1921a). In the following week's edition of *The Performer*, a letter from the circus proprietor, James Sanger, confirmed the need to form an opposition (*Performer* 1921b). *The World's Fair* reproduced a letter to *The Daily Express* from 'Lord' John Sanger explaining that animal performances were picked on, rather than hunting or fishing, because they were a small area easily attacked. There was no cruelty in training, because 'animals cannot be trained to hide fear' from an alert audience, and prohibition would mean unemployment at a critical time (World's Fair 1921c).

*The World's Fair* reported that the SGGB would oppose the bill and published a letter from Fred Wilkins, a circus proprietor, advising that Ernest Bell of the PADL should not interfere with the work of the RSPCA and that the posters against his circus in Birmingham had been covered up because they were libellous (World's Fair 1921d). 'Captain' Woodward asserted in *The Performer* that slanderous propaganda would fail in Britain, as it had done in America, and 'Captain' Permane, the bear trainer, also confirmed that no cruelty in the training of animals was necessary and that his were long-lived. But E.J. Holland, editor of *The Anti-Vivisection Journal*, replied to Albert Voyce's article, saying it did not respond to questions of transport and accommodation or of prosecutions, indicating that he must be ignorant (*Performer* 1921d). During March, because the prohibition bill had been read a second time unopposed, *The Era* repeated that immediate action was imperative. It reported the popularity of the SGGB's petition against the bill, described the numerous 'challenges' issued by trainers to critics to find cruelty in special training demonstrations, criticized unsubstantiated claims of cruelty as by W.C. Pilley in *John Bull* magazine, and mentioned the request by the Entertainment National Industrial

Council to the government to oppose the bill (*Era* 1921d, e). The Council, at its meeting on 18 March 1921, representing various entertainments organizations and unions and chaired by Sir Oswald Stoll, had heard from Albert Voyce that opponents were against the theatre and music hall generally, and it resolved that the existing laws were sufficient. Meanwhile the VAF passed a resolution against the bill and called members to protest to their MPs about it. 'Captain' Woodward had also stressed the educational value of animal acts (*Performer* 1921f). Before the month was out, William Savage, general secretary of the SGGB, announced that F. Millar, its parliamentary agent, had been authorized to arrange to help defeat the bill (World's Fair 1921e). 'Captain' Woodward called for a protest meeting and evidence from trainers to refute 'faddist' accusations of cruelty and devices of torture, for testimonials, and for the lobbying of individual MPs (*Era* 1921f).

However, James Horace McKenzie criticized the division of interests in the SGGB that had resulted in animal trainers not giving their full resources as members to fight the bill and relying too late on others in the Guild when unity was needed. Fred Wilkins appealed to all sportsmen to join in the campaign against the bill: 'An animal is like a wife: she is only trained by love and affection' (World's Fair 1921g). 'Captain' Woodward sent out a letter to the professional periodicals about arrangements to resume parliamentary defence work, calling for any available letters of approval from humane societies, Watch Committees, and so on to be supplied to the select committee as evidence. Dandy George advocated the licensing of animal trainers, to prevent job losses. He noted that with prohibition, no animals could be trained for the screen either, and the picture business would be hit (*Performer* 1921h). While in Parliament Brigadier-General Cockerill said at Standing Committee 'D' that animal performances were an advantage to the community because they showed what could be done by kindness, 'Captain' Woodward believed the bill had gone so far because of the scattered organization of the profession and the lack of professional unity against it (World's Fair 1921i). *The Encore*, a music hall and theatrical review, soon praised his activism: 'Under the leadership of 'Captain' Woodward, the animal trainers were rallied and press and public alike were kept informed as to the rights and wrongs of the question' (*Encore* 1921c). *The World's Fair*, describing itself as the first paper to draw attention to the need for action, continued to emphasize the need for a protection association, with more support from members being given to those already in opposition (*World's Fair* 1921m). *The Daily Herald* and Northcliffe Press[42] were especially opposed to performing animals, but 'months of voluntary work undertaken by Mr Joseph Woodward and latterly by James Sanger, culminating in the effective organizing of the opposition in the House of Commons by the VAF' helped to defeat the bill (*Performer* 1921q). Albert Voyce still complained of the dangerous general apathy in the face of it (*Performer* 1921s).

The meeting of MPs organized by O'Grady at the request of the VAF about the possible government Committee of Inquiry into the training of animals was

[42] Its owner, the Hon. Esmond Cecil Harmsworth, later 2nd Viscount Rothermere, was to become a life member of the PADL.

attended by 60 MPs and, representing the amusement industry, Sir Oswald Stoll, R.H. Gillespie (managing director of Moss Empires), James Sanger, Percival Percival (representing the Provincial Entertainments Proprietors' and Managers' Association as secretary) (Era 1921j),[43] Joseph Woodward, Albert Voyce, E.H. Bostock, G. Tyrwhitt-Drake (representing the SGGB), George Harrop (Tower Circus, Blackpool), and Bertram Mills, all of whom opposed the bill (*Performer* 1921m). It was claimed that evidence against performing animals was based on the reports of four or five hysterical people, offering old, limited, and dubious evidence. Stoll pointed out that 'Out of the 288,000 convictions obtained [by the RSPCA] over 80 years ... only 21 cases were connected with places of entertainment'. Referring to the 'green circular' of the PADL, it depended on 'imaginative, hysterical, probably neurasthenic witnesses'. Percival Percival called for no legislation without enquiry (*Era* 1921i). The VAF arranged for James O'Grady to represent its interests (*Performer* 1921p), and during the meetings of the select committee, a VAF representative was present to take notes of any unsubstantiated evidence (*616*).

Following the failure of the prohibiting bill, the select committee was appointed and met between 19 July and 11 August, 1921, and the following year between 2 March and 9 May. *The World's Fair* would regularly reproduce some of its minutes (*World's Fair* 1921n) and, while it was convened, complained that some of the Press were treating the controversy as a 'stunt', accepting the accusations of critical witnesses and with biased, and partial reporting of some who, often under a *nom de plume*, claimed an entertainments connection. It was wrong to say that any critical employees or professionals feared boycott if they did not preserve anonymity, and so the lack of attributable criticism must reflect an acceptable status quo. Although there were few animal trainers in the SGGB, they deserved support (*World's Fair* 1921o). In 'The Cruelty of Crank Legislation', Albert Voyce attacked a variety of restrictive legislation. The VAF had a 'deep rooted objection to the unfair principle of allowing people who have no knowledge of an industry to control the destinies of those who are engaged in it': they were victims of the 'mawkish and lying propaganda of the cranks' (*Performer* 1921t). The Executive Committee of the VAF, on 15 September 1921, thought the threat so serious that it was desirable to have a direct representative of its own in Parliament. Voyce did not want to stand, but Bayly was briefly considered (VAF Archives). However 'Captain' Woodward told *The Era* he thought the select committee had been a good idea because it showed how satisfactory was the management of animal performances (*Era* 1921l),[44] and Bertram Mills confidently invited select committee

[43] In a letter to *The Manchester Guardian* (which had regretted the failure of the bill) of 9 June, Percival Percival called for a select committee of enquiry instead of prohibition.

[44] At its executive meeting the following day, the VAF decided that the balance of funds donated by animal trainers and others in defence of animal performance (£193.9s.1d) be handed to Woodward for disposal (VAF Archives).

members to attend the International Circus and Christmas Fair at Olympia (*World's Fair* 1921r).

Because the work of the select committee had not been completed in 1921, it met again in 1922, and arrangements for the defence of professional interests were intensified. The SGGB called for more animal trainers to join it in order to help in the opposition (*World's Fair* 1922a), and its secretary invited members to put themselves forward as witnesses for the committee (World's Fair 1922b). Meanwhile, the VAF angered Brigadier-General Colvin, chairman of the select committee, by questioning its impartiality. The 'Variorum' column of *The Performer* by 'Artifedera' reported Colvin's objection to the letter he had received from Monte Bayly, VAF organizer, concerning the supposed distortion by some members of the select committee of past witnesses' evidence during the later questioning of new witnesses, thereby reinforcing unsubstantiated allegations criticising the training and performance of performing animals, and concerning the 'selection' of witnesses to this end, excluding VAF-sponsored trainers and so on, who could speak with authority. Bayly had asked whether the select committee was therefore eager to seek the truth. Colvin described this as a most improper letter because it reflected on the impartiality of the select committee, and the Speaker agreed. 'Artifedera' continued that last Session the VAF organizer had officially handled the case for the defence:

> It was to him the House of Commons officials referred as to who the witnesses on behalf of the accused were to be. All briefs, statements, names and addresses, and particulars were typed on VAF official notepaper and supplied to each member of the Committee. We were consulted in every way [with] utmost consideration. This Session the attitude is a complete *volte face* ... we have been absolutely ignored ....

... and witness nominees were rejected by the clerk of the select committee on 6 March. The VAF was told 'unofficially' on 7 March that no more witnesses from it were desired, as 'independent' witnesses were needed. A strong protest letter was therefore sent, because the 'accused' had to be allowed to defend their livelihoods, as was their right. On 9 March, three VAF spectators were only by luck called as witnesses, without advance notification by the clerk. 'Artifedera' confirmed that the VAF was not at all sorry about the letter. E.J. Hedenborough, clerk to the select committee, in a reply to Bayly on 10 March, said the committee unanimously found his letter offensive, and the Speaker thought it most improper: 'The Chairman wishes me to point out that this letter may ... prejudice, in the eyes of the Committee, your organization and the attitude which it represents'. It was also denied that trainers had been excluded from the witness chair, but the VAF remained unrepentant (*Performer* 1922c). This episode was referred to in the minutes of the select committee of 23 March (2499ff), when Voyce was challenged about its reporting in *The Performer*. He agreed it was discourteous but had then (rather than currently) thought its contents true, reflecting a 'smashing injustice' [that] 'got us moving'. But only the day before, he was maintaining the offensive in an article called 'Our fight against the faddists': 'Even Parliament itself is overrun with faddists—crank legislators who seek to impose irritating restrictions' because of the cruelty of a few 'black sheep'. He gave the assurance that present-day artistes would not tolerate on the same programme as themselves the appearance of a trainer who was guilty of cruelty: 'Reform, like charity, should begin at home. We would rather have our industry reformed from within than permit those who have no knowledge of it to make its laws'. A special conference of British animal owners and trainers was to be held in the VAF Board Room on Sunday, 26 March, with select committee members invited to agree a measure to improve the protection not only of performing animals, but also of trainers against the 'mawkish propaganda of certain crank societies'. Elimination of certain tricks was possible, but not general prohibition.

> We wholly agree with Captain Joseph Woodward that the public is the best censor in a matter of this kind, and that their undoubted appreciation of selected animal shows proves their innate popularity with the general run of variety patrons ... [above all] we object to the attempt of an outside minority to run any section of our industry. (*Performer* 1922d)

After the submission of the select committee's final report, there was some delay in the framing of new legislation, and in the time before the new Act was passed in 1925, the profession monitored the progress of intermediate revised bills. *The Performer* referred to Colvin's bill and the possible restriction on the use of some animals, to the use of mechanical appliances in conjuring, the introduction of registration, and to the unwelcome idea of court prohibition orders on complaints not just by the police but also by officers of any authorized society. It was confirmed that E.H. Bostock, James Sanger, Joe Woodman, Percy Broadhead, Bertram Mills, Fred Ginnett, and others had formed a plan of campaign to oppose this bill and that Monte Bayly as organizer of the VAF and acting as parliamentary agent had the matter in hand (*Performer* 1923a).[45] B.A. Mason in *The World's Fair* attacked the 'spoof and hypocrisy' of the bill, calling it discriminatory and regretting that opposition to it had been 'deplorably negligent': it was necessary for all to protest to their local MP before the second reading. All should get involved, not just animal trainers and exhibitors. 'What We Think' on the back page of this edition said the same about fighting the bill (*World's Fair* 1923c).

During a debate on the second reading on 23 March 1923 of the Performing Animals (Regulation) Bill,[46] O'Grady moved for its rejection.[47] *The Era* printed an editorial calling its readers to beware the third reading of the bill (*Era* 1923c), but *The Encore* (referring to the bill's success at its second reading, when it was carried by a vote of 169 to 35) published an accommodating editorial by A.L. Bayley, although he described the bill as unnecessary, because the charges made before the select committee had been unfounded. Slum children were a more legitimate object of concern, and this was:

[45] Bayly had been appointed as the VAF's parliamentary agent at the Executive Committee meeting of 2 June 1921 (VAF Archives).

[46] *Parliamentary Debates* (Commons) 1923, vol. 161, columns 2961–3020.

[47] O'Grady was blocking the bill on behalf of the VAF (*Performer* 1923b).

> One of those busybody pieces of legislation which annoy those whom they are intended to cure . . . but when all that has been said we still think that a great deal of the hectic agitation against the Bill put up by the professional and vested interests is somewhat beyond the mark of common sense. If it is true that animal training does not entail any cruelty, then the Bill cannot do any injury.

Only the guilty could suffer, and there was no sympathy for them. The bill did not threaten the vast capital invested in circuses and menageries:

> Although the opposition to animal performances might have come from cranks, the circus shows stand in danger of losing public confidence if these cranks get to work and get it into the heads of the public that there is cruelty exercised in their shows. Under supervision such accusations cannot be made, so the capital invested in animal entertainment organizations will be safer than ever before.

Opponents of the bill should therefore only concentrate on the committee stage in order to remove annoying amendments: wholesale opposition would not have public sympathy (*Encore* 1923).

With the prospect of new legislation that would require no more than registration, licensing, and restricted inspection, the contribution to the defensive campaign of political representatives was commented on. Pat Collins, president of the SGGB and MP for Walsall, at the time he was standing for Parliament again, told *The World's Fair* that he did his best to get support to take the sting out of the original Performing Animals Bill (*World's Fair* 1923j) (Fig. 3.12). Monte Bayly, organizer of the VAF, then wrote to emphasize that the full-time work in the select committee by O'Grady and de Frece saved the day from opponents to animal performance: over 16 pages of amendments had been tabled by them and they had been assiduous in attendance. Joseph Woodward was described as the animals' KC, with his Herculean work in preparing briefs and supplying all necessary witnesses and information, in which the VAF also took an active part (*World's Fair* 1923k) (Fig. 3.13). Bayly's comments looked as though he was ensuring that Pat Collins did not win undue credit for the success of the campaign, and just before the Act was passed at the end of 1925, E.H. Bostock announced that he was collecting for a presentation to 'Captain' Woodward, supported by *The World's Fair*, for his work in defence of performing animal acts (*World's Fair* 1925c, d). At the time of his death in 1945, *The Performer* recalled of Woodward, a member of the Circus Fans Association, that he 'was prominent in the defence of animal trainers and exhibitors whenever they were threatened by attacks from anti-performing animal groups or coteries', being 'particularly active in organizing defence against a threatened measure to restrict still further the training and presentation of certain animals . . . when finally an agreed Performing Animals Act came into existence' (*Performer* 1945). It was also reported in the local press that, having been called to give evidence to the select committee at bill stage, he had caused a sensation at Westminster by arranging for 'Carl Hartz [*sic*] to smuggle one of his disappearing canaries into the room and demonstrate its performance. Outside the House were bears riding bicycles and scooters' (*Thanet Advertiser* 1945, *Isle of Thanet Gazette* 1945). This account was an exaggeration, since Carl Hertz was allowed to perform the disappearing canary trick in front of the committee to attempt to show that no

**Fig. 3.12** Patrick Collins, by Bassano Ltd, 1922. © National Portrait Gallery, London. Ref. x122114. Reproduced with permission

**Fig. 3.13** Joseph Woodward in stage uniform, c. 1910. From the collection of Mrs Ann Rogers

cruelty or physical damage occurred (*1547–1551*).[48] There is no record that Joseph Woodward gave evidence himself, but Albert Voyce, as chairman of the VAF, reported to it:

> I have seen Captain Woodward train his animals over and over again when I have gone down to the theatre in the morning for my letters. Again, here is a man with an excellent reputation, and also a man who is greatly respected in religious circles, apart from the music-hall profession (2425).

The select committee had been interested to discover what role might be played by the professional associations in suppressing or dealing with any cruelty that might arise in the activities of their members. It was told by R.H. Gillespie that there were ten or a dozen professional associations across the country, often at loggerheads. However, the Joint Entertainers National Industrial Council represented all employees and the Actors' Association actors of the legitimate stage, and he would ask for suggestions from them to eliminate cruelty. It was decided that the chairman of the Council be invited to give evidence and

[48] Hertz, who was a conjurer and not a trainer, also worked with rabbits and ducks.

suggestions about the control of cruelty at the second select committee session, although this did not happen (*2004–2005*).

A reference was made during the select committee hearings to the VAF members' 'publicly stated' assurance that they would protest at any cruelty in performers to animals, and in 1914 at Olympia, the VAF had used a reassuring poster to declare that members found guilty of cruelty would be expelled (*215, 2891*). E.H. Bostock told the committee that it was best left to circuses and to the VAF, which represented nearly all the acts on the music-hall stage, to demarcate on 'undesirable' acts and to advise trainers to eliminate them: an independent regulatory body was not necessary. However, although he was an honorary member of it, he thought the SGGB not ideal to control standards because it had itself been criticized over the exhibition of animals (*3201–3202, 3255ff*). Arthur Valentine, an amateur conjuror, believed that magic clubs also, perhaps including the Magic Circle, would probably boycott cruel apparatus or tricks (1639f). *The Animal World* had earlier reported that:

> It is gratifying to know that some of the leading magicians of the day have decided to omit all tricks with animals—not merely on account of the actual tricks which, with a few exceptions, are harmless enough, but because of the conditions under which the stage animals are forced to live. (*Animal World* 1921a)

Some doubt arose as to the likely effectiveness of the VAF in a regulatory role. Albert Voyce claimed that Dandy George, although convicted for neglect of his dog, Rosie, had not been deliberately cruel and had left the VAF for reasons other than cruelty. Voyce confirmed that he would not necessarily expel someone from the VAF for a conviction for cruelty: it would 'depend on the circumstances' (2473, 2476). Samuel James told the committee that it was the role of the VAF to protect trainers and that it would ostracize complainers: there was a code of silence among members, and it was not seen as right to complain about colleagues (*720ff*). Sydney Arthur said that, in the same way, the National Association of Theatrical Employees had never reported complaints of cruelty, because such complaints had not been made by its members for fear that their names would be revealed (*732ff*). It was therefore not surprising that private advertisements in *The Era* and *The Performer* for witnesses to cruelty among former trainers were unsuccessful (*315*).

Criticism of the 1925 Act arose as soon as it had been passed, and the frustrations of opponents of animal performance were largely attributable to a well-organized professional opposition. Mitchell and others, in their joint letter to *The Times*, noted that the Act included only three out of the eight recommendations of the select committee report, 'the omission of the other five being due to the determined opposition of the interests involved'. Even those three were in a weaker form than the originals (*Times* 1927). Another letter to *The Times* was sent by Sir Walter de Frece from the House of Commons, in reply to one of 7 December 1928 from Sir Robert Gower and Lord Danesfort. He asserted that the 1925 Act protected all performing animals from cruelty, and their reference to a *Times* leading article of 4 December 1921 was therefore belated. He hoped the public would not be stampeded into thinking that there was any need for further legislation (*Times* 1928b). E.G. Fairholme, chief secretary of the RSPCA, then joined in to comment on de Frece's claim of belatedness, pointing out once more that in the Act only three survived out of the eight recommendations of the select committee, whose first report on the leading article in 1921 had been dealing with: the Act was therefore not fully protective (*Times* 1928d).

Although in the decades following the 1925 Act animal performance remained popular among some sections of the general public, it was subject to continued campaigns from animal rights activists and corresponding professional resistance that was representative mainly of the circus. These campaigns would intensify in the late 1970s and in the 1980s: although the PADL assumed a lower profile from the end of the 1960s, new organizations took on its mission. Animal performance increasingly became a politically sensitive issue at local authority rather than central government level; and the development of 'New Circus' in the 1980s represented a type of circus performance with a stance against animal acts.[49] A successor to the PADL, the Captive Animals' Protection Society (CAPS), organized demonstrations outside circuses in the 1970s, and in the 1980s it helped to finance Britain's first animal-free circus, set up by the daughter of the late Coco the Clown, in accordance with his dying wish. In the mid-1980s it started a successful campaign to stop animal acts at Blackpool Tower Circus. In 1987, vets Bill Jordan and Dr M. Woodford, an elephant specialist, criticized cellar accommodation at this circus, which used the veterinary consultancy of television-celebrity zoo, circus and dolphinarium vets David Taylor and Andrew Greenwood (Johnson 1990, pp. 110–112, 216). Between the late 1980s and early 1990s, the number of local authorities refusing applications from circuses presenting animal acts doubled to 200.[50]

## 3.6 Commercial Pressures and the Manager's Perspective

To retain public interest and ensure profit, animal acts in the music hall had to become more sophisticated, keeping pace with and pleasing a demanding audience. A reviewer had noted this trend in 1899:

> It need hardly be said that the dog, as the most sagacious of animals, continues to be most frequently seen on the stage, and it is impossible to enumerate all the troupes of canine performers which are 'on the road'. In regard to dogs, however, the public, it is said, has become much more exacting, and at the more important places only exceedingly clever animals will be appreciated. Sir John Lubbock's success in teaching his dog to read has

[49] In the United States, circuses without performing animals are called 'boutique circuses' by the traditionalists (Regan 2004, p. 133).

[50] Other aims of CAPS remain to end the captivity of animals in zoos and the exotic pet trade, as well as the use of captive wild animals in advertising and films. Undercover investigators seek evidence, and educational and campaign materials are produced. The Society lobbies politicians at all levels for legislation to end animal use in entertainment and has established an educational charity particularly to work with teachers and students.

probably heightened general expectations concerning these stage performances. (Dolman 1899)[51]

The increasing commercial pressure was believed to result in greater cruelty, as animals were coerced into ever more difficult routines (*259, 484*). Meanwhile, the value of highly trained and commercially successful animals was clear. In June 1906, Joseph Woodward had sold his seals and sea lions to a certain Berzac in New York for $9,000 (*World's Fair* 1911b), and much later, in 1934, Roy Huling would earn £50,000 from his seal act (*World's Fair* 1934). Wild animals in the circus had remained attractive before the war if they were exotic or apparently ferocious, and in seeking as a novelty to employ a female wild animal trainer, Frank Bostock noted that 'a good beast controller, male or female, can earn the salary of a Cabinet Minister in these jungles of mine...' (*World's Fair* 1911a).[52]

It was claimed by some witnesses before the select committee that the commercial importance of exhibited animals after the First World War had begun to vary between the circus and music hall and between wild and docile or domesticated animals. Sydney Arthur, the music-hall booking manager, thought that removal of animal acts would be no commercial loss to the music hall or to public enjoyment, because people were more enlightened and had become bored even with the more complicated acts recently developed. He thought there would be little unemployment, and other types of act could be turned to (*748ff*). James Sanger, on the other hand, claimed that performing animals were essential to his circus business and to employment in it, because animal acts were the key attraction and the rest 'padding' (*834, 872ff, 900, 1036*). George Harrop stated that although he no longer included lions and tigers in the programme at Blackpool Tower Circus because they frightened children, wholesale abolition of animal acts would close the circus: it had different customers from the music hall, which was not for children and families, and the circus could not become like the music hall, with its range of acts for a variety stage. Children liked the comedy of animal turns in the circus arena, and because these were essential for business, he regarded the select committee's work as of crucial importance (*1141, 1204, 1237*). Bertram Mills confirmed that animal acts were the most popular and that he received more public enquiries about them than for any individual performer: they were essential for business, and with prohibition, he would close his circus, being left just with music-hall turns in a circus ring (2191–2192). R.H. Gillespie said that although there would be no general public outcry if animal performances in the music hall were prohibited, some of them were a great public attraction: they were important for variety, although not all were suitable to attract the public. Some turns lasted 20 weeks,

[51] Lubbock was a pioneer student of the scientific study of animal behaviour in Britain, but he never taught dogs to read.

[52] It was reported the following week that he had received over 100 letters of application, and then on 28 October that only a Miss Gwendolyn Murray had remained brave enough to be appointed. But by 1933, only three circuses were exhibiting apes, big cats, and bears (*Parliamentary Debates* (Lords) 1933, vol. 87, column 727).

**Fig. 3.14** R.H. Gillespie, by Elliott & Fry. © National Portrait Gallery, London. Ref. x89403. Reproduced with permission

and therefore he needed only limited turns to fill the bills for his halls: he had booked only up to 50 animal turns in the previous 2 years (*1818*, *1873*, *1875*) (Fig. 3.14).

Bensusan wrote in 1913 that although managers of the variety theatres could not be expected to depreciate their own wares, Charles Morton, veteran manager of the Palace Theatre of Varieties in Shaftesbury Avenue (and described as 'father of the halls'), wanted all animal performances abolished because of his 50 years' experience of the associated cruelty, which he regarded as normal practice (Bensusan 1913a, p. 115). However, the RSPCA informed *The Times* that it had a reassurance from the management of variety halls that if there was public disapproval of animal acts, managers would not book them, and the matter was therefore up to the public (*Times* 1913). Ernest Bell said that the Performing Animals' Defence Committee had 'letters from 18 managers, representing a considerable number of theatres and music halls, expressing their disapproval of all animal turns and their practice of declining to allow them on their stages' (Bell 1918). But according to *The Animal World*:

> The manager of a modern music hall is keenly alive to the question of £ s.d. He is usually an astute individual, with his eyes wide open for the demands of his patrons. If they wish for animal turns, then animal turns they shall have. If they will have none of them, then his programme varies accordingly. Whether 'the turns' are in themselves good or bad is a secondary consideration. He does not occupy his position to discuss any question of ethics. His main business is to fill his theatre, and to earn dividends for his shareholders. (*Animal World* 1919a)

The RSPCA reported that seven protesters against a performance by an elephant at the World's Fair, Islington, got the performance stopped. They rose in their seats and charged the French trainer with cruelty from an instrument concealed in his hands. They informed the manager who then withdrew the turn to avoid further allegations of cruelty, which demonstrated the effect of even a small protest on the 'managerial mind' (*Animal World* 1922). Frederick Trussell assured the select committee that he would investigate complaints to keep the good name of the London Hippodrome (*3036*). Samuel James was a manager who took a more active interest and went on stage to see what was going on, but he pointed out that because of contractual variations there were differences in the attitudes of legitimate theatre and music-hall managers. Much depended on whether a booking was for one theatre or many music halls. Managers of a large group of music halls would be less likely to cancel the contract of an unacceptable act in the way he had done in his single theatre (*709ff*).

However, many managers were unaware, deliberately or otherwise, of questionable treatment of animals in their theatres. Fred Trussell assumed simply that animal training by cruelty was not possible (*2985*).[53] R.H. Gillespie said he had not seen cruelty and in up to 24 years' experience had received no complaints; however, he had not seen training other than rehearsals (*1764ff*, *1813*). Lily Grahame claimed that Stoll and Moss were never seen behind the stage (*2629*). Samuel James knew that music-hall managers were often unaware of cruelty, an exception being John Lawson of Walthamstow Theatre who had dismissed the owner of Consul the baboon [*sic*] (*712ff*). And Everard Calthrop, the amateur trainer, said the reason why those who had convictions were usually booked again was that the manager concerned remained ignorant of the conviction or even of the effect on his audience of performances resulting from cruelty (1883ff).

Many managers appeared indifferent to the possibility of cruelty, felt it was not their business, or were reluctant to invite trouble. Mollie Lawrence complained that managers turned a blind eye and did not go behind the stage (*2820*), while Jessie Hales thought it no use complaining to an indifferent management (2063f). R.H. Gillespie said he would be delighted to order reports of cruelty (but it was the performer who should be penalised, not the music hall). He had not done this before because he was busy and the matter 'hadn't come his way'. He was reluctant to demand admission of any conviction from prospective performers—this was best done through the VAF—but he would agree to consider not booking for a few years someone whose conviction was known. He could not speak for others because music halls were a competitive business and they might not do likewise (*1861*, *1994ff*). Frederick Trussell, who had limited witnessing of rehearsals, did not think it a manager's job to enquire as to whether there was cruelty. He was reactive only and would not invite trouble. However, he would not be mercenary in the face of cruelty, and he said his company would not object to his stopping an act (*3059*, *3099ff*).

Some witnesses nevertheless suggested that managers were both aware of cruelty and, being mercenary, took no action. Margaret Martin claimed that Frank Allen, general manager of the London Hippodrome, where she alleged cruelty took place, went behind the scenes but apparently with no result (554). Everard Calthrop admitted that managers might be mercenary as well as ignorant in booking cruel trainers (1840), and Jack Thompson said that after he had made a complaint against Zertho, the manager interviewed him, but the show continued. The theatre was also in a dirty state because of his dogs (1524ff). Lily Grahame was told by the manager

[53] He said he did not know why there had been convictions and misremembered giving evidence on behalf of the accused in a court case (*2902*, *2956*).

to stop accosting performers. The theatre manager allegedly allowed Consul to be beaten, and she feared being called hysterical by the management, who told her to mind her own business (*2598*, *2634*, *2677*).

Although Bensusan wrote that a manager had told him that shifters and carpenters had instructions to report any case of cruelty and that cruelty might occur in training, not in the performance (Bensusan 1913c, p. 119), it became widely suspected that employees were reluctant to speak out for fear of losing their jobs. Based on his experience of select committee proceedings, Lord Danesfort made this point clearly once more in a House of Lords debate in 1933.[54] Charles Wilson realized the vulnerability of employees when he told the select committee that RSPCA inspections were routine at theatres, not made just because of complaints, and therefore the management need not be suspicious that an inside report had been made (*2861*). Lily Grahame was reprimanded by management after her report to the RSPCA (*2564*), and Charles Haverly said that objections to cruelty would lead to boycott by managers and therefore loss of engagements (*575ff*). Managers interviewed by the committee gave a different impression. R.H. Gillespie undertook to put up notices in his theatres ordering reports to him of any cruelty; Frederick Trussell declared he would welcome reports of cruelty from employees; E.H. Bostock confirmed he would encourage employees to report cruelty; and George Lockhart said he would have authority to cut out cruel acts and would be supported by his directors (*2026*, *2972*, *3195*, *1319*). According to Bertram Mills, in the previous 2 years, around 230 animals had appeared at Olympia: all performances were personally assessed, and the RSPCA had passes for all hours. He also usually saw the training of horses for turns he booked, including those from abroad (2150, 2162–2167).

A few managers were actively opposed to animal performance. *The Era* reported in 1921 that Milton Bode, manager of seven theatres, was against it, not believing it necessary to a successful music hall. He did not make such bookings even if he knew they would be popular (*Literary Digest* 1920, *Era* 1921h). A legitimate-stage manager said to Athole Stewart that animals hated being on stage (*2316*), and Sydney Arthur confirmed that some music-hall directors were against animal acts (*740*). Arthur was a salaried booking manager rather than a variety agent dependent on commissions, and although, because he was not retired, he declined to name others, he told the select committee that what he had seen had persuaded him not to book any more animal acts; but his sincerity was questioned closely by the committee (*726*, *2366ff*). The critical witness, Mrs Albert Bradshaw, admitted the practical measures taken by some theatre managers against cruelty, by cancelling acts (*217*). E.H. Bostock had relieved a baboon trainer for striking his animal (*3227ff*), and Samuel James had dismissed a dog and pony act appearing towards the end of a pantomime at the Grand Theatre, Islington in 1906, because he had seen cruelty while unobserved. The performer had threatened to sue, but did not for fear of the bad publicity. (It would appear that the wording of contracts did not include

[54] *Parliamentary Debates* (Lords) 1933, vol. 87, column 744.

conditions regarding the treatment of animals by performers.) James pointed out that he had only once engaged an animal turn and had remonstrated with others about much cruelty he had seen elsewhere (*703ff*).

## 3.7 Racial Prejudice in the Controversy[55]

Racial prejudice and selective, usually inarticulate, racial discrimination influenced attempts to conduct an objective examination of charges of cruelty in the training and exhibition of performing animals. As the debate intensified, and following the appointment of the select committee, one explanation often given by both sides for shortcomings in the treatment of performing animals in training and exhibition was the alleged cruelty particularly or exclusively attributable to the 'alien enemy', 'foreigners', and distinct racial groups. The committee faced the problem of assessing attributions of real cruelty as opposed to unproven charges that may have resulted from irrational or emotive prejudice or from strategic prejudice intended to serve the interests of each side in the controversy.

Those critical of animal performances complained that existing laws failed to control cruelty which was considered especially characteristic of foreign trainers and their acts (whether presented by the foreigners themselves or by British showmen who had purchased their ready-trained animals). Meanwhile the British professionals attempted to deflect criticism of their own methods by attributing almost innate cruelty to certain foreign nations, races, and individuals and by emphasizing their wish, especially through the VAF, of barring foreign animal performances in the British circus and music hall. In sympathy with this, Frank Somers, the Leeds veterinary surgeon, told *The Yorkshire Evening Post* that when odd cases of cruelty occurred, 'almost invariably the offender [was] a foreigner, and a vicious foreigner at that' (cited in *The Era* 1921d). George Lockhart said to the select committee: 'It seems to me that the very fact that the great majority of the convictions [for cruelty to performing animals] are against foreigners must of itself be some sort of defence of the English trainer' (*1487*). Captain O'Grady, as a select committee member sympathetic with the professionals' cause, addressing the critical witness Lily Grahame about her experiences at the London Hippodrome between 1900 and 1909, reminded her that there had been proportionately very few convictions and these had been against 'alien enemies' (*2753–2759*). George Henry Harrop told the committee that before the war most of the animal acts had been foreign, principally German. VAF members would not now work with such ex-enemy turns from the European continent, the main pre-war source of animal acts: since the war, he said, the VAF had prohibited German and enemy alien acts because cases of cruelty had involved foreigners (*1164*, *1167–1168*, *1254–1256*).

[55] Much of this section was first published in Wilson (2009a).

Foreign, especially German, involvement in British animal acts had been considerable before the war, but during hostilities and in the 1920s, 'enemy aliens' were usually ostracized, only to reappear in greater numbers in the 1930s.[56] At the same time, however, ready-trained animals continued to be imported, giving strength to opponents' calls for the prohibition of all animal acts because of the impossibility of regulating foreign training methods. Jack Russell Vokes advised the committee that cruelty could be prevented in Britain by 'eliminating' some of the foreign trainers, because they were out of reach in a foreign country (*2086*). Two years after the passage of the Performing Animals (Regulation) Act in 1925, a letter to *The Times* from Galsworthy, Mitchell and others pointed out that many performing animals were still trained abroad with no legislation and with more general cruelty; and further cruelty resulted from confinement and travelling. They therefore asked: 'Will the public help to abolish this painful form of amusement by refraining from patronizing exhibitions in which performing animals have a part?' (*Times* 1927). Also dissatisfied with the effect of the Act, Mrs H.M. Lennox later wrote that discouragement of animal performances was necessary because many animals were foreign-trained (*Times* 1928c). The RSPCA continued to call for a ban on foreign-trained animals and demanded that, in the meantime, their exemption from quarantine should be abolished (RSPCA 1931). Its Performing and Captive Animals Special Committee decided in 1937 to support the current Performing Animals (Regulation) Amendment Bill and discussed the need to acquire evidence of cruelty to performing animals abroad, if necessary by sending at great expense special investigators who were familiar with the countries concerned and spoke their various languages (RSPCA 1937).

Whether or not the prejudice against foreign and especially 'enemy alien' acts on the part of the professionals was justified on grounds of cruelty, underlying it was without doubt a concern about domestic employment levels, about public perception of the nature of their own livelihood, and a fear that their activity might be curtailed by new legislation unless public concern was strategically deflected and concentrated in this way. The 'alien enemy' became an easy scapegoat, vulnerable because of the aftermath of the war, and foreigners, and other races generally, became victims of xenophobia, racism, and beliefs in British moral superiority. For example, Charles Butler declared: 'I know some very fine men in the circus world—Englishmen—and I would like to add that most of those trainers I have been speaking about [critically] are foreigners' (*515*). O'Grady said of ex-enemy Continentals: 'They have ideas that are not altogether in accordance with our British ideas'. George Henry Harrop replied:

> I do not know whether it is correct or not, but it has always been understood in the profession that the Continental people have been the greatest trainers in the world, especially Germans ... I have been in the habit of going to Germany for almost thirty

[56] Because of deportation and repatriation, the overall size of the German community in Britain declined from 57,500 in 1914 to 22,254 in 1919 (Panayi 1994, pp. 106 and 107).

years, but I do not think I shall ever take the trouble of going again—not to book any animal turns (*1257–1258*).

George Lockhart thought that only since the war had very few trained animals come from abroad and that any resumption would depend on VAF policy. There was no restriction before the war because 'we lived together in amity', and he admitted that the profession had not barred German-trained animals on grounds of suspected cruelty (*1435–1440*). Frederick Trussell, ex-manager of the London Hippodrome, explained to the select committee that lion acts were currently not so much in evidence because most of them belonged to German houses (*3047*).

The VAF's policy reflected a combination of post-war prejudice and protectionism rather than a concern for the welfare of animals, although such a concern was now often newly given as a reason for the boycott of foreign acts, when the characteristics of German trainers were stereotyped and criticized.[57] O'Grady claimed that while English artistes abhorred cruelty, they left it to the courage of the VAF to boycott the foreign acts responsible for it (*2760–2764*, *3098*). Bertram Mills and music-hall managers like Sir Oswald Stoll would seek VAF approval before engaging foreign acts, because it had more opportunities for knowing about the artistes (2194–2195). Albert Voyce confirmed that his VAF members had been boycotting enemy alien acts: 'We kept them out; we refused to work with them', even at the risk of artistes losing their employment if they refused to appear on the same bill.[58] Before the war, he said, the country was overrun with very cheap German and Austrian acts, which he had witnessed as undesirable. One such act was 'Zertho's Dogs', but it appeared before 1907 and before the VAF had been founded.

In those days [Germans and Austrians] were brought over here by a man called Barrasford, who came into our business. He flooded the country with German and Austrian acts, and a lot of these acts came over—perhaps a little terrier is made to stand on one foot or two feet on the palm of a man's hand, which was not a very nice thing, and I think myself it was painful to the dog. There were other acts, which I cannot remember now. We are going back 20 years now, and it is difficult for me to remember. But there came a time when the German act died out, and was not wanted.

Voyce added that these German and Austrian turns were, generally, unquestionably cruel (2434, 2461–2470). Soon after, *The World's Fair* reported a statement by Monte Bayly (VAF organizer) to *The Westminster Gazette* that in 1914 (well before the parliamentary interest) the VAF had itself suggested licensing, registration, and

[57] The VAF imposed a ban on ex-enemy aliens also because of the very low value of the German currency—German acts were prepared to work for starvation wages and would threaten those of domestic performers. The ban was lifted in 1924 largely because of an improvement in the German economy and the resulting wish of British performers to work in Germany (Wilmut 1989, p. 50). A Special General Meeting of the VAF lifted the ban on ex-enemy artistes on 28 March 1924 (VAF Archives).

[58] On 6 December 1922, Voyce wrote to *The Daily Mail* explaining the VAF's ban on ex-enemy aliens, and on 2 December, he had written to *The Evening News* about the introduction of German films into England (VAF Archives).

more severe penalties: 'The majority of objectionable exhibitors are foreigners, and they are not allowed here' (*World's Fair* 1922d).

Comparisons of British with foreign behaviour and morality in this area had begun in the late nineteenth century:

> For the most part foreigners are the purchasers of animal troupes, and the majority of them are idle, vicious and cruel . . . Although we hunt foxes, hares, and rats, and shoot pigeons from traps, we are, perhaps, the best-hearted nation on the earth—ready to kill everything, but reluctant to torture anything. (Bensusan 1913a, pp. 110 and 114)

Some of those on the select committee who aimed at legislation to eliminate or control animal performances were themselves enthusiastic hunters, notably Kenworthy (Kenworthy 1933, pp. 244 and 247). Writing in 1919 in support of the aims and methods of the emergent Jack London Club, Captain E.G. Fairholme, chief secretary of the RSPCA, charged alien owners with earning a dishonest living by forcing animals to perform unnatural and degrading antics: 'If we had strong laws to keep the aliens from our shores we should also, to an enormous extent, purge our public entertainments of these degrading spectacles' (Fairholme 1919).

Compounding a supposed moral inferiority among foreigners might be a variation in their intelligence, which could even be paralleled in performing animals themselves: 'From India, it may be said, all the clever elephants come; the difference between the brains of the Teuton and the Negro is not greater than between those of the Indian and African elephants' (Dolman 1899). James Sanger described why lions could bite a trainer:

> Some of these trainers, mostly these black chaps that for appearance have been put in for training lions, sometimes get a little bit too much treated in some of these villages and get over the mark, and instead of their knowing the place to walk back, they have a collision with the lion. If the man can only keep his head and know his position, a lion never thinks about attacking him (*812*).

Fred Ginnett referred appreciatively to a lion casualty from his circus staff as a 'good-looking fine nigger', when drink had not apparently played a part in causing the accident (738).

Assumptions were also clearly made that Germans, Slavs, Blacks, and Asians were particularly susceptible to perpetrating cruelty. For example, Joseph Kenworthy asked James Sanger: 'If you found one of your men—a negro or any of your men—being harsh and cruel to an animal, you would sack him?' (*1014*) When asked if he knew that the German animal dealer and trainer, Carl Hagenbeck, had been convicted for allowing an Indian mahout to goad an elephant so that it bled down its legs, Ginnett said that Hagenbeck was the greatest dealer in animals in the world but could not be responsible for 500 staff in every country, 'niggers, Africans, Indians and all sorts'. While the mahout was 'next door to a monkey', Hagenbeck was described as a nice old gentleman, even though a German, and had not been convicted of cruelty personally (819–827).[59] It is ironic that British interests should

[59] For an account of the Hagenbecks, see Rothfels (2002).

claim German trainers were the cruellest when humane techniques had been given such prominence by Hagenbeck well before the war and accepted as effective.

Trevelyan wrote of an unnamed former circus hand's assertion that as a rule lion trainers in Britain had been foreign and used brutally cruel methods and that the breaking-in of ring and school horses for equestrian work was also cruelly carried out by foreigners (Trevelyan 1936, pp. 95–99). Throughout the 1920s, and only after the performing animals issue had come under discussion in Parliament, the attribution by the specialist trade press of problems of cruelty was consistently aimed at foreigners. The only cruelty might be from 'recently arrived alien trainers, who we can do without anyway' (*Encore* 1921b). The horse trainer Arthur Aitchley had seen only one example of cruelty and that by a 'vicious foreigner', in his just-completed 17 years in the United States (*Era* 1921e). Joseph Woodward, as an active organizer of defence against any new legislation harming the interests of his profession, said that problems of cruelty belonged to the past or with foreigners with unpronounceable names (*Era* 1921l).

Specific instances of cruelty were given by witnesses to the select committee. Mrs Albert Bradshaw had seen 'a foreigner, a little stumpy woman with very big fat hands' beat performing cats (*214, 224*). In 1911, Basil Tozer, the former theatrical publicity agent, had infiltrated a place where dogs were being trained by a foreigner who hung them by their back legs and used a spiked collar to train them to balance on their front legs (*2409*, *2420*). John Oscar Mansley, the journalist who had been encouraged to offer himself as a witness by the PADL, spoke of a fearsome Russian keeper who in 1911 beat the 8-year-old chimpanzee, Consul, daily with an iron bar at his theatrical lodgings, where the animal's cage was so small it could not lie down (843, 864). Harold Laurence Hilliard, the manager-handler acting for the owner of Consul, explained that this charge was untrue and that the trainer was no Russian, but a Swiss named Derwoski (964). Jack Thompson, the composer and musical performer, described how a trainer thrashed another chimpanzee in a dressing room at the Palace Theatre, Oldham in 1919, and 'of course he was a foreigner' (1470–1491). Sidney Davis, the stage carpenter, reporting that Lipinski made performing dogs appear drunk on stage by fitting them with neck weights and by making them fear the whip, commented: 'I do not suppose you ever will see the show again, because he was an Austrian' and therefore boycotted (1257).

*The World's Fair* reported that although the select committee had been offered much evidence of cruelty in the training and performances of animals within a recent period, there was a growth of the humanitarian spirit in the country. It cited the comment in the report of the select committee that: 'The number of trainers of British nationality in 1914 was only about 20 per cent of the whole. These trainers enjoy the reputation of being more humane in their methods than those of foreign origin, and evidence has been forthcoming to that effect' (*World's Fair* 1922e). Charles Wilson of the RSPCA confirmed to the committee that training often took place abroad. Of performing animal turns on tour between 1 December 1913 and 30 June 1914, 110 out of 147 had been those of foreigners (19 of the remainder being British and 18 of doubtful nationality) (*276*). The theatre electrician, Albert

Hill, said that from 1914 there were very few animal turns in Britain, because they were mostly foreign acts affected by the war (1181).

The profession's journals[60] never failed to report cases of alleged cruelty that involved foreign trainers and exhibitors, such as when the RSPCA took action against Cossacks for cruelty to horses ridden through flames (*World's Fair* 1925b), but special attention was reserved for the 'alien enemy'.[61] The context had been set: for example, with the sinking of the *Lusitania* in 1915, Horatio Bottomley promoted Germanophobia in *John Bull* magazine. After the war, he said, 'no shop, no factory, no office, no trade, no profession' should be open to Germans. As a Member of Parliament in 1919, he advocated 'Britain for the British socially and industrially', and after the passing of the Aliens Act in 1919, liberal attitudes to foreigners disappeared altogether (Holmes 1988, pp. 112–114). During the war, Germans had faced social animosity, official racism, popular hostility, and boycott in the professional environment, and after its conclusion, they suffered an enhanced version of the employment discrimination traditionally suffered by minority groups (Panayi 1994, pp. 110, 111 and 121).

At the select committee hearings, Germans were frequently described as the most cruel, but at other times during the hearings, they were also described by their British counterparts as the best trainers. This association of attributes undermined the frequent assurances given by trainers and exhibitors that any effective training was only possible by kindness. Charles Haverly confirmed that a great many trainers were Germans and that they were also the cleverest, perhaps because they were more cruel than the English ones and so achieved better results. (Americans were fairly good trainers, he said, but the French and Italians were no good at all.) E.H. Bostock described (presumably white) American training methods as similar to British ones (*3245*). Perhaps Americans were considered favourably because they were allies and, typically, Anglo-Saxon and English-speaking. Haverly described the method of subduing lions with close confinement, trident, whip, and starvation, as used by a German at the Balham Empire 15–20 years previously; and also the ferocious beatings by the American John Cooper 30 years ago. Very cruel men came mostly from Germany, and while on the Stoll circuit, he had seen gratuitous use by one of them of an iron-tipped whip at the Leicester Empire in 1904 (*582, 548–555*). Samuel James had engaged in 1906 at the Grand Theatre in Islington a 'German-American looking gentleman' with a troupe of dogs and ponies. Because he thrashed his dogs when training them in the back somersault

[60] Principally the VAF's *The Performer*; the SGGB's *The World's Fair*; *The Encore*; *The Era*, representing the Provincial Entertainments Proprietors' and Managers' Association, and closely associated with the SGGB; and *The Magnet*, which described itself as the oldest penny professional journal published, devoted to the interests of the music hall, theatrical, and equestrian professions.

[61] For example, at its meeting on 14 April 1921, the Executive Committee of the VAF, chaired by Albert Voyce, resolved 'that this Committee vigorously protests against the insertion in "The Performer" of articles as letters directly against the VAF Policy as exemplified by recent letters in its columns supporting ex-enemy aliens' (VAF Archives).

and to jump from heights, he cancelled his contract. When at the King's Music Hall in Poplar in 1909, he saw a 'German-American-Jewish-looking gentleman' hitting baboons on a short leash across the snout with a log of wood, to make them jump (*703*). George Lockhart said that the best animal trainers came from Germany, and they were the best because they probably used methods that would give British trainers a very bad name indeed. He admitted he had never seen them at work, but they had trained horses to perform in ways not attempted or achieved by English trainers, so cruelty was suspected (*1430–1434, 1467*). Mrs Mollie Lawrence, the estranged wife of an artiste, declared: 'The German is the most cruel brute to animals' (*2821*). Mrs Margaret Martin, the ex-music-hall artiste, described a German woman at the Hippodrome who shouted at and beat cats with a dog whip (512), and Jack Thompson reported that he had complained to the manager of the Grand Theatre in Birmingham about the condition of a very tightly muzzled and bleeding dog kept in the wings by the German, 'Zertho' (1505–1526).

The music halls had originally given birth to jingoism, and they then encouraged anti-German sentiment before and during hostilities, but once it had started, some patrons found their treatment of the war trivial or distasteful and turned away, especially to the cinema (Cheshire 1974, pp. 54–55). Some representatives of the music-hall stage and circus seemed reluctant to endorse the negative stereotyping of foreign, especially German, trainers or their audiences. Frederick Trussell claimed that when he was manager at the London Hippodrome for nearly 20 years, he was aware of only one small conviction, that of a Frenchman, about the use of a spur, and that he did not understand that the method of treatment in training by 'late alien enemies' was particularly cruel (*2903, 3098*). Everard Calthrop, the amateur trainer, told the select committee that the foreign performances he had seen showed that training abroad was not necessarily cruel and that he had seen perfect performances with willing, eager animals (1900). After extensive visits to continental horse shows and circuses, Bertram Mills had only seen cruelty once, in the treatment of horses in the ring in Sweden; and a month previously in Berlin, he had disapproved of a display of a dog balancing on one leg on a man's hand. In both cases, he said, there was disapproval expressed by the local audience. He added: 'We quite appreciate the position between Germany and this country now, but I do not think the German public like cruelty any more than we do' (2159–2160).[62] Arthur Aitchley stated that German trainers were effective because the German was more pig-headed and persistent than the English, not more cruel in his methods (*2144*). It had long ago been claimed that phlegmatic English and German trainers were effective because of their patience and judgement (Anon. 1869, p. 209).

Ernest Bell stated that 62 foreign animal troupes came into England in 1913, and:

[62] Later, under the Third Reich, laws on animal protection were passed in 1933 and 1938. It was forbidden to use an animal for demonstrations, film-making, spectacles, or other public events to the extent that these events caused the animal appreciable pain or appreciable damage to health. In 1938 the Minister of the Interior also issued a Decree on Wandering Animal Shows and Menageries (Sax 2000, pp. 175–182).

> The largest capturers and importers of animals, Jamrach and Hagenbeck, were Germans, and we think we may say that the majority of the trainers and showmen are foreigners from countries which do not usually show much solicitude about the welfare of the animals in their hands. (Bell 1918, p. 95)

However, frequent favourable mention was made of Carl Hagenbeck during the select committee's hearing of evidence, and he was the foreign personality most commonly referred to, generally in unusually complimentary terms. Jack Russell Vokes found Hagenbeck's establishment acceptable because it was open and it was possible to pay to see good training (*2086*). E.H. Bostock could not speak too highly of Carl Hagenbeck and his mode of training or keeping animals (*3233–3235*). However, in Appendix I of the select committee's report of 1922, among 'declarations from persons in foreign countries who were unable to appear before the Committee', a Dutch painter, Paul Schultze, reported cruelties to horses he had seen while permitted to sketch at the training sessions of Carl Hagenbeck's circus at Scheveningen, near The Hague; and in the same year, 1920, Miss E. des Tombe, founder of the Dutch Jack London Club, saw cruelty in the training of polar bears and elephants at the circus of Wilhelm Hagenbeck (Carl's nephew) at The Hague. In these accounts, there is no indication that Carl Hagenbeck was personally involved, but Wilhelm Hagenbeck carried out some of the training himself. On 28 March, Wilhelm forwarded to Bertram Mills a letter from a close family acquaintance, Professor Dr Heck of the Zoological Gardens in Berlin, in which the humane methods of the Hagenbecks were confirmed and praised. Heck concluded by saying that the allegations made in the English newspapers gave him the impression that they had been prompted by 'untechnical and anti-German motives'. Mills had also kept Carl Hagenbeck briefed on the evidence presented to the select committee, and he had replied on 24 March, pointing out the humane methods of his family's business, and his and his family's membership of the Humane Society for the Prevention of Cruelty to Animals in Hamburg (United Kingdom Parliament 1922, Appendices I and II). Peter Chalmers Mitchell, probably the select committee's most convincing independent expert witness, confirmed that Hagenbeck was good with animals. He respected him and suggested that conditions in Germany were not worse than in Britain (2254–2256).

Disagreement and concern about the foreign factor continued from the 1920s into the following decades. T.H. Gillespie, then director of the Zoological Park, Edinburgh, wrote: 'I cannot conceive that my own country has a monopoly in kindness and common sense' (Gillespie 1934, p. 113); but in her book, published to counter his, Helen Trevelyan complained that:

> The great majority of animal acts are trained on the Continent, where the feeling for and treatment of animals is notoriously low, practically absent in many places, and by men whose names alone would hardly inspire confidence in the treatment of things less sensitive than animals. (Trevelyan 1936, pp. 105–106)

John Swain held similar views:

> We cannot stop the vast initial cruelty of the training of animals [abroad], as this is done outside the reach of our laws; but there is no reason why these senseless, brutal, disgusting exhibitions should be allowed to befoul and pollute the stage and music-halls of this country or our circuses and menageries, and thus deprive actors and actresses of their livelihood. (Swain 1933, pp. 296–297)

In her autobiography, Vesta Tilley (Fig. 3.15),[63] the actress and male impersonator, regretted that Variety was 'dead' by the mid-1930s, one of the reasons being the banning of foreign performers, because:

> In the old days, to make up a real Variety programme it was absolutely necessary to introduce foreign acts. The Germans were supreme in acrobatic and strong man acts, and many entertaining dumb shows. Performing animals were a strong item on the programme before public taste veered round against them, on the accusation [with which she did not agree] that such shows were only possible by exercising cruelty in the training of the animals. Foreign acts were barred largely owing to the War, but the Variety profession itself was not altogether blameless for their banishment, owing to the introduction of trade union methods in a profession that could prosper only by welcoming the largest measure of free trade in entertainment. (de Frece 1934, pp. 271–272)

In the House of Commons, well after the Performing Animals (Regulation) Act had been passed, Sir Robert Gower[64] asked the Home Secretary to cancel and stop further issue of permits to alien exhibitors of performing animals, but was told in reply that the few permits were for strictly limited periods and that withdrawal or refusal would be unjustifiable.[65] Then Peter Freeman (president of the Vegetarian Society between 1937 and 1942) asked the Home Secretary for details of the number of foreign performing animal turns touring in the country, the species of animals involved, the names of the owners, and whether they were domiciled. In reply it was stated that the information was restricted to certificates of registration, which showed that 205 trainers had been registered since the passing of the Act, of whom 125 were British and 80 foreign, but that there was no way of knowing how many foreigners were domiciled or engaged in training.[66]

During the second reading in the House of Lords of the Performing Animals (Regulation) Amendment Bill in 1930, Lord Auckland expressed his support, and as one who had owned and trained practically all kinds of carnivorous animals as shown in circuses and music halls, he had listened to trainers boasting of the brutalities which they found it necessary to inflict on their animals.

> Most of the trainers in this country are foreigners, and the abuse of their animals is dreadful. I think they might just as well be prohibited from exhibiting their animals in this country, and go back to their own countries and have their exhibitions over there.[67]

---

[63] Vesta Tilley was married to Sir Walter de Frece, who took the leading part on the select committee, opposite Kenworthy, in resisting new legislation. He was managing director of a company controlling up to 20 theatres and in parliamentary business connected with the performing animals' controversy; he represented the interests of the Society of West End Managers, the Theatrical Managers' Association, the VAF, and the music hall and circus generally.

[64] Gower was chairman of the NCDL from 1920, Conservative MP for Gillingham and chairman of the Animal Welfare Committee in the House of Commons between 1929 and 1945, and chairman of the RSPCA between 1928 and 1951. In 1931 he was presented by the president of France with a Sèvres vase in recognition of animal welfare work in that country.

[65] *Parliamentary Debates* (Commons) 1928–1929, vol. 224, column 951.

[66] *Parliamentary Debates* (Commons) 1929–1930, vol. 237, column 67.

[67] *Parliamentary Debates* (Lords) 1930, vol. 76, columns 907–908.

**Fig. 3.15** Vesta Tilley (Matilda Alice (née Powles), Lady de Frece) and Sir (Abraham) Walter de Frece in the 1900s, by Marcus Guttenberg. © National Portrait Gallery, London. Ref. Ax160045. Reproduced with permission

Lord Jessel then referred to Home Office records showing that 102 of 148 exhibitors and 206 of 314 trainers were British, participating in a sizeable domestic industry, but Danesfort pointed out that foreign involvement was greater if lions and other large carnivores were considered alone.[68]

Gower unsuccessfully asked the Home Secretary how many alien exhibitors of performing animals had been given a licence during the previous 12 months, how many there were currently, and whether he would consider withholding licences in the future.[69] Soon afterwards, he also unsuccessfully asked the Home Secretary to consider discouraging animal performances by refusing to grant permits to alien exhibitors, because there was no discrimination in the Performing Animals (Regulation) Act between British subjects and foreigners.[70] Questions along these lines were to continue. Sir Cooper Rawson complained about permits for circuses like Gleich from Germany, one that depended more than British circuses on performing animals.[71] Following his question to the Home Secretary in a later Session, Frederick Messer was told that 51 aliens currently had certificates of registration.[72] During the debate on the second reading of the Docking and Nicking of Horses Bill in 1949, a comment from Thomas Braddock demonstrated that wartime experiences and enemy status would again affect perceptions of Germans as trainers:

> We know that, before the war, the chief source of performing animals for circuses was Germany, and that the German people who were engaged in this particular line of business in producing performing animals stopped at no cruelty in order to train animals for their diabolical purpose. It is not without significance that the same race of people was capable of inflicting on a certain section of their own nation—the Jews—cruelties almost equal to those inflicted in the course of the training of performing animals.[73]

In the 1950s Peter Freeman was still expressing concern, asking the Secretary of State for the Home Department to provide information to animal welfare societies about foreign training conditions that would not be tolerated in Britain.[74] However, Edmund MacMichael thought it 'a base act of hypocrisy to point the finger of scorn at other nations for atrocities and at the same time to make the secret infliction of atrocities—especially for the sake of amusement—a source of revenue' (PADL 1945).

Addressing the question of the possible effect of a legislative prohibition of animal performances on employment, Sydney Arthur, the booking manager, had

---

[68] *Parliamentary Debates* (Lords) 1930, vol. 76, columns 911, 917 and 918.

[69] *Parliamentary Debates* (Commons) 1930–1931, vol. 247, column 819.

[70] *Parliamentary Debates* (Commons) 1931–1932, vol. 259, column 1040.

[71] *Parliamentary Debates* (Commons) 1932, vol. 264, columns 1248 and 1249.

[72] *Parliamentary Debates* (Commons) 1936–1937, vol. 318, column 1995.

[73] *Parliamentary Debates* (Commons)1949, vol. 462, columns 763 and 764.

[74] *Parliamentary Debates* (Commons) 1954–1955, vol. 535, column 1097. At this time central Europe was the main source of wild animal acts in the American circus, but such acts were becoming hard to maintain because horse meat was costly and scarce (Carlisle 1956).

informed the select committee that there were at the time more English trainers and exhibitors following the boycott of German acts, but that these were not many, and so there would be no significant unemployment problem, and other specialisms could be turned to in order to obviate any unemployment in the profession (*754–761*). By the 1930s, the presence of foreign trainers had increased, and on 4 May 1933 Lord Danesfort pointed out that there was no obligation to preserve the employment of trainers in Germany, most trainers and exhibitors then being foreign. He reported that 113 foreign trainers were currently registered, as opposed to 77 British (net numbers would be less because individual trainers were sometimes registered for more than one animal), and so domestic employment would not suffer significantly.[75] In the meantime, animal performances appeared to Helen Trevelyan to continue to guarantee large sums of money to foreigners (Trevelyan 1936, p. 91).

Circus and music-hall professionals were long used to working with foreign colleagues, having much opportunity to develop appropriate professional respect and familiarity, although there is a suggestion from the select committee evidence that individual performers generally kept themselves to themselves when on tour. As the performing animals controversy took shape, perhaps a new, contrived opinion of these foreign colleagues was sometimes expressed as part of a defensive strategy. In this controversy, in addition to any genuine concern about known instances of real cruelty, there would have been concern to protect domestic professional employment interests at a time when the music hall was in decline and when anti-German sentiment after the war was not just commonplace but also convenient. The reputation of German trainers in such circumstances was not helped by the fact that they were often acknowledged to be the most effective in their work. Those critical of animal performances might also share the anti-German sentiment common in society at the time, but they were less concerned with foreign threats to domestic employment in the music hall and circus than with the impossibility of regulating foreign training methods. For that reason, their concerns were directed less at the 'alien enemy' than to foreign sources in general.

In the absence of total prohibition of animal performances, it was fair that those genuinely interested in animal welfare should express concern about the application of acceptable standards in the circus and music hall; and because Britain had well-established animal welfare legislation[76] that barely existed in some countries and which was entirely absent in others, it was understandable that within this general area of concern, uncertain foreign practices were open to suspicion. Exactly the same type of concern continues today, for this same reason, regarding other areas of animal use, but it is not often couched in those xenophobic terms which were more readily used and accepted in the early twentieth century; and no strategic attempt is

[75] *Parliamentary Debates* (Lords) 1933, vol. 87, column 726.

[76] For example, the Wild Animals in Captivity Protection Act, 1900, and the Protection of Animals Act, 1911.

now made by the negligible domestic animal training interest to shift blame for any shortcomings on to foreign sources.

## 3.8 The Employee's Environment: Job Security and Tipping

An absence of reports of cruelty from employees was cited by those with animal performance interests as evidence that there was little cause for concern, but critics alleged that reports would not be made because staff believed their jobs would be jeopardized if they exposed cruel practices serving the commercial interests of the music hall and circus. There was little faith on the part of employees in the confidentiality offered by the RSPCA or the unions, not least because if a court case resulted, identities on both sides would be revealed. An internal culture of 'minding one's own business' also prevailed.

The evidence presented at the select committee therefore revealed, on the one hand, an alleged desire on the part of management to encourage and welcome the reporting of irregularities so that proper standards could be maintained and, on the other, allegations that the truth was being suppressed because of a fear of retaliation or boycott if complaints were made either within the profession or to external sources such as the RSPCA. The select committee heard statements of reassurance from those defending the status quo alongside claims of repression from critical bystanders within the industry, as well as some limited evidence of assertiveness from those who had spoken out, while employed, against the cruelty they alleged existed.

Bensusan claimed that stage hands had intervened and roughed up a cruel trainer during a rehearsal and that they would not allow cruelty during rehearsals at the Empire Theatre of Varieties, Leicester Square (Bensusan 1913a, p. 112; 1913b, p. 117). During the select committee hearings, Charles Jesson advised the witness Albert Hill, a theatre electrician who had described cruel practices, that the National Association of Theatrical Employees would provide protection for its members if they reported instances of cruelty, but he seemed unconvinced (1163ff). Charles Wilson of the RSPCA said that most prosecutions followed failed warnings to establishments that had resulted from tip-offs, but many staff did not realize the confidentiality provision (*2852*). While Lily Grahame had made reports to the RSPCA and protested to performers, but had been told both by these performers and by management to mind her own business (*2573ff*), Margaret Martin, the former music-hall artiste, had reported cruelty to elephants at the London Hippodrome to the Working Girls' Institute, but she feared for her job and noted 'a kind of freemasonry' among the artistes, who did not care to give each other away (514f).

*The Performer* criticized allegations at the select committee that employees were afraid to report cruelty because they feared the loss of their jobs (*Performer* 1922e). The fear of boycotts or sackings was described by Albert Voyce as a fabrication, and he believed the National Association of Theatrical Employees would protect its members in matters outside the province of their own duties (2452ff). R.H. Gillespie said artistes would probably report any cruelty to the VAF or RSPCA, and he had known no victimization of anyone for doing so (*1980ff*). He himself would welcome reports of cruelty (*1792ff*). He said he got complaints about some non-animal acts because they were cruel to the audience, but none from employees or the audience about performing animals (*1906ff*). George Henry Harrop regarded it as the duty of his staff to report cruelty, and he knew of no reprisals elsewhere. Furthermore, rehearsals at his establishment were open to view (*1146ff*). Arthur Aitchley denied there were boycotts, and reports could be made in confidence to the RSPCA (*2186*). Frederick Trussell did not believe any respectable management would sack a complainant and had not heard of that happening (*3094*). As for the circus, James Sanger suggested that because, unlike the music hall, he used his own rather than hired animals, he would be especially ready to listen to stage-hand complaints and he knew of no boycotts (*807*, *855f*). George Lockhart referred to the confidentiality of reporting to the RSPCA, and he had never heard talk of boycotts occurring (*1493f*).

But at the start of the select committee hearings, Charles William Rodgers of the NCDL stated: 'Employees have often privately protested to me against the cruelty which goes on behind the scenes, but they dare not come forward to give evidence for fear of losing their employment' (*3*). Sydney Arthur said that RSPCA and union confidentiality was not realized by those who might otherwise complain (*767ff*). Charles Wilson had heard hundreds of trainers' condemnations of each others' training methods, but only off the record because of their concern for their livelihoods, and for that reason, he did not want to divulge names (*310ff*). Manual staff were in fear of giving evidence in court, although they constantly did so in private conversation with him (*2842*). He was unwilling to force witnesses because he needed their support; otherwise, nine-tenths of his work would be spoiled; consequently, prosecutions were rare (*2845f*).

Basil Tozer, the former theatrical publicity agent, said that although the theatrical profession, scene-shifters, and stage managers were all very much against exhibitions by performing dogs, he felt he could write about cruelty because, unlike them, he had his own business to fall back on. Following an article, he had been contacted by a music-hall person who had witnessed cruelty to a dog but who feared boycott and so wanted no publicity (*2418–2419*, *2554*). John Oscar Mansley, the journalist encouraged to offer evidence by the PADL, had not earlier complained about what he knew because his struggling newspaper depended on theatres for its advertising revenue (857).

Mrs Albert Bradshaw believed that theatre managers themselves might also be fearful for their jobs (*268*), and Samuel James explained artistes' silence as a result of fear of boycott (*719*). According to Charles Haverly, as well as a fear of boycott, variety artistes tended to mind their own business (*620*, *674ff*). Charles Butler said there was a reluctance to speak out against animal acts, because proprietors like Sir Oswald Stoll might take reprisals against those who compromised their business interests (*530ff*). Jessie Hales said she had become afraid of boycott after giving

**Fig. 3.16** (Walter) Trevelyan Thomson, by Bassano Ltd, 1922. © National Portrait Gallery, London. Ref. x122078. Reproduced with permission

evidence of cruelty to the select committee, because to 'gossip' was not etiquette, and her own mother was annoyed at her for offering herself as a witness. She claimed that although the profession was generally against animal performance because of cruelty, there was the fear of boycott for speaking out. Like Margaret Martin, she also thought that a form of freemasonry existed, requiring individuals not to give each other away, disapproving of 'nosey parker' behaviour, and representing a reluctance to threaten another's livelihood. Stage hands, too, would fear the sack (2069ff, 2144, 2130ff).

Lily Grahame had her own fear of being boycotted. If the RSPCA were to prosecute, it would need witnesses, and therefore confidentiality would be removed. Hundreds would come forward to the select committee if their jobs were not under threat (*2604*, *2681*, *2775f*, *2778*). Trevelyan Thomson, a committee member (Fig. 3.16), confirmed that up-to-date evidence was difficult for the select committee to acquire because of the boycott fear (842), and Mollie Lawrence, the former wife of an artiste, said it kept critical artistes from the committee (*2820*). She said she had needed to consider her husband's position, and so had once cut short a remonstration with a cruel trainer (*2817*). Margaret Martin reported that working-class employees especially feared for their jobs, although she had no evidence of actual reprisals. When employed herself, she would have kept quiet, because she had children to support (527f, 561). Albert Hill, the theatre electrician, revealed that a performer had threatened to tell the manager of his theatre about his remonstration with him, but did not because Hill had said if he did he would go to the police. He added that he did not believe the trade union would offer protection against victimization, and not all were in a union. Employees were unaware of any union protection and believed a complaint would mean the sack (1084, 1113ff, 1156, 1198). Sidney Davis, the stage carpenter, also believed that stage hands would be sacked if they protested against cruelty to performing animals (1293).

If employees' fear of reprisals was a disincentive to reporting cruel practices, it might be that tipping was also an incentive to keep quiet. Albert Hill said that electricians such as himself, and stage hands, were disinclined to make any protest, not only because of a fear of being sacked, but also because they would lose the tips they relied on to supplement their income (1105f). Sidney Davis confirmed the important role of this 'touch-me-on-Saturday' from the 'animal people', who gave tips as a bribe not to squeal (1295f). Sydney Arthur said that tips were an important consideration for stage hands, and they could receive up to five shillings from a big turn after one week (*742*). Charles Haverly said that stage managers would not be straightforward in dealing with a complaint because they, too, were bound by an expectation of a tip from the person whose conduct was being questioned (*580*). Arthur Aitchley agreed that if there were no practice of tipping, reports of cruelty would be encouraged, but he did not in any case believe that training through cruelty was practicable (*2260*).

## 3.9 The Question of Dignity and of Effects on Children

In the nineteenth and early twentieth centuries comparatively, little was written in the popular press on wild animals' beauty, their natural behaviour repertoire, the particular features of their natural habitats, or their interests, but much on their ferocity, their candidacy for subjugation, their supposed anthropomorphized 'characteristics', and—through training to mimic human behaviour—their difference and inferiority. And, as Susan Nance points out, there was limited understanding of the breadth or function of species-typical behaviours because most wild animals had yet to be systematically observed in their natural habitats (Nance 2012). Instead, the lion became a 'pitiful figure, lord of all the jungle hunched upon a little stand, like a dunce in school' (Fox 1912, cited by Joys 1983, pp. 41–42).

Harvey Sutherland contributed a very sardonic article about the value and meaning to society of both lion taming and—more generally—animal training as an occupation (Sutherland 1902). Although at the time intended to be entertaining if not amusing, the article contains material which then and now many would find upsetting and provocative. Sutherland presents the material as simple factual elements of his discussion and on which he makes minimal comment, so reinforcing the effect of his journalistic technique. This effect is facilitated by his avoidance of attributing to the reader any assumed reaction to what he has to say—his article serves as a polemic by ignoring the normal conventions of writing such articles. It contains information from William Snyder that the young apprentice lion or tiger should be lassoed to draw it to the front of the cage, to have its claws clipped close, a muzzle fitted, and a collar added (for chaining up as needed later). Training can start with a rawhide whip, stout club, and a pistol with blank cartridges. Corporal punishment is allowed on the 'No lickun, no larnun' principle, continues the advice from Snyder[77]:

[77] William Snyder was Head Keeper at New York's Central Park Zoo, a post he held for 30 years after training elephants at RBBBC. In 1903 Snyder acquired Hattie, an elephant from Carl Hagenbeck, and trained her for the Zoo. Of Hattie, Snyder said: 'I never knew an elephant to pick up tricks so quickly. With most elephants it is a matter of time and beating it into them' (*New York Times Magazine* 1904).

> The lion is the king of the beasts, if we are to believe all we are told, but it is the experience of most lion-tamers that after he has made six or seven rushes at you with his clawless paws, has been knocked out of his senses as many times with the club, yanked around by the collar whose chain is in the hands of your assistants, stung by the rawhide whip, and has experienced the nervous shock of having blank cartridges fired of at him, he gives up and sees that resistance is useless.

Then the first 'fool trick' he learns is to sit on a pedestal. Little courage is actually needed, and the idea of the power of the trainer's eye is a myth.

Subjugation could be spiced with mockery. Madame Batavia was a Himalayan bear at Hengler's Royal Italian Circus near Oxford Circus, who offered an 'amusing and realistic presentment of a gorgeously dressed lady of fashion ... in picture-hat and brocaded dress' and caused a commotion and traffic jams in Piccadilly and the Strand. Her 'interesting and piquant entertainment' involved her passion for drinking stout (*Illustrated London News* 1904). The RSPCA quoted an unnamed paper's comment about its court action against the owner of a chimpanzee at Salisbury who admitted threats with the whip. There was no conviction, but the bench agreed the case should have been brought: the exhibition was regarded as pitiable and degraded the sentiment of the spectator, and it would have been no more cruel to train a man to behave like a chimp. 'These exhibitions serve no good or elevating purpose whatever, and are at variance with all modern teaching of the proper relationship between man and the dumb creation' (*Animal World* 1918). And 'an edifying sight, is it not, to see a dog, dressed up in all the glory of evening clothes, act the part of a drunken "man about town"?' 'What right have we to find pleasure in the antics of a monkey, garbed in the absurd evening dress of a "lady of fashion", forced to bicycle across a wire stretched high over our heads, when we know that the performance means hours of mental and physical torture, and years, perhaps, of solitary confinement in artificial and unhealthy surroundings?' (Fairholme 1919). A correspondent wrote to *The Animal World* that a chimpanzee 'had been trained to dress itself in evening clothes, to eat with knife and fork at a table, to drink out of a glass, to light a cigarette, to simulate drunkenness, to undress and go to bed. The hurried, strained attention of both man and beast was painful to witness. Yet this exhibition drew and amused hundreds' (*Animal World* 1919b).

In addition to opponents who argued generally that trained animal performances were undignified both for animals and their human audience, some witnesses who were trainers limited their objection to certain acts, such as the dog presented as a soldier or a drunk (188ff). Kenworthy objected to the humiliation of the circus lion because the animal was a national symbol.[78] This symbolism had been borrowed by the circus itself. When Sanger's Circus was at Weymouth in 1872, 'one of the lions used to be brought out of a morning and driven in a fancy car round the town with a lady beside him, the pair representing Britannia and the British lion' (Ex-Lion-King

[78] When the Home Secretary said he could not stop a Christmas circus performance at London's Olympia of 70 lions in a cage with an unarmed man, Kenworthy asked: 'Is the Home Secretary not aware that the lion is the national symbol, and that this is bringing the lion into contempt?' (*World's Fair* 1925e).

1872). For Mundy's wild animal show in Tennessee in 1901, a particularly docile, captive-bred lion was similarly carried for the street procession in the passenger's seat of a car, unfettered but with two people ready to shoot to kill if the lion showed any restlessness (Velvin 1911). During the interwar years in Germany, another animal, the eagle, was put to a more approved political use. The circus, its animals, and their trainers have sometimes fulfilled political roles that reflected the regime and its expectations. Both the Soviet and Nazi governments used the circus and its animals to give messages to mass audiences. Otte discusses the appearances in the Dresden circus ring of Erich Rudolf Otto Rosenthal (1885–1954), also known as 'Billy Jenkins, the first German cowboy'. He trained and used the nationally symbolic eagle, controlling the 'uncontrollable' and bonding with it, to impersonate the emblem on military standards and to pose above oversized German flags (Otte 1999).

George Bernard Shaw, in attendance at the PADL protest meeting of 10 February 1921, described animal acts as degrading (reported in *World's Fair* 1921b). He had earlier said that 'a performing animal trained by kindness would be just as intolerable and unnatural a spectacle as one trained in the usual way' (quoted in *Literary Digest* 1920). And in 1921, *Punch* published 'Algol's' sympathetic poem about a 'majestic' hippopotamus kept in the boredom and cold of captivity by a 'repulsive brute' and subjected to the indignity of the stage in order to entertain 'coarse' audiences ('Algol' 1921).

*The World's Fair* (1921l) cited a *Times* leading article on performing animals that asked why they amused. The article suggested that a psychological explanation was needed:

> There is something heartless, and even a little ill-bred, in our pleasure in performing animals ... forced to amuse us without being amused themselves. [The elephant was] a captive, a Samson making sport for the Philistines, who exult a little vulgarly in their triumph over her strength.

Use of animals in this way was a frivolous waste of man's valuable time, the article continued, and he did not know how to use his ingenuity properly. A genius like Darwin could spend months examining the natural behaviour of earthworms, but the rest of us would only have interest in them if they could be taught tricks.

> If all animals did tricks by nature we would never notice them, but he [Darwin] would watch and watch to discover why they did their tricks; and he would get all his pleasure for nothing and win fame by it [without tyranny over the objects of his curiosity] ... The worms are not incommoded by Darwin; but we must humiliate even the king of beasts and our own national emblem before we can be interested in him. (*Times* 1921)

During the debate on the second reading of the bill in the House of Lords in 1924, Viscount Knutsford, describing himself as experienced in training animals and also a conjuror by profession, expressed dislike of the way of life of performing animals rather than the nature of their training[79]; and he felt there was less of a problem with cruelty than with dignity:

---

[79] *Parliamentary Debates* (Lords) 1924, vol. 57, column 147ff.

> Personally, I dislike very much seeing beautiful animals, either day after day or night after night, in uncongenial and unnatural surroundings, performing tricks, and I am always struck when I see them with the much greater dignity that there is in the animals than in the man whose commands they obey.[80]

Earl Russell expressed similar views in 1930:

> For my own part I should, so far as I am personally concerned, deprecate the exhibition of these animals altogether, and I cannot understand the taste of an audience that desires to see them. . . . The more educated taste of audiences nowadays derives no pleasure from seeing animals in unnatural captivity, and performing unnatural tricks.[81]

In 1933, during the second reading of the Performing Animals (Regulation) Amendment Bill, Lord Carnock qualified such concerns for animal dignity:

> It may be undignified for a lion to ride round a circus ring on the back of a horse, but it is much more undignified for a poor man with his wife and children to go from doss house to doss house unemployed and penniless, and that is what this Bill would mean.[82]

Helen Trevelyan objected to T.H. Gillespie's description of training a sea lion: '"Let him realize (he does not say how, which is the most important part of the whole thing) that by holding a lamp on his nose for 3 min, a fish will come to him as certainly as if he had hunted and caught it, and your sealion is practically ready trained to perform." (The sublime idiocy of it is staggering.)' (Trevelyan 1934, p. 29)

The possible indignity of animal performances was matched by a voyeuristic interest in freaks. Tyrwhitt-Drake declared that in around 1924 he had the best collection of freak animals he had ever seen in Britain. He volunteered:

> I know to some people such exhibits are, to put it mildly, distasteful, but from experience I have found to a great number they are of very great interest, and if freaks are carefully selected—avoiding those with a limb short and other disabilities—I am bound to say they interest me, not only from their curious but from their scientific point of view. (Tyrwhitt-Drake 1936, p. 51)

Not long after he wrote these words, it would appear that political disapproval of performing animals was, ironically, encouraging the freak show. *The Times* reported:

> Owing to the performing animal acts we are rather short of circuses', said Mr Harry Gray, a leading showman, at the opening of the 300-year-old Mitcham Fair on Saturday. He was making an indirect allusion to the banning by the borough council of all performing animals. He said that the showmen could not entertain the children as they would like to do. A bigger variety of roundabouts and freak shows makes up for the absence of circuses. (*Times* 1939)

Concerns linking the exhibition of trained animals with a loss of dignity have persisted: 'Wild animals cannot be trained to perform tricks without assaulting their integrity as the wild creatures they are' (Regan 2004, p. 130). William Johnson

---

[80] *Parliamentary Debates* (Lords) 1924, vol. 57, column 154.

[81] *Parliamentary Debates* (Lords) 1930, vol. 76, column 914.

[82] *Parliamentary Debates* (Lords) 1933, vol. 87, column 741.

describes the abuse of dolphins: Sir Harmar Nicholls owned Pleasurama's London Dolphinarium in the early 1970s, featuring dolphins as part of a striptease show, as could also be found at the Moulin Rouge and, briefly, at London's Royalty Theatre under Paul Raymond (Johnson 1990, pp. 220–221). Many trainers have been eager to offer reassurance in this area, and so it was reported that John Chipperfield's policy was never to make fools of the animals but to show their ability through display (Cooke 1974). However, as recently as 1990, Marthe Kiley-Worthington (1990, p. 129) could suggest that 'Representing animals as slightly handicapped humans elicits ... care-giving behaviour from the human audience'.

Arguments about the implications for children of entertainment from animal performances were seen as having more immediate and practical effect than the more abstract concerns about matters of dignity. It was suggested by opponents that animal performances corrupted children's respect for animals. The profession countered that the performances were educational (morally and zoologically) and actually promoted respect for animals in children (*Era* 1921c).[83] However, teachers' unions were especially critical of children's attendance at animal performances and therefore, by extension, the circus. A witness conceded that lions and tigers were no longer included in the repertoire at the Blackpool Tower Circus, but only because they frightened the children. However, his circus depended on other, less sensational animal performances in order to maintain the interest of families with children, and his audience was different from that of the music hall, which was not aimed at children and their families (*1141*). Another witness believed that children, who were on the whole naturally cruel, would be encouraged to try to torture their pets into performing acts seen on the stage (*226f*): and tyranny may easily succeed where coaxing fails (*Literary Digest* 1920). But another witness believed the performances deserved the utmost encouragement because they were uplifting to the public and 'good for all boys and girls' (1827).[84]

In 1921, following teachers' objections to a bucking-mule act, Evan Good, the manager of the Palace, East Ham, contradicted claims in a resolution of West Ham National Union of Women Teachers against his use of performing animals as cruel and unnatural, when it had been urged that children should not attend music-hall animal performances which defeated attempts to teach children kindness to animals (*Era* 1921b). The Palace then challenged the union to witness and assess the performance of the horse Beauty, promising that if problems were found and cruelty

[83] 'Captain' Woodward said that animal performances augmented many natural history lessons begun at school. E.H. Bostock also believed that the performance element of his menagerie show had educational value (*3204*). Everard Calthrop argued that travelling live wild animal shows were good for the study of zoology: not all could go to the zoo (1998–1999). (Like Woodward, he was a Fellow of the Zoological Society.) Later, Hermann Dembeck (1970, pp. 139 and 140) claimed: 'The great animal trainers of the last 100 years have contributed more to the understanding of the more intelligent species of large beasts of prey than all previous books put together'.

[84] At the same time the witness, Everard Calthrop, an amateur trainer, did not approve of performances with wild beasts in cages: they were not elevating to the public and were likely to lead children to practise cruelty on animals at their homes (1812).

were proven, animal acts would be discontinued (*Era* 1921d; *Performer* 1921e, when the union was referred to as the National Union of Teachers). The teachers did not accept the challenge, saying that any kindness to Beauty could not affect the general principle. Meanwhile, it was reported in *The Era*, the Essex National Temperance Cadets went instead to see Beauty in her stables and were duly impressed with how they found her (*Era* 1921e). Commenting on this episode, an ex-cavalryman in the veterinary section suggested that the protesting teachers should learn about discipline and kindness from animal trainers and use the cane less at school (*Era* 1921b). The following year, *The World's Fair* asked in a front-page report: 'Are schools being used for propaganda purposes?' An exercise had been set at Hazlerigg School, Clapham, that promoted the idea that a circus boycott would reduce animal suffering, but when asked to comment, an official from the Board of Education said that 'every lesson is propaganda of some sort' (*World's Fair* 1922c). In one school in Hampshire, the headmistress gave each child a copy of 'A Letter to Boys and Girls about Animals used for your Pleasure', issued by the PADL, and essays were set on the wrongs of performing animals. In one essay, a child wrote that she and her sisters had asked not to go to the circus, and their mother had given them the money, which they took to the cats' shelter. In three villages, schoolchildren of the Band of Mercy canvassed for signatures for petitions against performing animals, and they helped to distribute literature (FHS 1922). Ernest Bell regretted that the Annual World's Fair at the Agricultural Hall in London would show wild animals cramped and confined in small cages when large numbers of children attended and appealed to school teachers to protest (Bell 1926).

In 1931 the Home Secretary and the President of the Board of Education were asked by Peter Freeman, Labour MP for Brecon and Radnor (in that year general secretary of the Theosophical Society in Wales), to consider legislation to prevent children attending trained animal performances; and reference was made to the recent unanimous resolution of the Royal Society of Teachers.[85] In 1933 Lord Banbury of Southam, during the second reading of the Performing Animals (Regulation) Amendment Bill, said that it was:

> . . . far better that the children should be brought up to consider kindness to animals than get their amusement from seeing those animals forced to do things contrary to their nature . . . [the bill] may do some good to children.[86]

Edmund MacMichael complained: 'As regards the effect on the mind and character of the community, and especially of the community's children, an overt act of brigandage is far less insidious than the training and exhibition of these miserable prisoners and captives' (MacMichael 1930a). Miss Laura Bransom, representing the London Head Teachers' Association at a public meeting in the early 1930s, gave a speech on suffering in performing animals and referred to witnesses of it at the earlier hearings of the select committee. She believed that the cruelty of animal trainers was

[85] *Parliamentary Debates* (Commons) 1930–1931, vol. 244, column 1676; vol. 247, columns 615 and 616.

[86] *Parliamentary Debates* (Lords) 1933, vol. 87, column 733.

the fault of their parents and teachers. Animal performances violated 'that sympathetic imagination which we [teachers] are at such pains to develop' through Nature Study in the curriculum, 'Animals' Weeks', and the work of Humane Education Societies in London and Manchester. The National Association of Head Teachers at its annual conference had passed a unanimous resolution:

> That in view of the training in taste and in the humane treatment of animals, which is an essential factor in the education of children, this Association is of opinion that the attendance of children at entertainments which include performing animal turns should be discouraged, and that the use of the school organization in connection with any such entertainment should be prohibited by Local Education Authorities. (cited in Trevelyan 1936, pp. 21–23)

Meanwhile Lady Cory had written to *The Times* to ask the public not to patronize performing animal acts, especially in the company of children (*Times* 1928a); and the Bishop of Bristol observed: 'It is quite intolerable that children should be taken to see lions sitting on stools begging for buns, the finest animals in creation thus being made to look fools' (quoted in Trevelyan 1936, pp. 93–94). A letter of 26 January 1932 was sent to Edmund MacMichael from Miss Ruth Young, secretary of the Association of Head Mistresses, saying its members were too busy to provide a representative for the council of the PADL, but its president, Miss E. Strudwick of St Paul's Girls' School, Hammersmith, wanted to consider support for League resolutions. 'The members of this Association are deeply sympathetic with the work which you have in view.' Bertha Edelbaum wrote to MacMichael on 15 February 1933, suggesting an enthusiastic response from heads of Jewish Schools to his offer of talks (MacMichael Papers).

Enid Blyton's children's circus stories first appeared in the 1930s and present the good circus as one where performing animals are well cared for by the heroes in the story and happy to perform. But she does not favour the use of big cats: 'Tigers hate doing tricks, and it's a shame to make them. But bears just love being clowns'; and she refers to the use of iron-tipped spikes by those not using the kindness she prefers (Blyton 1939, pp. 38 and 145). A successful circus meant as many performing animals as possible (except big cats). The hero, Jimmy, has the secret of handling with love and kindness every live creature there is. Although tigers, lions, and cats are not good in the circus, she told her readers it was unusual for animals not to like circus people (Blyton 1942, pp. 6, 9, 13–14, 52, 84).

Newsletters of the PADL in the 1940s recorded that the National Union of Women Teachers had helped to promote League talks in schools and that the National Association of Head Teachers and London Head Teachers' Association had denounced performing animal shows as bad for the character of children.[87]

---

[87] Richard Vincent Hughes Collection, Theatre Museum (THM/131). Records of the National Union of Women Teachers in the University of London Institute of Education Archives (UWT/D/127.1/3 and 1/15) contain pamphlets, reports, newsletters and memoranda of the PADL and of the Rodeo Protest Committee, and correspondence, mainly with MacMichael, dating from 1927 to 1955. The material includes correspondence with the Ministry of Education about visits of school children to the circus as part of the 'Victory celebrations' in 1946 and press cuttings and typed extracts about children attending circuses and performing animal shows.

MacMichael thought that 'the evil [was] aggravated and not lessened by the fact that the average child has no reason to suspect the presence of evil' in animal performance (PADL 1945). Things certainly went wrong in the United States a few years later. In 1951 at the opening performance of the Shrine Circus, the *Detroit Free Press* reported that 3,000 children had witnessed a lion rip a tiger to death (*Detroit Free Press* 1951, cited by Joys 2011, p. 401). The wild animal trainer Gunther Gebel-Williams looked back in 1999: 'Once there was a time the circus came to town and the children followed it dancing, but these days the children are more likely to be given coloring books by the animal rights people that feature a circus elephant weeping, and the trainer might be singled out for abuse' (Wadler 1999, cited by Joys 2011, p. 447). That year, RBBBC played to 27 million people, with an annual revenue of $600 million. Gebel-Williams had come to the United States in 1969 and rejected the 'fighting' style of big cat act, but kept the flaming hoops and the air of authority. He retired in 1990 and in 1999 Sara Houcke led a new change to 'sympathetic gentleness' (Collins 2000). She was referred to in circus publicity as the 'Tiger Whisperer' but the mildness of her act failed to secure lasting popularity, perhaps because the audience missed the old-fashioned excitement (Davis 2007), assuming it was not turning against use of captive wild animals. On Gebel-Williams's death and the installation of a memorial statue, PETA (People for the Ethical Treatment of Animals) described him as 'a man who spent his entire life terrorizing animals with whips, bullhooks and food deprivation' (Quinlan 2005, cited by Joys 2011, p. 448).

## 3.10 Human Casualties

The most frequent cause of human death or injury from captive wild animals has been attack by big cats or crushing by elephants. Such casualties have in themselves provided an argument against the keeping of wild animals for popular entertainment purposes, although the keepers of the animals concerned have not accepted such arguments. On the contrary, the risk from which the casualties resulted has been an integral part of the perceived attraction of this dangerous form of public human-animal interaction. When fatalities have occurred, the simple short-term response has been the (sometimes ceremonial) destruction, or at least severe punishment, of the 'guilty' animal. The underlying causes—unnatural and frustrating confinement, mistreatment, or biological factors (such as the elephant musth or age-related aggression in the chimpanzee)—have been misunderstood, ignored, or accepted by keepers as an occupational risk. The net result over the years has been occasional human catastrophe and continuing animal suffering. Human fatalities have, of course, provided excellent copy for the press, and the regular occurrence of

**Fig. 3.17** 'Strong-minded Lady. "I think it is shocking cruelty your going in there with a whip". Deputy Lion-tamer (feelingly). "So do I, Mum".' (1921) Reproduced with permission of Punch Ltd, www.punch.co.uk

*Strong-minded Lady.* "I think it is shocking cruelty your going in there with a whip."
*Deputy Lion-tamer (feelingly).* "So do I, Mum."

such accidents has improved the marketability of staged human-animal confrontation. Inherent dangers have traditionally not been explained scientifically, but promoted as part of the process of the just exercise of the right of human domination.

In the Roman amphitheatre, domination was attributed to those who sponsored the spectacles, and those who were dominated were equally animals and subservient humans. In due course animals alone filled the latter category, and the humans at risk began to serve a new master in the form of commerce. The motivations of the wild animals remained the same, and so a fine balance had to be struck between risk and commercial viability. Too many dead trainers would be commercially unsustainable, as well as unattractive to the now autonomous human participants. Some decorum was also needed to package the later rituals and make them acceptable to a more polite audience. But the danger had to remain, and this often resulted in the continued waste of human and animal life beneath a veneer of legitimate popular entertainment.

Some fatalities were the result of a different kind of cause. Fred Ginnett told the select committee of an accident that occurred with an apparatus that had become familiar during the preceding 20 years: the revolving stage used to enable a horse to gallop while remaining relatively stationary. This was thought by some to present a danger to the animal, but in this case the victim was human. The revolving stage device was often questioned, and on this occasion, at the Coliseum, a pony escaped unhurt when thrown off the stage into the orchestra, but a boy was killed (719).

Women trainers became a novelty in the nineteenth century, and a well-publicized casualty was Ellen Bright, who was killed in 1880 by a tiger at the age of 17 (Bostock and Velvin 1903, p. 31). Then lions mauled Van Amburgh's daughter to death in 1883. Dolly Castle abandoned training after losing part of her hand to a lion, and Valecita, a leopard trainer, was killed by her animals at her Michigan training barn in 1924 (Joys 2011, p. 375). But the human casualties were usually men who were at some point in the production of their heroic role (and if they survived, they—alongside some of the women—would gravely boast of their scars) (Fig. 3.17). 'Captain' Bonavita had to have his arm amputated after a lion attack, but continued to perform with the same animal (Velvin 1910–1911). After

Hagenbeck had sold a lion to an American circus, it was returned, pickled in a cask, with a request for a refund. The trainer had been mauled as a result of a mistake and the lion beaten and killed by his rescuers (Hagenbeck 1912, pp. 121–123). More recently, it was possible that the execution of a tiger in similar circumstances would result in charges, but they were dropped: Graham Chipperfield had shot a tiger after it mauled his brother Richard at St Petersburg in Florida (*New York Times* 1998). It was inevitable that problems would also result from escapes. In 1914, Mme Andree, who with others was later charged with criminal negligence, had her lions escape after a vaudeville performance, during their transfer to moving dens. Although not vicious, having been supplied by Hagenbeck as cubs, one was shot, as was a policeman in error by his colleague (*New York Times* 1914).

Danesfort referred to the many cases reported by inspectors of the RSPCA in 1928 and 1929 in which lions attacked the trainer and injured him. In one case at the Tower Circus in Blackpool, a free fight developed among a large number of lions, bears, tigers, and other animals performing in a closed ring, as reported in *The Blackpool Gazette* (1928). The Czech trainer, Trubka, lost charge of ten lions, which posed on pedestals and jumped through hoops; seven bears, which were drawn round the ring in carriages by dogs; two tigers; one puma; and two dogs: they were all eventually subdued with hoses. Danesfort observed that this sort of exhibition was not popular with the general public, although it appeared in some cases to have an attraction for 'strangely constituted audiences'.[88] Letters in *The Spectator* describe how Herr Adolf Cossmeyer at Bertram Mills Circus was washing his polar bear and slipped and fell, or the bear put its paw on his shoulder and brought him down. The bear then bit off his nose and ear and mauled his leg, so that he died soon afterwards. MacMichael said that Mills should accept responsibility for keeping this animal in his circus, and the authorities would not have ignored the event if the bear had belonged to some wandering gypsy instead of to a first-rate show (Yeats-Brown 1930, MacMichael 1930b). Peter Freeman asked the Home Secretary to consider banning the importation of performing bears, because of recent accidents such as this one.[89]

Elephants frequently figured in accidents. *The Spectator* published 'The elephant inquest' in 1900 after a Sanger's Circus keeper had been killed at Crystal Palace and the elephant shot after unsuccessful poisoning. The elephant had already crushed a keeper who had been sacked by Sanger for cruelty but who was re-employed as a labourer. The elephant was not destroyed on the earlier occasion because its motives were understood. Although very docile, the elephant had also been assaulted, this time with a lance, by the second keeper, and broke its chains to trample him. *The Spectator* wondered whether this was justifiable homicide: it left a very unpleasant impression of the management of the 'elephant herd' in Britain. Circus elephants also had a record of 'accidentally' crushing people they remembered with hate (*Spectator* 1900). A later example in the United States was the

[88] *Parliamentary Debates* (Lords) 1930, vol. 76, columns 902–904.

[89] *Parliamentary Debates* (Commons) 1930–1931, vol. 249, column 1365.

**Fig. 3.18** George Lockhart crushed by his circus elephant, Sauce, at Walthamstow railway goods yard, 1904. The elephant escaped from the train which was transporting him to the next circus destination. Mary Evans Picture Library

elephant Boo, also violent because of her inhumane early training, and she killed a trainer who had retaliated with a hot iron bar and burnt her trunk. For 5 years she was kept in a cellar on a property used for disbanded circus animals, before killing another keeper in 1937 (Allen and Kelley 1941, p. 59, cited by Tait 2012, pp. 92–93). Charles Haverly told the select committee that one of the Lockhart family was killed by an elephant in reprisal for cruelty, because elephants would bear a grudge (*602*). However, in a reply to a rather clumsy question from Kenworthy, George Lockhart explained that his father was killed by a frightened elephant when trying to recapture it during transportation in London (*1308*, *1312*) (Fig. 3.18). This stampede at Walthamstow came about because the elephants were young and timid and easily took fright (*1379*).[90]

Any regular incidence of human casualties, especially fatalities, resulting from the training, transport, and exhibition of performing animals would serve the arguments of those aiming at prohibition. However, statistics of casualties were not specially gathered, and reference to them was usually made in Parliament following a well-publicized press article in a national daily, even though the less accessible trade press was the most reliable source of such news. When Peter Freeman (Fig. 3.19) asked the Home Secretary the number of accidents, fatal and non-fatal, caused in Britain by performing and captive animals in circuses, music halls, and travelling menageries over the course of 5 years, he was told that the information was not available.[91] Dr Somerville Hastings also asked the Home Secretary the number of persons killed by wild animals in circuses and menageries over 10 years, but again the information was not available.[92] Freeman retained a long-term interest in this matter, and 25 years later unsuccessfully asked the Secretary of State for the Home Department, Major Lloyd-George, to consider

[90] Accidents involving elephants have continued over the decades. For example, two people were crushed to death in 1985 and 1993 by elephants of the Clyde Beatty–Cole Bros Circus (Miller 1999).

[91] *Parliamentary Debates* (Commons) 1929–1930, vol. 237, column 582.

[92] *Parliamentary Debates* (Commons) 1930–1931, vol. 249, column 400.

**Fig. 3.19** Peter Freeman, by Bassano Ltd, 1931. © National Portrait Gallery, London. Ref. x84189. Reproduced with permission

legislation prohibiting the exhibition of wild animals and the turns in which they appeared, noting the number of deaths and accidents that had been caused. Lloyd-George thought the number of accidents small, perhaps only two fatal ones in the previous 25 years, but Freeman cited a very recent case of a man being mauled to death by four lions and another when a man only just escaped with his life.[93]

## References

### *(Articles in periodicals are anonymous or untitled unless otherwise shown.)*


Aflalo F G (1900) The ethics of performing animals. The Fortnightly Review March, pp 382–391

'Algol' (1921) To a performing hippopotamus. Punch 160, 23 March 1921, p 229

Allen E, Kelley BF (1941) Fun by the ton. Hastings House, New York

The Animal World (1918) Performing animals. October, p 117

The Animal World (1919a) June, pp 69–70

The Animal World (1919b) Performing animals. August, pp 87–88

The Animal World (1921a) Conjuring tricks with animals. April, p 43

The Animal World (1921b) Performing animals: legislation. July, p 74

The Animal World (1922) Public protest on behalf of a performing elephant. February, p 14

The Animal World (1923) The performing animals bill. September, front page

Anon (1869) Haney's art of training animals. A practical guide for amateur or professional trainers. Giving full instructions for breaking, taming and teaching all kinds of animals including an improved method of horse breaking, management of farm animals, training of sporting dogs; serpent charming, care and tuition of talking, singing and performing birds; and detailed instructions for teaching all circus tricks, and many other wonderful feats. Jesse Haney, New York

Baker FH (1995) A description of Blackwell "A" Winning Colliery, Alfreton, Derbyshire in 1933. Bulletin of the Peak District Mines Historical Society 12(6):39


93 *Parliamentary Debates* (Commons) 1954–1955, vol. 537, columns 2220 and 2221.

Beers DL (2006) For the prevention of cruelty. The history and legacy of animal rights activism in the United States. Swallow Press, Athens
Bell E (1918) The performing animal. The Contemporary Review 113:93–100
Bell E (1926) Christmas amusements Letter. The Spectator 18 December, p 1112
Bensusan SL (1913a) The torture of trained animals (reprinted from The English Illustrated Magazine). In: Trist S (ed) The under dog. A series of papers by various authors on the wrongs suffered by animals at the hand of man. The Animals' Guardian Office, London, pp 109–117
Bensusan SL (1913b) The sorrows of trained animals (reprinted from To-day, 11 February 1899). In: Trist S (ed) The under dog. A series of papers by various authors on the wrongs suffered by animals at the hand of man. The Animals' Guardian Office, London, pp 117–118
Bensusan SL (1913c) The case of performing animals (reprinted from Black and White, 1899). In: Trist S (ed) The under dog. A series of papers by various authors on the wrongs suffered by animals at the hand of man. The Animals' Guardian Office, London, pp 118–121
Billboard (1924) 13 December, p 106
Birmingham HR, Garvie AE, Harvey JM, Hawtrey C, Henderson A, Welldon JEC (1922) Letter. The Spectator 11 August, p 207
The Blackpool Gazette (1928) 18 October
Blyton E (1939, new edition 1973) Hurrah for the circus! Dean, London
Blyton E (1942, new edition 1973) Circus days again. Dean, London
Bostock EH (1927) Menageries, circuses and theatres. Chapman and Hall, London
Bostock FC, Velvin E (ed) (1903, reprinted 2009) The training of wild animals. Watchmaker, Seaside
Carlisle DT (1956) And in this ring, ladies and gentlemen...! The New York Times 8 April, p BR3
Cheshire DF (1974) Music hall in Britain. David and Charles, Newton Abbot
Chindahl GL (1959) A history of the circus in America. Caxton, Caldwell
Collins G (2000) Tiger trainer burning bright: circus artist spurns whip and chair. The New York Times 24 February, p E1
Cooke R (1974) A family circus. The Illustrated London News 2 December
The Daily Herald (1921a) 2 June, p 4
The Daily Herald (1921b) Clear the stage. Editorial 3 June
Davis CB (2007) Cultural evolution and performance genres: memetics in theatre history and performance studies. Theatre J 59:595–614
de Frece M (1934) Recollections of Vesta Tilley. Hutchinson, London
Dembeck H (1970, German edition: 1966) Willingly to school. How animals are taught (trans: Johnson C). Harrap, London
Detroit Free Press (1951) 3,000 children see circus lion rip tiger to death. 20 February
Ditmars R (1933) Toledo Blade 25 February
Dolman F (1899) Four-footed actors. About some well-known animals that appear on the London and provincial stage. The English Illustrated Magazine 21(192):521–526
The Encore (1921a) 10 March, p 3
The Encore (1921b) 5 May, p 3
The Encore (1921c) 9 June, p 3
The Encore (1923) 29 March, pp 2, 3
The Era (1921a) 23 February, p. 16
The Era (1921b) 2 March, p 16
The Era (1921c) 9 March, p 15
The Era (1921d) 16 March, p 16
The Era (1921e) 23 March, p 18
The Era (1921f) 30 March, p 12
The Era (1921g) 13 April, p 18
The Era (1921h) 27 April, p 12
The Era (1921i) 8 June, p 13
The Era (1921j) 15 June, p 14

The Era (1921k) 17 August, p 11
The Era (1921l) 17 August, p 12
The Era (1922a) 22 February, p 11
The Era (1922b) 22 May
The Era (1922c) 24 May, p 13
The Era (1923a) 11 January, p 25
The Era (1923b) Editorial 1 March, p 13
The Era (1923c) 28 March, p 13
The Era (1923d) 18 July, p 11
Ex-Lion-King (1872) Lions and lion-taming. Every Saturday 17 February, pp 173–176
Fairholme EG (1919) Trained animal performances. An insult to animals and human beings alike. The Animal World, March, pp 28–29
FHS (1922) The case of the performing animal Letter. The Animal World November, p 127
Fox EL (1912) When the jungle goes to school. Harper's Weekly 56, 30 November, p 18
Gillespie TH (1934) Is it cruel? A study of the condition of captive and performing animals. Herbert Jenkins, London
Hagenbeck C (1912) Beasts and men: being Carl Hagenbeck's experiences for half a century among wild animals (trans: Elliot HSR, Thacker AG). Longmans Green, London
Holmes C (1988) John Bull's island. Immigration and British society, 1871-1971. Macmillan, Basingstoke
Hornaday WT (1922) The minds and manners of wild animals. A book of personal observations. Charles Scribner's Sons, New York
The Illustrated London News (1904) The playhouses. 15 October, p 531
The Isle of Thanet Gazette (1945) 14 December
Johnson W (1990) The rose-tinted menagerie. Heretic Books, London
Joys JC (1983) The wild animal trainer in America. Pruett, Boulder
Joys J C (2011) The wild things. Dissertation. Bowling Green State University, Bowling Green
Kenworthy JM (1933) Sailors, statesmen—and others. An autobiography, etc. Rich and Cowan, London
Kiley-Worthington M (1990) Animals in circuses and zoos. Chiron's world? Little Eco-Farms Publishing, Basildon
Limon D, McKay WR (eds) (1997) Erskine May's treatise on the law, privileges, proceedings and usage of Parliament. Butterworth, London
The Literary Digest (1920) Cruelty charged in training trick animals for stage and movie. 25 September, pp 102–110
MacMichael ET (1930a) Letter. The Spectator 11 October, p 497
MacMichael ET (1930b) Letter. The Spectator 25 October, pp 583–584
MacMichael Papers (uncatalogued and unpublished material relating to the director of the Performing Animals' Defence League between 1929 and 1968, kindly lent to the present author by Edmund MacMichael's daughter, Mrs Esther Denham)
McKenzie C (2000) The British big-game hunting tradition, masculinity and fraternalism with particular reference to the "The Shikar Club". Sports Historian 20(1):70–96
Mighetto L (1991) Wild animals and American environmental ethics. University of Arizona Press, Tucson
Miller EK (1999) Animal rights groups protest circus visit. The New York Times 1 August, p LI3
Mitchell PC (1929) Centenary history of the Zoological Society of London. Zoological Society of London, London
Mitchell PC (1937) My fill of days. Faber and Faber, London
Nance S (2012) 'What is modernity for animals and us?' http://performinghumanity.wordpress.com/tag/training/. 12 November. Accessed 27 Nov 2012
The New York Times (1914) Five lions loose in theatre throng. 18 December, pp 1, 8
The New York Times (1931a) 23 April, p. 23
The New York Times (1931b) 25 April, p 13

The New York Times (1940) 31 March, p 124
The New York Times (1998) Man who shot tiger could face charges. 12 January, p A15
The New York Times Magazine (1904) Hattie of Central Park. Most intelligent of all elephants. Her cleverness is a revelation to trainers—why, she understands English. 19 June
Otte M (1999) Sarrasani's Theatre of the World: monumental circus entertainment in Dresden, from Kaiserreich to Third Reich. German History 17(4):527–542
PADL (Performing Animals' Defence League) Newsletter no. 27, winter 1945
Panayi P (1994) Immigration, ethnicity and racism in Britain, 1815-1945. Manchester University Press, Manchester
Peck GW (1906) Peck's bad boy with the circus. Thompson and Thomas, Chicago
The Performer (1921a) 17 February, p 9
The Performer (1921b) 24 February, p 13
The Performer (1921c) 3 March, p 17
The Performer (1921d) 10 March, p 21
The Performer (1921e) 17 March, p 12
The Performer (1921f) 24 March, pp 7, 9
The Performer (1921g) 24 March, p 10
The Performer (1921h) 7 April, p 15
The Performer (1921i) 14 April, p 9
The Performer (1921j) 14 April, p 13
The Performer (1921k) 21 April, p 8
The Performer (1921l) 28 April, p 13
The Performer (1921m) 2 June, p 8
The Performer (1921n) 9 June, p 3
The Performer (1921o) 9 June, p 5
The Performer (1921p) 16 June, pp 5, 9
The Performer (1921q) 16 June, p 9
The Performer (1921r) 16 June, p 10
The Performer (1921s) 23 June, p 9
The Performer (1921t) 11 August, p 7
The Performer (1922a) 18 January, p 3
The Performer (1922b) Cranks on the crawl. 22 February, p 5
The Performer (1922c) 15 March, p 6
The Performer (1922d) 22 March, p 7
The Performer (1922e) 8 June, p 4
The Performer (1922f) 26 July, p 3
The Performer (1923a) 21 February, p 11
The Performer (1923b) 28 February, p 5
The Performer (1923c) 21 March, p 18
The Performer (1923d) Invited protests against animal acts. 4 April, p 10
The Performer (1923e) 11 April, p 14
The Performer (1923f) 25 July, p 5
The Performer (1945) 13 December
Punch (1921a) 160, 16 March, p 213
Punch (1921b) 160, 8 June, p 453
Punch (1921c) 160, 29 June, p 512
Punch (1923) 164, 13 June, p 567
Punch (1924) 166, 27 February, p 221
Quinlan P (2005) Statue stirs emotions of bygone era. Sarasota Herald-Tribune 6 December
'The Recorder' (1921) The Magnet 1 October, p 3
Regan T (2004) Empty cages. Facing the challenge of animal rights. Rowman and Littlefield, Lanham

Rothfels N (2002) Savages and beasts: the birth of the modern zoo. Johns Hopkins University Press, Baltimore

RSPCA (1931) Minutes, policy and publications committee, 6 November

RSPCA (1937) Minutes, performing and captive animals special committee, 30 September

Ryder RD (2000) Animal revolution. Changing attitudes towards speciesism. Berg, Oxford

Sax B (2000) Animals in the Third Reich. Pets, scapegoats, and the Holocaust. Continuum, New York

The Spectator (1900). The elephant inquest. 24 February, pp 271–272

Sutherland H (1902) Training wild animals. Ainslee's Magazine 8(6), January, pp 518–525

Swain J (1933) Brutes and beasts. Noel Douglas, London

The Thanet Advertiser and Echo (1945) 11 December

Time (1934) A bully and his betters. 12 February

The Times (1913) 29 November

The Times (1921) Performing animals. 7 June, p 11

The Times (1922) 4 December

The Times (1923) 23 March

The Times (1927) 29 December, p 6

The Times (1928a) 10 December, p 10

The Times (1928b) 15 December, p 6

The Times (1928c) 18 December, p 10

The Times (1928d) 29 December, p 4

The Times (1935) Obituary, Lord Danesfort. 1 July

The Times (1937) 15 January, p 9

The Times (1939) 14 August, p 16

Trevelyan H (1934) Heaven's rage. C W Daniel, London

Trevelyan H (1936) Laugh, clown, laugh! C W Daniel, London

Tyrwhitt-Drake HG (1936) Beasts and circuses. Arrowsmith, Bristol

Tyrwhitt-Drake HG (1946) The English circus and fairground. Methuen, London

United Kingdom Parliament (various dates) Parliamentary Debates (Commons). HMSO, London

United Kingdom Parliament (various dates) Parliamentary Debates (Lords). HMSO, London

United Kingdom Parliament (1921a) Report from the Select Committee on Performing Animals, together with the Proceedings of the Committee and Minutes of Evidence. HMSO, London

United Kingdom Parliament (1921b) Report from Standing Committee D on the Performing Animals (Prohibition) Bill with the Proceedings of the Committee. HMSO, London

United Kingdom Parliament (1922) Report from the Select Committee on Performing Animals, together with the Proceedings of the Committee and Minutes of Evidence. HMSO, London

VAF (Variety Artistes' Federation) Archives, box VAF 5, held by Equity (British Actors Equity Association), London

Velvin E (1910-1911) Critical moments with wild animals. McClure's Magazine 36, pp 377–385

Velvin E (1911) Critical moments with wild animals, second paper. McClure's Magazine 37:155–163

Wadler J (1999) Still bringing out the animal, and animals. The New York Times 9 April, p B2

Wilmut R (1989) Kindly leave the stage! The story of Variety 1919-1960. Methuen, London

Wilson DAH (2008) Politics, press and the performing animals controversy in early twentieth-century Britain. Anthrozoös 21(4):317–337

Wilson DAH (2009a) Racial prejudice and the performing animals controversy in early twentieth-century Britain. Society Animals 17(2):149–165. http://booksandjournals.brillonline.com

Wilson DAH (2009b) 'Crank legislators', 'faddists' and professionals' defence of animal performance in 1920s Britain. Early Popular Visual Culture 7(1):83–101. http://www.tandfonline.com/doi/full/10.1080/17460650902775393

The World's Fair (1911a) 7 October, p 8

The World's Fair (1911b) 23 December

The World's Fair (1921a) 5 February, p 13

The World's Fair (1921b) 19 February, front page
The World's Fair (1921c) 26 February, p 10
The World's Fair (1921d) 5 March, front page, p 10
The World's Fair (1921e) 26 March, p 6
The World's Fair (1921f) 26 March, p 10
The World's Fair (1921g) 2 April, front page, p 7
The World's Fair (1921h) 16 April, front page
The World's Fair (1921i) 23 April, p 5
The World's Fair (1921j) 7 May, p 16
The World's Fair (1921k) 4 June, front page
The World's Fair (1921l) 11 June
The World's Fair (1921m) What we think. 11 June, p 21
The World's Fair (1921n) 6 August, front page, p 7, and later editions
The World's Fair (1921o) What we think. 6 August, p 13
The World's Fair (1921p) 27 August, p 5
The World's Fair (1921q) 3 September, p 6
The World's Fair (1921r) 3 December, front page
The World's Fair (1922a) 18 February, p 13
The World's Fair (1922b) 18 March, p 7
The World's Fair (1922c) 29 April, front page
The World's Fair (1922d) 10 June, front page
The World's Fair (1922e) 10 June, p 10
The World's Fair (1923a) 20 January, p 16
The World's Fair (1923b) What is a performing animal? 24 February, p 18
The World's Fair (1923c) 3 March, p 18
The World's Fair (1923d) 10 March, p 9
The World's Fair (1923e) What we think. 31 March, p 21
The World's Fair (1923f) 28 July, p 19
The World's Fair (1923g) 4 August, p 11
The World's Fair (1923h) 11 August, p 16
The World's Fair (1923i) Other people's views. 17 November, p 12
The World's Fair (1923j) 24 November, p 21
The World's Fair (1923k) 4 December, p 14
The World's Fair (1923l) 15 December, p 23
The World's Fair (1923m) 22 December, p 11
The World's Fair (1924a) 5 April, p 12
The World's Fair (1924b) 5 July, p 29
The World's Fair (1924c) 12 July, p 4
The World's Fair (1924d) 19 July, p 10
The World's Fair (1925a) 4 July, p 9
The World's Fair (1925b) 10 October, p 9
The World's Fair (1925c) 7 November, front page
The World's Fair (1925d). What we think 7 November. p 25
The World's Fair (1925e) 19 December, p 9
The World's Fair (1934) 22 September
Yeats-Brown F (1930) Letter. The Spectator 3 October, pp 438–439

# Chapter 4
# Means and Ends

**Abstract** This chapter discusses the historic means and effects of acquisition, movement and importation of animals for the entertainments industry, the various methods alleged to have been used in taming or training these animals (according to rearing, age and tractability) and the nature of the acts they were trained to perform. International examples are given, including the activities and methods of Frank Bostock, Carl Hagenbeck and the succession of American big cat exhibitors. Reference is made to evidence and opinion presented before, during and since the British parliamentary select committee on performing animals of 1921 and 1922, including other considerations such as standards of accommodation, the suitability or otherwise of the music-hall and circus environment, the characteristics and behaviour of particular trainers, and veterinary and specialist opinion regarding health and psychological welfare.

## 4.1 Capture, Transport and Captivity

The select committee highlighted concerns about the acquisition and confinement of animals that would gather momentum over ensuing decades. The dialogue frequently turned to the matter of capture and importation, transport and travelling as well as routine accommodation arrangements. Other more abstract issues that were soon to attract further attention in different contexts involved the nature and consequences of captivity and the significance of captivity in relation to evolution. The business of capture was also the ideal source for many published accounts especially of the nineteenth century that satisfied widespread interest in the heroic human exploration and conquest of untamed nature in far-off places. The more typical context for such accounts was the hunting expedition, but the capture of live animals for zoos or circuses provided the same storytelling opportunities, and indeed the two activities were often combined within the same expedition. The capture of wild animals was at least as wasteful, destructive and cruel as the hunting of them for sport. However, if sport hunting were not involved, one difference was that the western entrepreneurs frequently assigned the work to indigenous people who had the specialist know-how for seizing the animals required of their locality.

 
D.A.H. Wilson, *The Welfare of Performing Animals*, Animal Welfare 15,
DOI 10.1007/978-3-662-45834-1_4

It was also often the case that young animals were sought (Bostock and Velvin 1903, p. 113ff). The activity was no doubt less deserving of any heroic assignation than that enjoyed by the intrepid hunter of big game, although defeating logistical challenges frequently required heroic organization, commitment and financial risk.

The acquisition of various cubs usually involved the killing of fiercely protective parents. *Haney's* noted that bears are born blind, like puppies, and remain so for about 8 or 9 days, but with care could be raised even if taken when only 4 or 5 days old (Anon. 1869, pp. 137 and 138). Later, chloroform might also be used to secure bears (Bell 1918, citing Velvin 1914). Trauma and sometimes fatal injury could occur during capture, and animals died of fright or internal injury in pitfalls, or after refusing to eat. Captured animals could be subdued by starvation before transportation (Bostock and Velvin 1903, pp. 109, 110, 116, 117). Snares and nets were used as well as traps and pitfalls. Raymond Blathwayt visited Hagenbeck to find out more of the process by which he acquired animals for his own purposes and for others, like Barnum, who had visited in 1872. For the capture of lion, elephant, rhino and giraffe for his agent in Nubia, the parents were surrounded and killed by natives with sword, spear or gun and the suckling young taken and raised on goat's milk. Leopards and hyenas were caught in baited traps, and in Borneo and Java, trapfalls and pitfalls were also used for a variety of animals, most of which died as soon as they were caught, the cause being internal injuries, self-killing through excitement or failure to eat. Internal injuries, especially in rhinos, might not be apparent until death occurred a couple of months after capture. Young elephants in Burma and Ceylon were separated from the herd by dogs and noosed around the hind legs. The trailing rope was followed and eventually tied round a convenient tree, the elephant later dropping down in despair, when the rope was fastened round its other legs, and the animal left for several days until calmed down (Blathwayt 1893a). There was great demand from Barnum and Forepaugh circuses for elephants: 67 were taken from Ceylon in 1883 (Hagenbeck 1912, pp. 28 and 29).

Adult tigers in India were often trapped in pitfalls as well: 'A kid with a stone tied to an ear is secured to the middle of the frail [pit] cover. Tortured by the weight of the dangling stone and frightened because of loneliness, the unfortunate kid bleats piteously as the men withdraw. Far into the jungle the mournful sound penetrates almost incessantly—for hours, sometimes for several days ...' until a tiger arrives. Jaguars can be trapped in a large 'mouse trap' baited with a monkey or dog. A hippopotamus calf is caught by harpooning the mother and tiring and fighting it—Gustave Hagenbeck, brother of Carl, lost his life this way. The rhinoceros mother must be killed for its calf—an army of natives dodge and weave until the tendons in the surprised mother's rear legs can be cut with swords (Rolker 1905). In northwest Africa, if adult lions or elephants turned to defend their young, they too were killed by the traditional method of cutting the Achilles tendon so that they were hobbled and bled to death. Capture of most young hippopotami and crocodiles was effected by harpoon, when a careful aim ensured injury rather than death (Hagenbeck 1912, pp. 51–58). Hagenbeck knew that slaughtering the mother was also the usual way to obtain an infant walrus. One of his was caught in what he describes as a very ingenious though pathetic manner. 'The mother was first killed and drawn into close proximity with the boat, which was then allowed to remain still. Before very long the cub came looking for his mother, and having found her, climbed up on her back, where he was easily caught and secured' (Hagenbeck 1912, p. 248).

The performing animals controversy drew attention to newly growing concerns about the unnatural and unhealthy nature of often continuous travelling for performing and exhibition animals, starting with the import of those captured in the wild and ending with the routine of travel between exhibition sites.[1] Blathwayt had described a procession across the desert in 1870 of captured animals between Cassala and Suakim in Nubia, with 160 drivers and lasting 36 days. On arrival, elephants were transferred by sling from a barge to the steamer, very often being killed by the breaking of their necks or legs. Deaths of animals on board might result from seasickness and failure to eat. On one occasion, only 12 out of 40 reached Hagenbeck in Hamburg. On a voyage in 1881, 33 of 42 ostriches broke their legs in a storm at sea and were thrown overboard (Blathwayt 1893b). The experience of fear and seasickness on a rough sea voyage after capture would usually subdue a wild animal (Velvin 1911). Hagenbeck confirmed that half of the captive animals would die before arrival in Europe. He gives details of the transport of a large collection of animals across the desert to a Red Sea port, including 150 head of cattle and hundreds of sheep and goats for milk and slaughter and 100 heavily laden camels. The procession suffered from heat and lack of water and went by night. Some of the animals were driven, and others were in cages on the backs of camels. A hippopotamus would be regularly bathed and slung between two camels. Each day, 15–20 miles were covered for up to 6 weeks, with arrival in Hamburg after 3 months. Half died through sunstroke or, as also in the Mongolian and Siberian expeditions, terror or disease. The death rate in the less manageable adult animals was always higher (Hagenbeck 1912, pp. 51–58, 65–69, 89, 92, 94).

The role of the Hagenbeck concern as the world's leading wild animal importer and dealership was interrupted both by Carl Hagenbeck's death in 1913 and by the First World War, with the British blockade of German ports. One of those who stepped in was the American Frank Buck, and he attempted to minimize casualties on long journeys, partly by paying close attention to the animals' dietary requirements (Joys 2011, pp. 162 and 168). A British dealer based in London's docklands, John D. Hamlyn, was also active in the early twentieth century. Hamlyn imported more anthropoid apes than anyone, and these included the celebrated gorilla 'John Daniel'; and as trainers of chimpanzees, he and his wife were described as unequalled. After the war the German trade was embargoed, but Hamlyn died before he could benefit, and instead another did: Bruce Chapman started a travelling menagerie after the First World War, soon adding a circus. Until 1920 he was a bird

[1] The Government of Southern Rhodesia would later prohibit the catching of wild animals for circuses and menageries on account of the cruelty inflicted in the capture alone (*Parliamentary Debates* (*Commons*) 1947, vol. 438, column 108).

dealer and then came from Birmingham to London in 1929 to become the largest wild animal dealer in Europe before his death in 1935, when much of the trade went back to the Continent. He imported vast quantities of stock, large and small, supplying both the Continent and America (Tyrwhitt-Drake 1946, pp. 182, 187 and 188; 1936, pp. 62 and 63).

Before the controversy in Britain had fully got under way, *The World's Fair*, in an article of 1911, 'Wild beasts at sea', admitted the unpleasantness of transportation: 'It is a fact that a voyage thoroughly upsets wild animals, and it takes some time after reaching dry land ere they recover their normal condition' (*World's Fair* 1911a). The problem could get worse later in the century. Johnson alleges that in the late 1980s, elephants, bears, tigers and lions owned by the Chipperfield organization endured months of extreme confinement, injury and death during poorly planned sea voyages to and from the Far East and New Zealand, resulting in a conviction on return to Southampton (Johnson 1990, pp. 140–142). But routine transportation within the country of destination also caused difficulties, and it was this that earned regular criticism from those who were more easily able to observe conditions at home. Their attention concerned domestic as well as wild animals. The circus and zoo owner Tyrwhitt-Drake recounted his experience of gnus and a wild ass dying from the trauma of handling, putting into confinement and travel: 'In all these cases the animals were used to seeing human beings, and were in perfect health, but in spite of this just died of excitement and shock' (Tyrwhitt-Drake 1936, pp. 35–37). At the same time, he complimented the travelling menageries of the nineteenth century:

> The animals certainly looked well in their comparatively small travelling dens; on the whole a good deal better than those in the stationary collections and zoos. One reason for this was, I believe, that the journey from place to place, bumping along often over indifferent roads, gave them much-needed exercise, while the taking down of the shutters of the wagons gave them plenty of fresh air and sunlight (Tyrwhitt-Drake 1946, p. 168).

But he commented:

> I do not think anything is more troublesome and worrying than moving wild animals. It is certainly the most dangerous time, both for the animal itself, as its nerves are liable to be upset, making it scary, and also for the people (Tyrwhitt-Drake 1936, p. 27).

In July 1913, he had lent his entire menagerie to the Scottish Zoological Society (directed by his friend, T.H. Gillespie), which had just built its park in Edinburgh but had practically no animals to put in it. The animals were placed in packing cases in closed trucks of a train, and 'difficulty of packing' was experienced (Tyrwhitt-Drake 1946, p. 181). 'All my shifting dens are boxes with iron bars in front and a sliding door, which lifts up, behind; once the animal is enticed inside the slide is dropped, fastened with a bolt' (Tyrwhitt-Drake 1936, p. 35) Fig. 4.1.

Harold Nicolson's later review for *The Spectator* noted with some sarcasm that Jennison (1937) had given the assurance that all wild animals much enjoyed foreign travel, and that in the course of their journey they were so entertained by observing new sights and sounds that 'they do not really mind having an uncomfortable cage, and that the truly boring time only comes when they reach the capital and are

**Fig. 4.1** Sir (Hugh) Garrard Tyrwhitt-Drake, by Henry Walter Barnett, 1910–1920. © National Portrait Gallery, London. Ref. x45419. Reproduced with permission

**Fig. 4.2** Sir Harold Nicolson, by Howard Coster, 1935. © National Portrait Gallery, London. Ref. x24456. Reproduced with permission

housed in a great gaunt cage which never moves'. Nicolson said he was obliged to Mr Jennison for telling him that lions attached to travelling circuses were happier than those who were allotted spacious residential quarters in Regent's Park. 'It is nice being given new ideas' (Nicolson 1950) (Fig. 4.2).

The small size of travelling cages and crates was frequently criticized, in particular. Tyrwhitt-Drake explained that travelling crates had only to be small enough to be handled and to prevent an animal like a leopard being knocked about; the size would also depend on the length of the journey. There was no cost saving in using the smallest possible crate because charge was by the mile (265ff). Henri Rey had told the select committee that payment was for the animal by weight, not the size of box, which did not therefore have to be unnecessarily small, but committee member Trevelyan Thomson believed that the total consignment, not just the animal, was charged by weight (1804ff). Fred Ginnett said that tight travelling spaces were necessary, on the horsebox principle. He had sold and despatched to Barnum and Bailey across the Atlantic a lion and lamb in their accommodation cage, of ample size at 16 × 7 ft, but having a special small transit cage within for the lion (629).

T.H. Gillespie was critical of the cramped travelling boxes where animals could live except for an hour or two on the stage: he recognized this as the most cogent objection to performing animals. But he went on to claim that animals preferred cages, because after weeks of confinement they would want to return to them,[2] and animals like lions and bears slept a good deal of the time in any case: if accustomed to it from an early age, the size of cages in a circus or travelling menagerie was no problem. As for pacing, and repetitive and rhythmical movements, this was a pleasing way of amusement and of passing the time, being equivalent to the pleasure in human dancing or pacing a room: 'Do not pity the lion, therefore; he is merely enjoying himself!' (Gillespie 1934, pp. 76ff, 85, 144 and 145). Garrard Tyrwhitt-Drake later acknowledged that 'In these days [1946] the public likes to see animals and birds exhibited in cages and dens as large as possible and under something like natural conditions'. But

> A travelling showman's wagon cannot be more than 20 ft long, 6 ft high, and 6 ft wide, and yet—and I speak from personal experience and from my knowledge of other travelling shows—the old-fashioned wagon cage is the healthiest and best to keep the inmates fit and well in. They are always dry, can be kept warm when necessary, and at other times provide plenty of fresh air and sunshine. (Tyrwhitt-Drake 1946, p. 183)[3]

The testimony of an unnamed former circus hand referred to by Helen Trevelyan and dealing with the treatment of performing animals before the First World War had described the suffering during travelling occasioned by exhaust fumes and cramped conditions (Trevelyan 1936, p. 95ff). For the select committee witness, Miss Jessie Hales, the insanitary nature of travel arrangements was a result of exclusive concern with commercialism (2120). E.J. Holland, editor of *The Anti-Vivisection Journal*, criticized sea lion accommodation and travel as witnessed himself, noting that there seemed a reluctance to justify the resulting unnatural existence: 'I must confess I have no great desire to hear once again the sound of a sea lion crying in the night at the back of a theatre stage or when packed for travelling. It is not a happy sound' (*Era* 1921d). At first, the Woodwards' tricks had been based in tanks of 40 × 20 ft, largely to compensate for the seal's clumsiness and tendency to accidents on land, but this arrangement proved difficult for travelling. Although allegedly less intelligent than seals, sea lions with their versatile hind flippers were able to gallop rather than leap on land, and with a mixture of the two types of animal on stage, as in the 'sealskin band', soon only small, 6-ft tanks were used for accommodation and travel (Mallabar 1902).

As for music-hall animals, Peter Chalmers Mitchell described their travelling conditions as variable. The use of confined boxes was safer but unpleasant because

---

[2] A similar argument has been given more recently to defend the battery hen system. The animal has probably become agoraphobic.

[3] The select committee had heard that Tyrwhitt-Drake had faced a court action brought by the RSPCA regarding the size of cages for his wild animals. During an adjournment, the case was withdrawn after an agreement to improve the cages and an acknowledgement by the RSPCA that the animals concerned were in excellent condition (224).

of frequent travelling, which was therefore unnatural and unhealthy; on the other hand he noted that many menagerie animals were extremely healthy (2257ff). However, Everard Calthrop argued that transit conditions for performing animals were probably better than for many other animals, such as those going to market, because they were valuable and must be fit to perform (1956/9); and Bob Anderson, owner of an 'educated' horse, expressed similar views in a letter to *The Era*: 'Why convict one because it's food for the mind, and not the other because it's food for the stomach?' (*Era* 1921a). Tyrwhitt-Drake claimed that no attendance, feeding or water was needed for wild animals on a long domestic journey, say between his premises at Maidstone and Edinburgh (17 hours' travel), but that performing dogs were attended (265ff). Mitchell admitted that he had not seen definite cases of suffering occasioned by regular transit (2281), but within a few years, a committee of the Zoological Society of London prepared a full report into necessary standards for all stages in the transport of living animals; and with the help of the RSPCA copies were distributed by the Society to zoological gardens, dealers and shipping agents all over the world (Zoological Society of London 1928; Mitchell 1929, p. 242). Much later, in 1946, the problem of travelling in close confinement remained, as revealed by Peter Freeman's unsuccessful request to the Secretary of State for the Home Department that he remove the licence of a circus proprietor convicted for forcing a lioness to travel in a cage so small that she could not turn round.[4] A very recent study of the effect of travelling compares the reactions of farm and circus animals in current conditions: 'The transporting of unhandled naïve animals from the farm and the transport of circus animals is very different: circus animals (similar to some competitive horses) become very familiar with being transported', and there is little evidence from pulse rates of distress or trauma (Kiley-Worthington 1990, pp. 41 and 219).

Although the witness Jessie Hales admitted she had not seen cruel training, she saw cruelty in the whole life of a performing animal, because of the confinement and lack of exercise, as for dogs (2125/7). Mrs Albert Bradshaw also considered accommodation and transit conditions were enough reason for abolition:

> I have seen animals on station platforms in very tiny boxes where they could hardly breathe; and in the theatres I have seen them, in the dressing rooms, simply left there the whole of the day. I have also seen them under the stages of several theatres. (*260f*)

The transport of dogs was frequently mentioned in the select committee hearings. E.H. Bostock had seen small travelling boxes for music-hall dogs: they were necessary for quarantine requirements and to cut down the expense of travelling, he thought, but exercise was given after a journey (*3225*). 'Captain' Devereux wrote in *The Era* that his dogs travelled in luxury, thanks to Melox dog food (*Era* 1921e), but Charles William Rodgers of the NCDL told the select committee that the boxes that Leonidas's dogs travelled in were the smallest his veterinary surgeon had ever seen (*5f, 89*), and Albert Hill said that Lipinski's dogs had the smallest possible boxes for

[4] *Parliamentary Debates* (*Commons*) 1946–1947, vol. 439, column 234.

travelling (1192). There were specific instances of neglect or cruelty in transport, including lack of food, water or exercise. Mrs Albert Bradshaw described how in 1908 a dog that was part of 'Jimson's Pets' had 'drinking diabetes' and was denied water, even on a long journey from London to Ireland, when the dogs were housed in exceedingly small boxes (*176*). Rodgers said that in 1919, his League prosecuted Dandy George for leaving a dog in a railway station locked in a box for 24 h. He believed that performing dogs generally suffered from continuous travelling in confinement (*5f & 89*). Miss Jessie Hales, the music-hall and theatre performer, said it was usual to pack up on Saturday night as soon as a turn was over and either send the animals to the railway station or keep them at the theatre overnight for a very early start on often long journeys, to reach the new theatre by Sunday night, or on Monday morning after an overnight stay at the station: there was little chance for a meal for people or animals (2009ff). Athole Stewart, the actor, saw dogs left about on platforms in unsuitable baskets—he could not imagine anyone sending a dog to a dog show in the same way he had seen dogs carried about for business in the music halls (*2334f*). Jack Thompson, the composer and musical performer, told the select committee that Zertho would come to and from the Grand Theatre in Birmingham with a very small suffocating basket of seven or eight howling dogs on an open trolley, in the snow and rain of winter (1518).

Transport arrangements aside, many have condemned permanent captivity, especially of wild animals, as unnatural and cruel. The inadequate size of cages and an impoverished environment made the matter even worse. Of course, the plight of zoo animals as well as those used for popular entertainment has earned attention, and in spite of detailed improvements of conditions in most mainstream operations since the early twentieth century, confinement still provokes philosophical objections. Tait suggests that the 1960s book and film *Born Free* helped to persuade public opinion that for wild animals, zoos, circuses and other captive situations were a poor substitute for a life of freedom in the wild (Tait 2012, p. 164). Such objections are not new. Joys records Diana Donald's observation that

> Rousseau, Buffon and Stubbs saw animals always losing to man's tyranny. In captivity, their natural society was destroyed and they no longer could choose their own mates, live in their natural social organization, or express their intelligence and courage. The effects of civilization were either death or an atrophying of natural powers. (Joys 2011, p. 122; Donald 2007, p. 166)

Citing Altick, Joys refers also to Leigh Hunt, the poet and contemporary of Byron, who

> ... questioned why animals should be kept in captivity at all if conditions could not be improved. Why, he asked, would a great people, 'think themselves warranted in keeping any set of fellow-creatures in a state of endless captivity—their faculties contradicted, their very lives, for the most part, turned into lingering deaths?' (Joys 2011, p. 119; Altick 1978, p. 318)

Henry Salt wrote of menageries:

> A speedy death may, for all we know, be very preferable to a protracted death-in-life ... To take a wild animal from its free natural state, full of abounding egoism and vitality, and to

> shut it up for the wretched remainder of its life in a cell where it has just space to turn round, and where it necessarily loses every distinctive feature of its character—this appears to me to be as downright a denial as could well be imagined of the theory of animals rights. . . . it cannot be maintained that the establishment of wildbeast shows is in any way necessary for the advancement of human knowledge. For what do the good people see who go to the gardens on a half-holiday afternoon to poke their umbrellas at a blinking eagle-owl, or to throw dog-biscuits down the expansive throat of a hippopotamus? Not wild beasts or wild birds certainly, for there never have been or can be such in the best of all possible menageries, but merely the outer semblances and simulacra of the denizens of forest and prairie—poor spiritless remnants of what were formerly wild animals. To kill and stuff these victims of our morbid curiosity, instead of immuring them in lifelong imprisonment, would be at once a humaner and a cheaper method, and could not possibly be of less use to science. (Salt 1894, pp. 39 and 40)

It is ironic that Carl Hagenbeck admitted the problem, one that he had found so lucrative:

> Captivity, indeed, involves an enormous change in all the animal's ordinary modes of life. From being free to roam on the deserts or the steppes, and compelled to exercise cunning or swiftness to secure its food, it is now confined in a comparatively cramped space and not called upon to exercise any of its normal activities whatever. The giraffes, elephants, and gazelles, which are accustomed to wander in large herds through the plains or forests, find themselves suddenly isolated and condemned to drag out their existence in solitary confinement. It is not to be wondered at that weakness and enervation should ensue as a result. Among all animals a sort of mental depression seems to take place when they have been only recently captured; and in the case of the more highly evolved and nervous animals, such as the anthropoids and especially the gorillas, this mental depression often terminates in death. (Hagenbeck 1912, pp. 203 and 204)

Hagenbeck said that his animals were given space and liberty that alleviated the distress caused by captivity and that they should be enabled to play: it was very important to combat suffering from tedium (Hagenbeck 1912, pp. 233, 236, 241, 287). But there were fatal self-mutilations which he did not understand and inevitable problems of physical health: all his five Black Sumatran rhinoceroses died because of enteritis (Hagenbeck 1912, pp. 167, 228, 229). A correspondent of *The Spectator* asked: 'Can any forced withdrawal of a wild animal from its natural conditions, however effected and however "kindly" it be treated, be anything but cruelty?' (Cather 1931; Estcourt-Oswald 1931) Others complained that the menagerie was like 'a prison for the blameless with absolute boredom and distress . . . the onlookers of the dullest intelligence. Has not the crowd enough with its abominable cinema?' and 'It has been said that these circus animals are in "good condition". So are prisoners, and good bodily condition is no criterion of happiness . . . Many would consider that the space provided for them should be measured, not by square feet, but by acres' (Bayne 1921). In reply to Margaret Bradish, an active member of the RSPCA, who had asked Thomas Hardy for messages of sympathy to be sent to a public meeting of protest against keeping animals and birds in lifelong, close captivity, he confirmed he had already written in complaint to the press (Purdy and Millgate 1984, p. 12). And Francis Yeats-Brown wrote to *The Spectator*: 'I hope all of us who have been prisoners [he had been held by the Turks as a prisoner of war between 1915 and 1917] will help to end the suffering entailed by zoos and

*Elder Sister.* "DON'T FRIGHTEN THE POOR ANIMAL. IT ISN'T FAIR."

**Fig. 4.3** 'Elder Sister. "Don't frighten the poor animal. It isn't fair".' A cartoon of 9 May 1923, when the confinement of animals and the symbolism of the British lion had come under public scrutiny as a result of the performing animals controversy. Reproduced with permission of Punch Ltd, www.punch.co.uk

circuses (Fig. 4.3). A bird is meant to use its wings and voice in the air . . . a cage is a lazy or blasphemous attempt to hold more beauty for ourselves than we have earned' (Yeats-Brown 1929).[5]

The menagerie as a destination for imported wild animals began to decline after the First World War. James Sanger had sold off his flesh-eating menagerie animals during hostilities and seemed not to want to re-establish it (*876*). Wombwell's menageries had travelled the country for more than 120 years, since 1807, but in 1932 the collection was sold to London Zoo and thereafter there were no more proper travelling menageries (Tyrwhitt-Drake 1946, pp. 171 and 183). However, an interest remained in displaying unusual animals, for profit and without due regard to conservation. Tyrwhitt-Drake wrote of the rare okapi, discovered around 1900: 'It must be borne in mind that they are not considered good livers; by this I mean that they are delicate and difficult to keep in captivity, and consequently a return for its attractiveness to the British public must be obtained quick!' And even London Zoo became involved in breeding blue budgerigars for high profits (Tyrwhitt-Drake 1936, pp. 64 and 65).

Although Bostock worried about depriving wild animals of their freedom, he decided that captivity had saved them from hunger, thirst and the hunter and that his captives had educational value (Bostock and Velvin 1903, pp. xiii–xvi). Chindahl also claimed that the circus menagerie had high value in natural history education, but cages tended to be inadequately labelled. In two large circuses, a cage label for an African gemsbok read 'James Bock, India', and for a mandrill, 'Blue-faced Mongrel' (Chindahl 1959, p. 214).

The predicament of the big cats had long attracted critical attention. The lion, Napoleon, had been caged for 18 years in 1833:

> You cannot but observe the weary feeling with which his fierce and burning nature yields to this his uncontrollable destiny... till some prying visitor pokes him with a cane, or the keeper rouses him. Then again his restless soul rages... like the living coals of fire his eyes

[5] Yeats-Brown was a decorated pilot and accomplished author and until the previous year had been assistant editor of *The Spectator*.

> glow and flash, and the ground trembles with his mad roar... A rugged, state-prison-looking fellow, thrust at him with his stick, and when his eyes, inflamed with sullen horror at the past and deep despair for the future, were lifted towards him, he spit in his face. (*New York Mirror* 1833, cited by Joys 2011, p. 105)

Lions were frequently born in captivity in the mid-nineteenth century, but few of these reached maturity, many dying at the time of shedding their milk teeth (Anon. 1869, p. 127). *Haney's* advised in 1869:

> All wild animals when captured, after they have reached an adult age, display at first a passionate resistance to confinement and all efforts to soothe them. While this lasts it is usual to keep them without food. The exhaustion induced by this deprivation greatly aids in quelling their rage, besides teaching them the hopelessness of resistance. With cubs this is scarcely ever necessary; though they sometimes display anger, they are so easily overpowered or restrained from mischief, that it is hardly worth while needlessly to make them suffer hunger. As soon as the old ones become quiet they are fed by the tamer, who thus lays the foundation of their future good will. (Anon. 1869, p. 136)

An 'Ex-Lion-King' knew that 'It takes a lion well into 10 years to come to his full growth; and when this is once attained, confinement seems to bear uncommon hard upon them' (Ex-Lion-King 1872). Captive lionesses would often devour their cubs out of nervous distress, and fighting in lions lead to serious injury so that 'sewing and patching' had to be undertaken while they were tied to the cage bars (Bostock and Velvin 1903, pp. 40 and 44). In rehearsals for Alfred Court's mixed act in Madison Square Garden in 1940, an Indian leopard killed a snow leopard and Court killed animals in self-defence, one being a polar bear and another, curiously, a Great Dane (Joys 1983, pp. 159–161).

In 1911, Frank Bostock gave an account, which would soon after have been considered unguarded, of fatal fights in his menagerie between lions, tigers and other wild animals some years previously. 'I saw one lion seize a smaller lion by the throat with his teeth and dash its head against the bars of the runway, cracking it like an eggshell.' This was 'child's play' compared with separating a lion and tiger which generally had an instinctive hatred for each other: 'The tiger lay dead in the corner of the cage with his throat half torn away'. He said that wild beasts feared man 'just as we would fear some unknown power if such a one were to come among us who could apparently deprive us of our liberty and happiness.' He had to kill troublesome animals, even if they were performing tigers worth £1,000 (*World's Fair* 1911c). Troublesome animals could also be made profitable. Fights in Clyde Beatty's performance at Detroit Shrine Circus in March 1951 resulted in the killing of two tigers by a lion called Prince. 'Following the attacks at least 6,000 customers had to be turned away each performance ... As the disruptions continued, he quietly replaced Prince with another lion, given the same name' (Joys 1983, p. 198; 2011, p. 402).

On the pretence of seeking an autograph, Helen Trevelyan got backstage where 'amidst din and darkness' she was shown lions and tigers kept long term in cramped cages. She was told how convenient these 'prisons' were:

> ... since they were simply run on to a lorry at one end and off at the other, without the animals ever coming out ... These creatures, to whom liberty must be the breath of life, to

> whom the sky should be the limit, are caged for *life*, and no action is, or can be taken. (Trevelyan 1936, p. 102)

Joys (1983) has assembled a range of opinions concerning the condition of wild animals, especially the big cats, in American circus captivity. In the prologue to her book, the trainer Dave Hoover advises:

> Don't pity the lion or tiger because he is caged. All the big cats in circuses today were born and raised in cages. This is their safe and secure world where they receive all their creature comforts. The caged cat no more longs for freedom than a native of the Amazon jungle longs for a TV set, simply because he has never experienced it. (Joys 1983, p. xv)

Consequently, retired animals are reluctant to leave their cages. An angry Clyde Beatty expressed similar views: 'My lions and tigers are better cared for, in captivity, than they could possibly be in their native jungle' (Beatty and Anthony 1933, p. 130, cited by Joys 1983, p. 276). In the 1920s, publicists like Courtney Ryley Cooper explained that pacing the cage was merely for exercise (Cooper 1923, pp. 113–116, 123, cited by Joys 1983, p. 93) and that the trainer was guardian of all aspects of the animal's welfare, 'communion and understanding' having developed over the previous 50 years (Cooper 1924, pp. 7, 8, 17–20; 1925, pp. 87 and 89, cited by Joys 1983, p. 94). According to another publicist, Al Priddy, animals loved the travel, the excitement of milling crowds and even the music associated with circuses (Priddy 1930, pp. 43–50, cited by Joys 1983, p. 98). But a trainer had written in 1904 that tragedies occurred because a caged animal, like an imprisoned man, was unbalanced: some faculties were dulled, others sharpened, and he was dealing not with his own kind but with the domination of an alien, whom he regarded with terror and smouldering, unpredictable rage (Adams 1904, pp. 188–197, cited by Joys 1983, p. 250). The trainer Roman Proske believed that prolonged, unnatural confinement produces increasing frustration in a big cat until there is an opportunity to express its pent up, explosive force in one murderous onslaught (Proske 1956, p. 175, cited by Joys 1983, p. 259). Phillipe Diolé argues that confinement leads to psychological imbalance and sexual disorders, animals being psychologically demolished because unable to flee from man; and attempted alleviation through training represents only servitude and degradation (Diolé 1974, p. 20, cited by Joys 1983, p. 270). But Joys is more optimistic: 'The tamed wild animal has no flight tendency—unlike the newly captured animal it has become emotionally stable' (Joys 1983, p. 279). Hediger states that boredom in confinement resulting from absence of enemies and freely administered food can be alleviated by training as 'occupational therapy': rather than oppose it, campaigners should promote 'biologically suitable' training, which is also aesthetically pleasing (Hediger 1950, pp. 158–161).

Joys blames recent cruelties in the United States on ignorance, negligence, amateurism, inadequate help and cheap, inexcusable cage conditions, especially in the small 'mud shows' (Joys 1983, p. ix). This view had been expressed since the 1900s by circus proprietor Al G. Barnes (Joys 1983, p. 68) and mixed fighting act trainer, Peter Taylor, who made the common and favourable comparison of conditions in the circus as opposed to the zoo (Taylor 1925, p. 62, cited by Joys 1983,

p. 97). An example of a critical news item about 'mud shows' appeared in the *Circus Report* in 1977, 'a type of story that is becoming all too frequent'. It referred to four tigers being confined in a $4 \times 8$ ft box; the confiscation of two lions, one dead and half eaten, the other having to be destroyed; an infant elephant being poked with a penknife, drawing blood; and other cases of cramped housing, starvation and beatings witnessed by the Humane Society of the United States (*Circus Report* 1977).

Peter Chalmers Mitchell understood that different animals reacted differently to confinement, and that, surprisingly, menagerie lions did not seem to do any worse than those in a zoo, were very fit and bred well (2223). But concern about the unnatural confinement of animals has since continued to grow and soon rivalled concern about other related aspects of the treatment of animals, whether in entertainment, experiment or agriculture. Speaking of captivity, Mitchell warned that

> It is necessary to dismiss from our minds the fallacy of reading human emotions into animal psychology. I am sure that for them *Stone walls do not a prison make, Nor iron bars a cage*, if we are to read into 'prison' and 'cage', the human emotions aroused by these words. (Mitchell 1937, pp. 361 and 362)

Gillespie also noted that captivity was often interpreted as 'imprisonment', with its connotations of historic human experience and with the significance of human imprisonment. In the question of captivity, it was necessary to consider what the animal requires rather than what critical human observers would require for themselves (Gillespie 1934, pp. 26, 31, 38). Anthropomorphism might be compounded by a misunderstanding of wild animals' physical requirements.

The very size of the elephant, however, made its captivity both difficult and subject to special criticism. The elephant was a 'poor wretch' shut up 'days, nights, and years in this dark low prison, where the blessed sunshine and the breath of flowers never come!' (Joys 2011, p. 105, with reference to *New York Mirror* 1833). When they became too difficult to control, they were destroyed with varying levels of competence. Hagenbeck describes the killing of his rogue or sick elephants, and captive elephants and other animals were also said to have been killed by rats (Hagenbeck 1912, pp. 153–157, 163). An elephant called Tip was killed in 1894 with potassium cyanide in his bran, on the orders of the Board of Park Commissioners of New York Central Park. He had grown to hate his trainer, Snyder, and had 'murdered several keepers'. The *New York Times* gave a titillatingly gruesome report of his death: 'for more than 9 h the poison was corroding his vitals', then in a paroxysm he burst his cage, trumpeting in agony and 'spouting big drops of blood' (*New York Times* 1894). Tait (2012, pp. 92 and 93) cites sources describing the cruel treatment of unruly elephants. During his musth, Tusko was 'covered with chains, with martingales on his tusks, cross-hobbles on his four legs and a chain-basket affair that controlled, to a certain extent, the movement of his head and trunk' (Lewis 1955, p. 57). In 1994, an African elephant named Tyke was executed in public on the streets of Honolulu, gunned down by 87 bullets which caused a slow death over 2 h. An alternative to execution might be banishment: when an animal permanently 'goes bad', 'he is sent to the lonely cage to spend the rest of his life in comparative obscurity, disturbed merely by the passing crowd and his daily meals' (Bostock and Velvin 1903, p. 98). Attempts had been made to control other animals that might become threatening simply by restricting their movement:

> The narrow limits of the cage in which he [a rhinoceros] was shut up did not allow him to manifest much intelligence. The object of the keeper was to make him forget his strength, or forego its exercise . . .to open his mouth, to move his head to the right or left, or to lift his leg were the usual acts by which he was required to testify his obedience. . .and he was scrupulously rewarded for the least thing he was required to do. The distinction he made of persons, and the great attention he paid to everything which passed around, demonstrated that, in more favorable circumstances, his intelligence might have been more strikingly manifested. (Anon. 1869, p. 143)

Mitchell recalled that soon after 1903 Hagenbeck (who died in 1913) called on him: 'I have known no more attractive personality', he said, one who had a genuine affection for animals, and arranged imports under the best conditions; and he described him as a 'pioneer in the housing and exhibition of animals in captivity', wanting more space for exercise, more fresh air and less confinement in small artificially heated houses (Mitchell 1937, pp. 121 and 122). Hagenbeck had supplied him with animals and had advised the Zoological Society of London on transport, housing, feeding and treatment (Hagenbeck 1912, pp. v–xi). Both men were very interested in acclimatization, believing in the value of open air for the health of wild animals, and that lower temperatures in new environments were not necessarily a problem. Mitchell pointed out that 'We are too ready to think that [warm-blooded] animals will thrive only in the climates with which we associate them' (Mitchell 1937, p. 364). But in Tyrwhitt-Drake's experience of working with wild animal dealers, he found that 'the mortality of animals travelling from their native climate to one often entirely different is naturally high' (Tyrwhitt-Drake 1946, p. 184). Bostock had found that polar bears suffered greatly and became exhausted in training because of their naturally very cold environment (Bostock and Velvin 1903, p. 53). The Moscow State Circus would later overcome training problems with bears: their motor cycling and boxing bear had been trained not to hibernate (*New York Times* 1964).

Joseph Kenworthy had told *The Sunday Chronicle* that his bill would end such practices as the confining of seals in tanks of dirty water (*World's Fair* 1921a). Lily Grahame described monkeys chained or in very small cages in the dark under the arena at the London Hippodrome (*2780ff*); and Margaret Martin pointed out that it would have been difficult to exercise animals like bears and seals (545). Jessie Hales reported seeing cats in close confinement every day except when performing (2098), while Alfred Roelgin, the bird trainer, said it was possible to keep birds 'packed up' for 6 months, after which they were still able to go straight into a trick (2365). Mrs Martin complained that performing animals were kept without daylight in a smelly run at the Hippodrome (513), but Frederick Trussell, former manager there and at other theatres, confirmed that there was good accommodation for elephants (*3130*). Lily Grahame had told the select committee simply that an elephant was better in the forest than on a cement floor (*2700*). In retrospect, some comments made by Tyrwhitt-Drake appear unguarded or naïve. He wrote unremarkably of a violent elephant that would kill other performing animals, and eventually apparently died of cold, having been stabled under an arch in Liverpool, and of the thrill of seeing a trainer enter a cage of five or six performing lions, 'in a space hardly big enough in which to swing a cat' (Tyrwhitt-Drake 1946, pp. 150 and 170). E.H. Bostock meanwhile informed the select committee that accommodation in music halls was better in the provinces than in the cities (*3258*).

Helen Trevelyan referred to Carl Hagenbeck on wild animals' depression and sometimes death after capture and to John Galsworthy on the subject of depression: he described eagles and condors in zoos as 'statues of winged grief' (Trevelyan 1936, p. 46). But was this again a matter of human perception? Mitchell claimed that 'in most cases animals gain far more by freedom from enemies, regular food and general hygiene, than they lose from captivity' (Mitchell 1937, p. 362). James Sanger told the select committee that he did not believe that keeping them under unnatural conditions was cruel to wild animals, and the unusual performance in the ring by the hippopotamus 'Nora of the Nile' was not a pathetic sight. He was not convinced she died prematurely because she was kept under such unnatural conditions (*1102ff*). In the same vein, E.H. Bostock did not believe Charles Haverly's statement about lions being confined and deprived of food and water for 5 days to make them docile (*3165*).

An unnatural captive state for wild animals was defended by T.H. Gillespie, director of Glasgow Zoo and sympathetic with the performing animals sector (the source also of some of his animals), and by G. Tyrwhitt-Drake, a friend of and supplier to Gillespie and a circus, zoo and menagerie owner in his own right. Aside from any concept of the right of a wild animal to be allowed the dignity of freedom in its natural environment, both men ignored the fact that the human intervention they promoted interrupted the natural process of evolution in the wild: necessary activities accompanying a wild state were dismissed only as inconveniences to the animal. Natural activities like hunting were carried out in the wild only out of necessity and not through choice, commented Gillespie, who commended the 'sweetness of doing nothing' in captivity. He rightly pointed out that birds fly to escape or hunt rather than from choice or for pleasure, this being the explanation of why in some secure environments, birds became flightless (and, incidentally, in the case of the dodo, suffered for it). It would seem, therefore, that Gillespie approved of sudden interference in long-term, gradual, natural evolution. This would be distressing interference because it would apply to an unrealized future state in an animal's possible development.[6] Gillespie attempts to reinforce his view by superficially noting, without exploring the reasons, that captive animals that escape prefer to return to captivity. In the case of lions, they are lazy and prefer living in cages than in enclosures or in the wild: they are roamers in the wild by necessity and not by choice. But Gillespie does not acknowledge that they are confined in cages

[6] Both Sir Julian Huxley (secretary of the Zoological Society of London from 1935) and the eminent ethologist W. H. Thorpe have argued that the frustration of natural behaviour in an animal is the greatest of cruelties (Wilson 1999, p. 228).

for human interests and not by their own strategic choice, the concept of choice being in any case an irrelevance, suggestive of conscious decision making. He continues: 'Captive animals are indifferent, within reasonable limits, to the extent of their boundaries, once they have become accustomed to them'. This process of acclimatization involves a dulling of instincts and cruel frustration, and Gillespie does not mention the natural stimuli of territoriality, hierarchy, competition, wariness, fear and so on, and the routine challenges of an existence in the wild that promotes evolutionary development (Gillespie 1934, pp. 55, 59ff, 70ff, 75). Webster believes 'The welfare of an animal must be defined ... not only by how it feels within a spectrum that ranges from suffering to pleasure but also by its ability to sustain physical and mental fitness and so preserve not only its future quality of life but also the survival of its genes'. He continues: 'It is seldom, if ever, in the welfare interests of a truly wild animal to be seduced into an increased sense of dependence on man ... In the natural environment where the rule is survival of the fittest, it is, I repeat, best to leave well, alone' (Webster 1994, pp. 11 and 227).

Tyrwhitt-Drake had taken an opposite view of the effects of confinement: 'Wild animals do much better in comparatively small, dry, covered-in cages and loose boxes than they do in very large open cages and paddocks'. He advocated an impoverished routine in captivity: lions only roam free in the wild to hunt or find a mate, and

> The secret of keeping carnivora in fine condition in captivity is to give them sufficient food to keep them content for twenty hours and hungry for the next four. They will then get up with the instinct to catch, and prowl up and down till the meal is brought to them, taking the exercise that is so necessary for them.

He continued:

> A properly fed animal will amuse itself for hours playing in a small cage, if only it is given something with which it can do so, a shelf or two to jump up and down on, a ball or even a log of wood. Everyone knows how very young children are amused for hours and days with the simplest things, and a young child's brain is far in advance of a mature lion's.

A comparison was in order. As to the question of the close confinement of wild animals, Tyrwhitt-Drake wrote:

> Take the case of a primitive native. I venture to think that if he was offered freedom with its attendant hunt for food and family worries against confinement in an enclosure with warm sleeping quarters, one or two wives and plenty of good food and drink, and no worries, he would choose comparative confinement rather than freedom and its accompanying struggle for existence, and still more so it is with an animal.

These unnatural conditions of daily existence promoted by Tyrwhitt-Drake would in principle disable the evolutionary process and frustrate instinct. His argument was also anthropomorphic because an animal could not understand and appreciate the arrangements. Any alleged physical comfort would certainly be outweighed by psychological stress and boredom, and his assertions that 'most of us prefer a fairly small cosy room', and that caged animals need a proper diet rather than exercise to keep fit, ring hollow (Tyrwhitt-Drake 1936, pp. 75–78).

There was also general criticism of the confinement of common domestic animals such as the dog. Ernest Bell warned of being deceived by a dog's behaviour and noted that the owners of performing dogs enjoyed official concessions that prejudiced the animals' welfare:

> If the performer is a dog, he may frisk about wagging his tail and looking pleased. But you do not know that he has been kept in a small box all day, and his delight at being free and able to stretch his limbs is quite real and unfeigned, but it is only the reaction of previous restraint and misery ... By one of the strange anomalies or absurdities of our law, while the dogs of private owners are kept for weeks in quarantine, the dogs of performers are allowed in without any quarantine, so frightened is the Government of interfering with any trade interests! This means that the dogs have to be kept indoors under strict supervision, and when not actually performing they are kept in boxes or hutches, in which they are carried to and from the theatre, or are, perhaps, left there day and night. (Bell 1918)

Arrangements for performing dogs attracted much attention in the select committee. Any shortcomings in accommodation for performing dogs could be exacerbated by the quarantine requirements. Charles Butler, the music-hall singer, said that the quarantine laws prevented outdoor exercise for dogs (*474*). Jack Russell Vokes showed the select committee the special quarantine bag he used to carry his dog. It had caused some controversy, but he wanted to present it as a reasonable device. He also reported that the Ministry of Agriculture and Fisheries checked on theatre accommodation, including facilities for exercise, for quarantined dogs before he arrived; if it were not satisfactory, he had to use quarantine establishments. In addition, he was allowed to exercise his dog under close supervision in the countryside if contact with other dogs was avoided (*2030*, *2066f*, *2111f*). Butler testified that when not performing, dogs were not allowed out of their travelling cages, remaining confined in dressing rooms, which new artistes would find in a disgusting state the following week (*428*). Mollie Lawrence had 18 years' such experience of dogs being kept all day, except for a few minutes' exercise, in dressing rooms that became squalid (*2833*), and Charles Haverly saw Peterson's dogs confined in cubicles in a hamper, except for a few minutes' exercise each day (*558*). But Jack Russell Vokes, owner of 'Don the Drunken Dog', said such stories were untrue (*2120*), and George Lockhart affirmed that exercise for his and other dogs was given in the circus and music hall (*1446ff*). Nevertheless, in May 1916, the manager of the Drury Lane Theatre had been heavily fined for failing to provide proper housing for dogs (*1849*). This resulted from a conviction obtained by the NCDL (Bell 1918).

In a recent major investigation into conditions in the late twentieth-century British circus, commissioned by the RSPCA, Marthe Kiley-Worthington claimed that measures of welfare could be made through indications of stress or suffering leading to disease, use of prophylactic drugs, surgery and failure to breed, noting that drugs and surgery were little used in circuses for these reasons, the majority of circus animals being in good physical condition. But to qualify these views, it could be said that stress does not always lead to physical problems; and veterinary treatment may sometimes be absent when it is needed. And although Kiley-Worthington advocates environments that enable the natural behavioural repertoire, they are surely impossible to provide in a travelling circus. She found much evidence of behavioural restriction, and no doubt such evidence was felt by the RSPCA to invalidate her later conclusions, causing it to reject them. Funding for the publication of her work was instead provided, ironically, by the Association of Circus Proprietors of Great Britain (ACP). She reported that at the start of her study, many carnivores were kept in beast wagons except for the 10 min or so of the performance each day. Some elephants were kept shackled for 24 h, except for performance. Circus animals were also often confined for long periods during transport. Such restriction remained a problem, even though in November 1987 the ACP began welfare inspections of its member circuses. The ACP's Veterinary Scheme for 1989 (subject to annual revision and improvement on advice of its consultant veterinary officers) as recommended by Taylor and Greenwood, the international zoo vets, was made a condition of membership and was largely concerned with standards of travel, accommodation and exercise (Kiley-Worthington 1990, pp. 15–17, 28, 32, 44, 52, 59, 224–226).

During her visits, Kiley-Worthington recognized no evidence for neuroses and behavioural pathologies in elephants, even though all elephants in zoos and circuses performed stereotypies. On the other hand, she identified frustration and conflict behaviour, especially when elephants were shackled without food or objects to manipulate. Neither did she see evidence for neuroses and behavioural pathologies in big cats. There was frequent pacing but in their case no stereotypies. They spent about two-thirds of their time resting (perhaps as in the wild, a characteristic that some keepers argued made them suitable for captivity). Bears did show evidence of distress through stereotypic behaviour and frustration through abnormal pacing and bar biting: they were abnormally active but remarkably unaggressive. Horses in the circus showed less stress than racehorses and horses kept stabled for teaching (Kiley-Worthington 1990, pp. 78–81, 87, 92).

To ensure an adequate environment for an animal, Kiley-Worthington identified a need to consider the species' evolutionary background by taking account of factors that she lists: social organization; habitat; the natural food, its nutrient content and textures; the performance of the entire behavioural repertoire; the species-specific characteristics of the receptors; the species-specific brain anatomy; the species' communication system; how the animals learn; and what that species might be expected to learn:

> Mothers should be allowed and encouraged to raise their own young. In order to achieve this it is often important to be able to handle the mother safely as it is often easier to help the mother, but not be a substitute for her, even though, as is the case with so many captive-born females these days, she has had no experience of being mothered (she was herself hand raised) and therefore possibly less likely to be a good mother.

Furthermore, assessing the time budgets of an animal in different environments can give an indication of the suitability for an animal's behavioural needs of a new environment in which, for example, it does not have to search for food (Kiley-Worthington 1990, pp. 74, 187, 209). It is difficult to understand how some of these factors and needs could be accommodated in a circus, but Kiley-Worthington had stated in 1989:

> As a wildlife ethologist I spent many thousands of hours watching and recording the behaviour of wild animals. I have to admit however that I have learnt more about how they respond and feel, what they can learn, how fast they learn and how individuals 'tick' from studying animals in the circus, and in particular from training animals myself. (Kiley-Worthington 1989)

In his introduction to Johnson's critical work on the captive-animal entertainment industry, Desmond Morris also emphasises the need for captive animals to be guaranteed an adequate physical and social environment, for educational as well as welfare reasons: but animal performance ('a travesty of nature') does not provide this, and outdated acts demean the wild animal as a caricature of humanity. Its environmental deprivation is also deplored by Prince Sadruddin Aga Khan in the preface (Johnson 1990, pp. 9–11). Johnson argues that removal from the wild to circus captivity of intelligent social animals such as the elephant (and dolphin) will cause much psychological stress resulting from the frustration of long-evolved natural behaviour patterns including foraging routines; mother, infant and herd interactions; bathing and grooming; and infant learning and development specific to the natural habitat. There is also the special susceptibility of captive dolphins to noise pollution, 'synthetic sea water', the requirement to eat dead fish and transport on irrigated stretchers (Johnson 1990, pp. 171, 180, 205, 217). Regan points out that dolphins swim up to 40 miles a day and can dive to over 1,600 ft; have a strong familial and social structure; and make use of echolocation in the expanses of the sea. Housing them in concrete tanks is therefore highly unsuitable, and their unnatural tricks, attributed to special 'communication' with their trainers, result only from a taught dependence on dead fish. Regan observes with satisfaction that the British public boycotted marine parks and they went out of business (Regan 2004, pp. 136–139).

Regan sums up the problems of captivity for performing animals. In the circus, wild animals suffer prolonged confinement, limitations of space, loss of familial teaching and social structure, and abnormal stereotypical behaviour denoting mental illness and psychological defeat. United States federal regulations governing cage size do nothing to relieve the deprivation resulting from an unnatural existence denying the freedom to roam a home range. In the case of circus elephants, for example, there are 'no matriarchs, no kin groups, no migratory routes' (Regan 2004, pp. 125–140). Webster believes that welfare of wild animals in circuses can be assessed according to the Farm Animal Welfare Council's 'Five Freedoms'[7] concerning hunger and thirst, discomfort, pain and ill health, behavioural restriction and fear and distress, although such animals are more highly motivated than domestic species to make an active contribution to the quality of their own existence, so may be more inclined to appreciate the 'sixth freedom', the freedom to be free. They face the psychological problems of boredom and frustration of

[7] http://www.fawc.org.uk/freedoms.htm.

patterns of active behaviour such as hunting and exploration and suffer anxiety through an inability to carry out strongly motivated patterns of passive behaviour such as hiding, especially the hiding of offspring (Webster 1994, pp. 11, 225).

## 4.2 Wild, 'Docile' and Domesticated Animals and the Age Factor in Training

Much attention was given before, during and after the select committee hearings to the relative suitability for training, performance and exhibition of wild, 'docile' and domesticated animals. The 'docile' category was accepted as including animals such as elephants and sea lions, ones that could be taken from the wild but became easily tractable. Another factor considered in relation to the type of animal was the age at which the animal should be tamed or trained. Jack London in *Michael, Brother of Jerry* had described the cruel subjugation of adult animals caught in the wild, but during the British controversy it was frequently confirmed by professionals such as James Sanger that only young wild animals were conditioned to perform later in life and that in some cases, because an animal (usually a male ape or large cat) grew unpredictable or dangerous in adulthood, it had to be disposed of. The American Cleveland Moffett had also warned that unless such animals are taken from the wild at less than 3 years, they could not be trained; and lions born in the menagerie tended to be stupid (Moffett 1894). The matter was complicated by further differences in the characteristics of young animals taken from the wild and those wild animals that were bred in captivity. Captive-bred animals could also be dangerous, as explained by William Snyder:

> We have learned how to breed these wild animals in captivity . . . The only advantage aside from the added information about their life history has been that they come cheaper to raise than to capture. They aren't a bit tamer. They aren't so easy to manage. Familiarity breeds contempt. A lion right out of the jungle is scared half to death by everything around him here, and can be cowed into doing what you want easier. (*Pacific Rural Press* 1902)

Finally, it was acknowledged that the right animal was needed for the right act, noting build and temperament.[8]

Concern was frequently expressed that wild animals were particularly unsuitable for training and performance or exhibition. Such concern lasted over time,[9] but Mitchell told the select committee: 'I do not think it is possible to set any limits to what you can get a wild animal to do, supposing you keep within the natural

[8] For example, Tyrwhitt-Drake 1946, pp. 133, 136 and 137. Tyrwhitt-Drake was unenthusiastic about the employment of domestic cats: they were too small, did not like the work and had no interest in it.

[9] For example, Sir Robert Gower unsuccessfully asked the British Home Secretary if he would consider introducing legislation making public performances by wild animals illegal (*Parliamentary Debates* (*Commons*) 1930–1931, vol. 247, column 631).

disposition of the animal' and if the trainer was kind and knowledgeable. He believed that wild animals had an even higher training capacity than domestic animals and were more intelligent (2219, 2245). Furthermore, some thought that domestic animals deserved at least equal consideration during the controversy. A leading article in *The Times* (quoted in *World's Fair* 1922) regretted the failure of the original prohibiting bill and criticized the distinction made by the select committee between the interests of wild animals and those of equally vulnerable domestic animals. And Everard Calthrop had told the select committee he did not agree that there was more scope for cruelty with wild animal training than with the training of dogs and horses (2000ff). But for some, it was easier to avoid cruel treatment in the training of domestic animals. Although she disapproved of the use of the tight bearing rein in Bertram Mills Circus, a correspondent of *The Spectator* noticed that there were 'performing dogs who seem to enjoy themselves immensely', unlike the weary, sad-eyed lions on their painted stools or the 'baby elephant with diamante collar and a huge pink bow, who rides a bicycle and stands on two legs with a flag gripped in his delicate trunk' (Barry 1927).

George Lockhart was not keen on wild animal acts because it was not possible to make the animals do very much: it was hardly worth the trouble, and he felt that domesticated rather than wild animals enjoyed their tricks (*1329*, *1419*). To begin with, Bertram Mills presented no wild animals at Olympia, but this was only because of the single ring facility which would bring the animals too close to the audience (2164). George Henry Harrop believed that domesticated animals were a greater draw than the wild ones (*1199*), but this would depend on their achieving special abilities, and Charles Butler remarked: 'I have never seen a household trick performed by a pet dog on the stage: it would not be exciting enough' (*483*); and Mrs Albert Bradshaw also pointed out that the training of domestic pets was entirely different: it was done as a game and was unreliable (*186*). In Bertram Mills Circus Christmas 1930 season, Yeats-Brown said that Alfred Loyal's dogs gave a delightful show, seemingly without cruelty:

> The brown poodle who turns a back somersault was wagging his tail for minutes before he did his piece: evidently he looks forward to the applause which greets his feat; as also does a glorious Alsatian, versed in juggling. Where are we to draw the line? It is at the point which divides wild from domesticated animals . . . the poor lions do nothing but slouch on to tubs or group themselves sulkily in various parts of the arena: it is an idiotic performance. We should do all we can to help these prisoners for life who are supposed to enliven us by the spectacle of their humiliation. (Yeats-Brown 1931)

Some wild animals were easily domesticated to perform equally readily as dogs, and Jacques Charles gave the example of the sea lion:

> Comme je manifestais un jour mon étonnement au Captain Woodward qu'aucun de ses animaux ne ressentît l'appel de la nature et ne tentât de reprendre sa liberté en filant vers la haute mer . . .
>
> – Pas de danger, me dit-il, ils ont trop peur.
> – Peur de l'eau?

– Peur de ne pas être capable de se suffice à eux-mêmes; ils sentent qu'ils ne pouvraient plus trouver leur nourriture et ils sont habitués a des repas à heures fixes, ils préfèrent leur vie bourgeoise à l'aventure. (Charles 1952, p. 149)[10]

Helen Trevelyan cited the cruelty of training as originally reported by some circus people, who said that wild animals such as tigers and lions had to be subdued by fear to perform or behave unnaturally and also that cage-bred beasts were more dangerous, being sly, untrustworthy and contemptuous of humans (Trevelyan 1936, pp. 94 and 95). Gillespie echoed this last point, which was a popular belief, that wild-born animals settled down in captivity more readily than captive-bred ones (Gillespie 1934, p. 34). A recent and uncritical platitude held that 'They [big cats] obviously feel quite happy in their friendly subordination to man' (Dembeck 1970, p. 140). Such routine docility would depend on conditioning from an early age. James Sanger told the select committee that only young wild animals were trained up, and he had not heard of anyone else starting with adults (*957*); and Fred Ginnett did not train forest-bred cats or other adult animals from the wild, nor, he said, did others (615/756). But in 1900, Samuel Hopkins Adams had thought that all trainers agreed that a wild-caught animal was preferable to a captive-bred one, which had not had experience of the nurturing back to comfort and compliance after the debilitating capture and voyage and which was a 'spoiled child' and had no fear or respect. If coerced it would attack, and then 'only the harshest measures, long continued, will avail' (Adams 1900). E.H. Bostock said that wild animals were lightly trained in simple acts only from being cubs: 'to attempt to train a fully-grown animal would be simply absurd—it is never attempted or thought of' (*3155*), and Henri Rey confirmed that it was possible only to train from the cub stage (1697). Similarly, Bertram Mills, approving of wild animal shows because they were instructive and popular, said that wild animal training was possible without coercion, from being cubs (2181/3). According to James Sanger, lion cubs, once trained, could be well managed as adults. When they grew up, 'If the man can only keep his head and know his position, a lion never thinks about attacking him' (*811*). In this connection, Henri Rey did not agree with Hagenbeck (1912, p. 134) that lions necessarily turned unpredictable and dangerous after taming from cubs (1741f). Tyrwhitt-Drake identified mature male wild and domestic animals as unpredictable and dangerous, adding that it was not possible to train (as opposed to tame) lions until the age of about 18 months (356, 363). Mitchell confirmed that wild animals for training were best taken young. Adult wild animals had to be made to fear their trainer, and that was when cruelty could come in. He noted the moodiness of wild animals, when they might become uncooperative. It was also difficult to expect them to perform to time, and so cruelty in training and keeping them up to the mark was a great danger in 'time performances'. He had seen such

[10] 'As I revealed one day my astonishment to Captain Woodward that none of his animals felt called by nature to attempt to take back freedom by making off towards the open sea . . .—No danger of that, he said, they are too afraid.—Of the water?—Of not being able to survive on their own; they know that they would no longer be able to find their food and they are used to meals at set times, they prefer their bourgeois life to adventure.'

cruelty: if a wild animal became badly behaved, it had to be severely controlled and kept in fear (2220).

The problem of unpredictable, hysterical and dangerous behaviour was a particular feature of the adult male ape, who, like the elephant, became less controllable when affected by sexual instincts. Gillespie downplayed any problem with chimpanzees, describing them as best fitted for training but threatened by legislation. He suggested that a certain eminent authority (perhaps Mitchell, in which case his suggestion would be wrong) had been quoted out of context to support such legislation: the only need was to ensure the use of young chimpanzees (Gillespie 1934, p. 136). Mitchell had told the select committee that it was wrong to train chimpanzees for public exhibition, because they became dangerous and almost hysterical. Apart from the chimpanzee, it was wrong to perform the gorilla or orang-utan, or any type of large manlike ape (2226ff, 2247ff). At this time, *The Encore* (1922) agreed with the problem of the temperament of the performing ape. Harold Laurence Hilliard had described chimpanzees like Consul as unpredictable and reported that E.H. Bostock's brother Frank had been savagely attacked by Consul (1018ff). E.H. Bostock himself confirmed that orang-utans, chimpanzees and baboons became increasingly savage and dangerous as they got older, especially males, when they were put in a cage and eventually given up (maybe for exhibition) (*3246ff*). But in an article in *The World's Fair*, Bostock criticized the current bill that called for the prohibition of the training and performance of the large carnivora and apes: he found this preposterous and unnecessary. He described anthropoid apes as the most human and therefore the most easily trained and claimed that the select committee had received no evidence of cruelty to these or to the large carnivora (*World's Fair* 1923). And Tyrwhitt-Drake later wrote that 'Chimps will easily learn to strike a match, light a cigarette for themselves, and thoroughly enjoy a smoke.' Although they were liable to become unpredictable when mature at 8–12 years old, they only required the same training as for a child (Tyrwhitt-Drake 1946, p. 142).

Heinrich Hediger, director of the Zoological Gardens, Zurich, and professor in animal psychology and biology in the University of Zurich, was one of the few animal psychologists later to examine such problems:

> The trainer works intuitively, like an artist, spontaneously using all sorts of biological distances and reactions as a result of his experience; the animal psychologist on the other hand measures, names and describes them.

With large, predatory wild animals,

> The process of circus training is not a simple one, but is a complex of several partly overlapping individual processes. Broadly speaking, three main divisions may easily be distinguished:
>
> 1. The aim that the desired exercise in training should be understood.
> 2. The overcoming of inhibitions and resistance.
> 3. The aim of physical adaptation.
>
> ... Very often the performance of an action already understood is rendered impossible through various inhibitions and resistances ... inhibitions conditioned by flight, or

> unfamiliar milieu. Resistance is the result of unwillingness rather than inability, that is, active opposition, without biological necessity ... sometimes requiring punishment, and also the assertion of the trainer's dominance in view of an animal's awareness of social rank, especially if it is an alpha male. The third facultative phase of the training process, aiming at physical adaptation, is necessary in those cases in which the animal is not physically able to perform the requisite actions [without training].

Hediger continues:

> With animals born in captivity, or accustomed to it, the flight distance is considerably less than under normal conditions in freedom ... In contrast to flight reaction, which may occur in any direction, radially, critical reaction has the advantage of being aimed in one direction.

Wild animals would identify

> ... three types of home—of the first order (= cage it lives in): of the second (= individual 'place' in the ring): of the third (= pedestal; pyramid structure)—as places of relative security which attract the animal ... Displays of this kind, based on fundamental biological knowledge, are in sharp contrast to such showy performances as bears riding motor-cycles, elephants sword-fighting, or animals firing guns, to quote only a few particularly grotesque examples ... Individuals of high social rank are less amenable, and thus more difficult to work with ... The socially organized animal must literally not be approached too closely, for it would immediately take it as an attack on its rank ...

– a special problem with older animals. The trainer must adopt and achieve the role of the leading animal if he is to maintain control. Hediger also notes that stereotyped movements in captive animals reflect the need for their continuous occupation to counteract boredom. Training, tricks and disciplined play answer this need (Hediger 1955, pp. 121–127 and 133–136).

Stokes criticizes the performative inadequacy of the wild animal act whose fate is sealed as a civilized theatrical entertainment. Audience applause has meaning for the trainer, but little if any for wild animals—they 'return to their cages alone and still feared. There is no reconciliation, no return home'. At least for domestic animals such as dogs or horses, 'the arena of public performance extends to the off-stage world that they occupy alongside us; for wild animals the off-stage world—jungle, desert, savannah—has never been shared; and today, at a time of environmental crisis, it may even be lost forever, leaving the animal quite literally with nowhere to go'. Performance requires freedom and intentionality, which is absent in wild animal acts, and the wild animals do not benefit:

> Whatever they may have gained from performance (other than release from imposed boredom) was of no use either to themselves (as well as extending its abilities a domestic pet trained in obedience may lead a safer life) or, unlike a guide dog, to others. After all, even the supposedly humane principle of Hagenbeck's 'gentling' is based on the idea that you cannot teach an animal anything to which it is not *already* 'naturally inclined', which it does not, in some sense, already know. There can, in that very precise sense, be very little that is truly 'performative' in any display by a wild animal. (Stokes 2004)

## 4.3 Unnatural and Terrifying Acts?

Hagenbeck explained that 'The real difficulty in the treatment of captive carnivores begins . . . when one attempts to train them to perform; for here one is endeavouring to make the animals do what is contrary to all their natural instincts. All the difficulties, however, can be overcome by patience [and] by a careful study of the brute mind' (Hagenbeck 1912, p. 117). In this way in 1975, Baumann trained his tigers unnaturally to roll over in a prone position, which the audience mistakenly interpreted as relaxation (Baumann and Stevens 1975, p. 18, cited by Tait 2012, p. 154). A longer established and profoundly unnatural display was the difficult training of mixed groups of incompatible animals. This was 'training animal nature against its instincts' and demonstrated man's power over it (Bostock and Velvin 1903, p. 33). *Haney's* describes it as one of the most entertaining and popular features of Barnum's Museum in the mid-nineteenth century: a bizarrely peaceful 'Happy Family' would be assembled in a cage for the audience, consisting of many varieties of animals that were predator and prey or at least enemies in their natural state. Van Amburgh had presented the spectacle of the lamb lying with the lion, and a contributor to *New Sporting Magazine* went to see his performance at Drury Lane, and then—'after the play comes the farce'—he visited John Austin's halfpenny exhibition at London's Waterloo Bridge, where a variety of familiar animals (cat, rabbit, mouse, owl, pigeon, dog, rat, etc.) of the 'violentest contrarities' could be seen in their 5-ft cage (Fig. 4.4). This cage looked like 'a London club, where a concourse of people, hating each other at heart, associate, and endure each other's

*Group of Animals of opposite natures living in the same Cage.*

**Fig. 4.4** Group of animals of opposite natures living in the same cage. From Ogilby 1829, p. 21 and borrowed by John Austin for use in his handbill, as reproduced in Selborne 1839, p. 73

society, in silent patience'. For 26 years, Austin had trained animals of opposite natures 'to live together in peace and contentment' so that 'birds sometimes perch on the head of a cat, eat, and frequently rest for the night there', and an occupant if released would return to the cage. This was achieved by feeding well and by socialization from a very early age (Selborne 1839; Ogilby 1829, pp. 19–21). *Haney's* explained that to prepare such presentations, the animals must be kept in small separate cages that were gradually drawn closer together, until two or more different animals could be introduced into a single cage. They should be kept fed to satiety, without drugs and under close supervision, any aggression being punished by 'poking and punching and stirring them up generally' (Anon. 1869, pp. 175–177). In a variation of this, Durov trained incompatible animals such as a fox and hen to feed from the same basin (Dvinsky 1956, p. 46).

There were also combinations in performances of wild with domestic animals, especially lions with horses. Henri Rey, who had worked at the Brixton training establishment for the previous 6 years, said that lions were trained from the age of 5 or 6 weeks to ride on a horse that became familiar and acquiescent (1779/1804). Of equestrian lions, Tyrwhitt-Drake wrote: 'It is a curious fact that harmless and domestic animals, if they are brought up with the big felines, not only have no fear of them but actually boss them' (Tyrwhitt-Drake 1946, p. 177). Mitchell told the committee that the lion-on-horseback trick was not necessarily cruel and indeed that van horses were not frightened of lions when arriving at his zoo (2233, 2311).[11] Critics considered some or all animal acts undignified and pernicious because the acts themselves or the circumstances in which they were prepared and performed appeared unnatural and often terrifying to the animals. They objected to the sort of performance that an audience, and especially its children, would not suspect. As observed by William Snyder: 'Whenever you see a monkey race in the circus remember that before the ponies submitted to the process they experienced the most agonizing fear. I never saw any signs either that the monkeys enjoyed the ride very keenly. However, if it amuses the children, what need we care?' (Sutherland 1902).

It was claimed that unnatural and difficult acts to timetable were enforced by conditioning through punishment, but those representing professional interests argued that the process involved only the exploitation of an animal's natural inclinations (such as greed) and of natural abilities (such as balance) and that performance achieved by such conditioned response through reward was neither cruel nor unnatural. Furthermore, they argued that both natural inclination and ability, and the attractiveness of the reward, met the need to perform to timetable. The truth about methods, performances and environment varied according to trainers, animals and location, and the difficulty of substantiating the claims of each side also fuelled the controversy, much as has been the case with that over vivisection. Ernest Bell thought that 90 % of turns involved cruelty. Examples of

[11] Perhaps this was because horses had evolved without overlap with lions in their home territories (Clive Phillips, personal communication 8 July 2014).

music-hall turns that he witnessed and which passed without adverse comment from the audience included a pony whipped to keep up on a revolving table which it fell off; a terrorized cat thrown by the hind legs into the air, to revolve and alight on the trainer's hand; and a horse stumbling on a dog that fell from its back and was whipped as punishment (Bell 1918). The need to introduce novel acts of increasing difficulty meant that any natural abilities were exploited for unnatural purposes. The sure-footedness of mules has enabled trainers to teach them to walk over a number of empty bottles placed upright on a floor or platform (Anon. 1869, p. 62). A Barnum and Bailey trick was for an elephant to pick up a greyhound and throw him over his back 'as punishment for being saucy' (Velvin 1910–1911). The elephant Bosco could sit himself at a table, pull the bell, allow himself to be served by a monkey, drink out of a bottle, eat off a plate, and 'would dine in the most orthodox human fashion' (Hagenbeck 1912, p. 158). Antony Hippisley Coxe spoke approvingly of Bertram Mills Circus at Olympia: 'I have never before watched baby elephants crawl on hands and knees, nor seen penguins partner seals. Yet such originality is in itself typical of the constant search for something new' (Hippisley Coxe 1953).

Disagreements and uncertainties arose during exchanges at the select committee over the nature and meaning of an animal performance. Should chimpanzees' zoo tea parties, military training and use of animals, and the use of horses in tournaments be considered as so different from animal turns in the circus and music hall? Did wild animals warrant different treatment from domestic animals? And was it appropriate to make wild animals perform with domestic animals? Mitchell admitted that some wild animals at London Zoo were effectively self-trained by the experience of their new circumstances, without any forcing of their natural instincts and only through their modification of their way of life in nature (Mitchell 1929, p. 243). In attempting to advise the select committee on what might be natural or abnormal tricks, he commented that all mammals were experimenters, modifying their instincts and habits under new circumstances (2237). At Hornaday's zoo, the animals were more than self-trained: mammals, birds and reptiles were trained:

> (1) to be peaceful, and not attack their keepers; (2) to not fear their keepers; (3) to do as they are bid about going here and there; (4) to accept and eat the food that is provided for them, and (5) finally, in some cases to 'show off' a little when commanded, for the benefit of visitors ... Every large zoological park is a school of wild animal education and training; and it is literally a continuous performance. Let no one suppose that there is no training of wild beasts save for the circus ring and the vaudeville stage. Of the total number of large and important mammals that come into our zoological parks, the majority of them actually are trained to play becomingly their respective parts. An intractable and obstinate animal soon becomes a nuisance. (Hornaday 1922, pp. 206 and 207, cited by Joys 2011, pp. 287 and 288)

Captain H.E. Hobbs, founder of the Tailwaggers' Club, believed that the training of sheep dogs and police dogs was based on natural behaviour, patience and kindness, but that 'the aim of the stage performer is to get dogs to perform tricks which are utterly foreign to their natural instincts and impossible to their physical powers'. Performance training was rushed and must therefore be based on cruelty, and Helen Trevelyan quoted Dr Johnson's comment: 'I'd rather see a man on four legs than a dog on two!' (Trevelyan 1936, p. 89). Indeed, James Sanger admitted that all tricks were unnatural, but did not see that as a wrong in itself (*943*). E.H. Bostock disapproved of the dog trapeze act and thought it worthy of elimination, but he believed that the somersault was not difficult to train in dogs. An example of a long-running dog act was that of the Duncan family. It was introduced by 'Professor' Duncan in 1879, when one dog capable of 52 tricks performed at a charity performance for widows and orphans of the Tay Bridge disaster. It became known as Duncan's Collies and played at the Palace Theatre, London, for 9 months at a stretch. It was regarded as one of the best animal acts in variety, and Duncan toured America, Australia, South Africa and most of Europe. After giving a performance for Edward VII at Buckingham Palace, the act became 'Duncan's Royal Scots Collies'. Victor Duncan took over from his father in 1926, and the act lasted until the late 1950s, featuring on the bill during the successful Danny Kaye season at the London Palladium in 1948 (Busby 1976, p. 47). When the act had returned from a foreign tour, it appeared at circuses only, since the Ministry of Agriculture agreed to approve accommodation and a special trailer to meet the quarantine laws. Standing dogs skipped with ropes and jumped onto the backs of kitchen chairs to balance on their hind legs: 'None of our tricks was ever copied—they were too damned hard ... you've got to get a good hind-leg dog—it takes about a month to teach a dog to walk on its hind legs. It would probably take another 3 months to get him on the back of that chair.' They were taught to balance, not walk—'when he falls forward he puts his paws on your chest, and you move backwards, and eventually he's walking'. For the chair trick, the dog would learn to jump up onto a board that would be raised gradually to the height of the chair back. 'The great thing was to make sure the dog didn't fall. If he fell and hurt himself you were back to where you started.' All 10 dogs were trained to do everything, so substitutes were possible; but they were less reliable in dress rehearsals because there was no audience (Wilmut 1989, pp. 114 and 115).

Meanwhile, sensational acts with wild animals on the music-hall stage had almost ceased by the 1920s (*3151/2*, *3188*), and George Lockhart had told the select committee that the audience was then more interested in comedy animal acts rather than sensational tricks (*1468*). As an early indication of disapproval of sensational acts, in 1904 Sydney Buxton MP had unsuccessfully asked the Secretary of State for the Home Department to end a performance called the Plunging Elephants at the London Hippodrome in which elephants were driven to the head of a chute to be precipitated into water.[12] Use of this spectacle as an example of unacceptable entertainment was made regularly thereafter, and Fred Trussell, for 20 years manager of the Hippodrome, told the committee that the 'chute business' had become inflated since: there was no problem at the time, when there was a simple misunderstanding in the Press of the comments of the Chairman of London County Council and when the RSPCA were satisfied (*2898ff*).

[12] *Parliamentary Debates* (*Commons*) 1904, vol. 130, column 538.

In May 1921, Colonel Burn presided over a House of Commons evening meeting in support of the Performing Animals Prohibition Bill, when evidence was heard from witnesses to training. A letter was read from Mr Milton Bode, who had connections with theatres and music halls in the Midlands, saying he had evidence of many acts of cruelty in the performance by animals of hazardous tricks. At that meeting Kenworthy said the aim of the bill was to abolish unnatural tricks (*Times* 1921). A former music-hall manager admitted to the select committee soon after that all unnatural acts had to be learnt by cruelty (*703*), and Tyrwhitt-Drake conceded that unnatural tricks were likely to be unpleasant (288). Reference was made before the committee to terrified trapeze dogs at the Victory Circus at Olympia in 1920 (*7f*), and to Sibbon's terrified trapeze cats, and rope-walking cats with others in perambulators at the Metropolitan Music Hall, Edgware Road in 1911 (*179*). Henry Trengrouse, magistrate of Hampton Wick, described cruelty at the Palladium when a dog frightened into a tremor walked on its fore and hind legs, and up a ladder, but there was no audience protest and the RSPCA was not called (2382ff). Helen Trevelyan referred later to a former circus hand who had made a sworn statement to a 'well-known league' about the cruelty of training dogs for somersaults and rhesus monkeys for wire-walking (Trevelyan 1936, p. 95ff). Mitchell noted that training could be taken much further, to tricks which were outside animals' natural instincts and also opposed to them but which drew on their natural or physical abilities. The Zoological Society of London itself refused both ready-trained animals and offers of training (Mitchell 1929, p. 244).

Those who represented the profession claimed that nearly all performances exploited natural behaviour and that the few unnatural tricks involved no cruelty. Although it was an unnatural spectacle, Mitchell did not deny to Jesson the apparent happiness of a chimpanzee on a bicycle at a demonstration before select committee members outside the Houses of Parliament (2278/9). According to James Sanger the strange positions and manner of walking of elephants in a performance was natural: it was done for their own amusement and could be seen in the zoo (*973*). Even an elephant walking over a horse should raise no concern (*2280*). Training a fit dog of a suitable type to perform somersaults by rewards of sugar or meat was a simple matter (*1010*, *2173*). George Lockhart said that training was based only on what came naturally. It was also acceptable to train dogs to walk on their forelegs because they enjoyed the work, and the trainer would not want otherwise (*1409ff*). Circus proprietor Fred Ginnett claimed that starting with rewards of titbits when a kitten, 'you can get a cat to walk on his hind legs as far as you like', and he attributed trembling in dogs on stage to excitement, not fear (707, 748ff). Tyrwhitt-Drake regarded a trained animal as behaving naturally but on command and that cruelty was unnecessary (228/333; Tyrwhitt-Drake 1946, p. 135). A dog walking on his front legs, on a trapeze or performing a back somersault did seem unnatural to him, but not cruel (286/294; Tyrwhitt-Drake 1946, p. 132). Bertram Mills also emphasised that tricks were based on natural behaviour: 'I have never seen a trick yet that an animal would not do more or less naturally when playing with other animals' (2197). Acts he had seen would not deserve elimination, and Reuben Castang, the professional animal trainer, was also unwilling to see certain tricks eliminated, because some animals were extraordinary (2189, 488). But Harold Nicolson did not like what he saw at Bertram Mills Circus at Olympia. He asked: 'Is the pleasure which we derive from a reversion to childhood quite as simple or as innocent as we like to believe?' He did not think that elephants, with their 'slow and massive obedience', were intended by nature to stand on tubs. In the same way, bearing reins, or standing on their hind legs, were unnatural for horses, even if no brutality were used. And 'it must indeed be a low type of humour which derives pleasure from seeing animals behaving in the guise of human beings. The thoughtless laughter, even of adults, which greets and encourages these performances fills me with sadness' (Nicolson 1948). In a letter of reply, Anthony Hippisley Coxe referred to Nicolson's 'extremely sensitive mind' and insisted that 'all the best tricks performed by animals show off their natural movements'. He thought that 'the circus in this country suffers from lack of serious criticism' (Hippisley Coxe 1948).[13]

'Captain' Woodward, interviewed by *The Era*, said that his sea lion acts were based simply on enlargements of natural behaviour (*Era* 1922). He told *Chums*:

> I endeavour to employ every one of their senses, even their reasoning powers, which are by no means inconsiderable... I keep them in actual sea-water, which is changed every day. In the summer I am obliged to add ice in order to keep the temperature as nearly as possible the same as that which they experience in their wild state... In the water seals are capable of performing some wonderfully clever feats, but the inconvenience of a suitable tank is too great for exhibition purposes... Their natural traits have been so far overcome that one animal would catch live fish and throw them out of the tank. (Smith 1898)

Gillespie noted that successful training and performances were based on what a sea lion did well naturally, instinctively and physically: catching and balancing (closely related to its fish-catching routine in the wild) and playing. The best-trained sea lion he had ever seen on stage belonged to Joseph Woodward, involving a motor cycle act developed as a new routine after Woodward had decided not to retire in the early 1900s. It was exhibited by Mrs Woodward, who sat in the sidecar of a motor cycle which the sea lion, 'Jellicoe', appeared to drive. The elaborate sketch involved a bang, a breakdown, an inspection of the engine, the setting-up of a picnic and balancing acts. Mrs Woodward also told Gillespie that Jellicoe for some time, perhaps mischievously, missed out a bit of the routine—not during rehearsals but only in front of an audience, when no correction to this sequence of conditioned responses would be made (Gillespie 1934, pp. 120–132). Gillespie believed that because all performance was based on natural behaviour (although dogs and chimpanzees could do more, and he disapproved of cat acts), it was no marvel and easy:

[13] Hippisley Coxe was an enthusiast and historian of the circus who donated his collection of circus memorabilia to the (Victoria and Albert) Theatre Museum in 1978. He is said to have trained his own troupe of performing cats, called 'Coxe's Catrobats', which could walk a line of champagne bottles, jump through hoops, count and ring bells (Matthews 2011; Northern Star 1983).

> One would teach a sea-lion to jump off a motor cycle at the sound of an explosion and look underneath by the simple means of firing a blank and at the same time letting the animal see you throw a fish under the cycle. Very soon the sea-lion has only to hear the report of an explosion to leap down and look under the cycle for the fish, which afterwards, may not be thrown there, but given to it by hand when it has made the routine movements. In similar fashion, the animal will be led, step by step and movement by movement to perform a series [of conditioned reflexes] which, in the aggregate may seem complex and difficult but which are, in detail, of the same simplicity and naturalness. (Gillespie 1934, pp. 114, 115, 126)

In a reference to Joseph Woodward's performances at Olympia and other places involving feats of memory and skill by these 'seemingly most lethargic and useless of creatures', it was noted that naturalists claimed that the seal (rather than the dog) had the largest brain in proportion to its size after man (Dolman 1899).[14]

## 4.4 Taming and Training: By Fear or 'Kindness'?

Between the time he started to train animals in the 1870s and the beginning of the twentieth century, there was, said Hagenbeck, 'probably no sphere in which the growth of humanitarian sentiment has been more striking than in the treatment and training of performing animals ... sympathy with the animal, patience with its deficiencies, has brought about a perfection of education which cruelty altogether failed to secure.' In contrast, 'Many years ago I saw at an auction in England four "trained" lions, whose whiskers had been scorched off and who were frightfully burned about their mouths' (Hagenbeck 1912, p. 118). That was probably around the time when Haney published in 1869 its *Art of training animals* in the United States. The methods advocated in it are mostly reasonable, and not too far removed from Hagenbeck's much vaunted approach. It might be that Hagenbeck's worldwide business, as well as his academic interest and the attention given him by the press, acted as specially influential and efficient vehicles for the promotion of relatively humane training techniques. It was no doubt the case that many trainers only gave them lip service, since harsh treatment of wild animals was already proven to be effective in the necessary subjugation of them; and violence was often still part of the display itself, perhaps reflecting worse behind the scenes. The methods advocated in *Haney's* depend on positive and negative reinforcement by rewards or the withholding of food; on verbal cues with praise and caressing, or admonition; and on light use of the whip or pin pricks as punishments or incentives. The reader was told that fear of severe violence in the animal would divert attention from lessons and inhibit responses to instructions, and 'he will have a sneaking look which will detract materially from the appearance of his performance'. Only one trainer was known by the author to use these 'cruel and unwise' techniques: he 'trains his horses with a club'. But the legitimate use of pins with visual or aural

[14] Elsewhere it was claimed that seals were less tractable and intelligent than sea lions (Todd 1918).

cues would enable controlled kicking or nods and shakings of the head in mules and 'talking horses'. As a general rule, the severity of teaching should be related to the disposition of the given species. For example, a thorough flogging of a celebrated monkey who refused to complete a tight-rope performance in a New York theatre had no effect—he only jabbered and howled; and 'Being well fed and well treated, when it behaves itself, even a flea will become tame ... As fleas are not well adapted for being flogged a new device is resorted to, a piece of burning charcoal, or heated wire, is held over them until they are subdued' (Anon. 1869, pp. 15, 39, 163, 172).

Hagenbeck developed his more gentle methods of training from 1884 and began to apply them through his trainer, Deyerling, and brother-in-law, Heinrich Mehrmann, from 1887. He described himself as more of a naturalist than a trader and disliked training with whips and red-hot irons—things that he maintained were of the past and ineffectual because they caused hatred instead of trust. He also recognized differences of character within as well as between species, leading to the need for careful selection. These 'modern methods' could be used to accomplish difficult and unusual tricks such as a lion riding on horseback (Hagenbeck 1912, pp. 30–32, 125, 126, 136). Hagenbeck's British counterpart was Frank Bostock, who also disapproved of cruelty in training, but qualified this assurance with the comment that 'kindness is the whip used to lead wild animals to obey'. Only in the last 20 years of the nineteenth century were elaborate tricks with wild animals developed, he said (Bostock and Velvin 1903, p. 30). He found it strange that many people believed that valuable animals, expensive to acquire, train and maintain, were drugged or treated cruelly.[15] For wild animals such as lions, Bostock explained that they should be made to forget their own power, and to punish an animal for natural aggression was counterproductive. The best animals were wild-born, young and new to captivity: captive-bred animals grew to know their strength and resent their confinement, until they reverted to just eating and sleeping. The wild-caught lion (if not killed on the journey from the wilds by fear) is a sick, weak, nervous wreck on arrival and full of hatred. One quiet trainer must then take over all aspects of the lengthy gentling and familiarization processes. The lion is subdued by countering aggression with a chair, a stick to its sensitive nose and continued routine familiarization, leading to the ability to touch it. Simple training in the cage follows, and then arena training, using exact routines with nothing unexpected or unusual, and remembering that 'no wild animal is ever tamed, only trained'. This training was best done at night because as nocturnal animals they were then more alert (Bostock and Velvin 1903, pp. xvi, 45, 46, 120–159, 166, 167, 233). But things could go wrong. Tyrwhitt-Drake let it be known that incorrigibles were dispatched sometimes incompetently by gun or poison and that Frank Bostock castrated four dangerous lions to quieten them down, when they lost their manes (Tyrwhitt-Drake 1946, pp. 150, 151, 178).

[15] Van Amburgh had supposedly called drugging despicable, 'and no fair proof of anything but the powers of physic' (Horne 1838, p. 38, cited by Joys 2011, p. 54).

In a book for Russian children, kind methods of training, only using food rewards, are also attributed to Vladimir Durov and are contrasted with the usual beatings. Durov opened an animal training establishment in 1912 and applied the same principles in training as Hagenbeck and Bostock. These included establishing in the animal a sense of the trainer's confident authority and being aware of features of the animal's natural behaviour, curiosity and fight-flight characteristics, which could then be modified through conditioned reflex. Durov's House opened to the public and under his descendants attracted a growing number of visitors to see over 35 species of performing animal—there were 15,000 visitors in 1940 and 250,000 by 1955 (Dvinsky 1956, pp. 12, 18, 22–29, 34, 35, 61, 116). Neirick explains how in the Soviet Union animal performance was put to political and symbolic use, resulting in alternating training methods. Animals were trained in order to ridicule enemy figures and systems and to reflect social and technological achievements in Soviet society. Advances in aeronautics and rocket science were illustrated by pigs, rats and bears flying above the arena by hot-air balloon, model aeroplane and parachute. There was also a message for domestic and, later, foreign audiences to be found in the way animals were trained. Except during the Second World War, the relationship of performing animals with their trainers was supposed to reflect those desirable features alleged to be promised in human life under the Soviet system: friendship and cooperation, hard work, understanding and best use of individuals' natural abilities, patience and a determination to succeed. With wild animals, it was all about teaching rather than taming. However in wartime, Neirick shows that the acts were designed to promote ideals of bravery, and so confrontation and patient domination, with a wild animal taking the part of the enemy, began to echo the more aggressive training and exhibition methods which trainers such as Durov and others had earlier criticized as characteristic of exploitative capitalistic societies (Neirick 2012, pp. viii–ix, 21, 45, 71, 72, 96–100, 102–106, 170, 178, 195–197).

Commenting on the possible use of drugs and alcohol in training, according to *Haney's* 'The statement that belladonna or the leaves of *datura stramonium* are put in the food of tigers to act on their nervous system and create hallucination and terror, is, we believe, unfounded; no hallucination equals the simple reality of a heavy iron bar.' But bears sometimes acquired a fondness for liquor, which could be used as an inducement or reward for performing. 'Many monkeys also had a great liking for strong drink, and this weakness is frequently taken advantage of by other trainers to induce them to perform; a bribe of a little liquor often proving a more powerful incentive than anything else. A mandril, who, at one time, created considerable excitement in London, where he was exhibited under the title of "Happy Jerry", was a remarkable example of monkey devotedness to the rosy god. Gin and water was his besetting weakness, and to obtain it he would make any sacrifice or perform anything within the bounds of possibility' (Anon. 1869, pp. 131, 139, 140, 163). According to Ellen Velvin, none of these procedures were used in the early twentieth century: drugging (because it would cause lethargy or savagery), declawing (because of its cruelty, and it could cause fatal infection) or defanging (it would be difficult, would lessen an animal's value, and affect its digestion) (Velvin 1910–1911). But in his letter to *The Times*, Thomas Hardy (1913) referred to the insidious cruelty of the use of drugs, and the select committee was told that lions and tigers were drugged with laudanum mixed in their water before a performance to make them sleepy (*421*). Fred Trussell claimed he had never heard of drugged lions, Henri Rey said doping would make a lion unusable and Fred Ginnett pointed out that a doped lion would not know what he was doing and be dangerous, like inbred 'mooney' lions (*3127*, 1707, 635). And while Mitchell noticed a curious absence of (dangerous) front teeth in performing chimpanzees (2228), Rey also said that lions' teeth and claws were not drawn (1695, 1711).

The methods of Hagenbeck and Bostock came to be referred to in the United States as the continental or European system, with different sorts of non-confrontational acts resulting from it. It was not adopted there as readily, and detractors (often with hypocrisy) accused some of its proponents of declawing and defanging their animals. In the prologue to Joys (1983), Dave Hoover, Clyde Beatty's successor, claims disapprovingly that although European wild animal performers were usually 'quiet and docile', many had been declawed and defanged. Roger Smith makes the same charge (as well as that of gelding) against Clyde Beatty's critics in the book's epilogue (Joys 1983, pp. xiv–xv, 281). But around the turn of the century, the Los Angeles SPCA had occasioned $50 fines for each of five people for attempting to goad to fierceness with revolvers, whips, a club and a fork an emaciated, declawed, toothless and harmless old lion at an amusement park concession (perhaps one of poorly regarded 'mud shows') (Bensusan 1913, pp. 124, 125). Beatty himself expressed continuing opposition to unnaturally declawing big cats because it frustrated them and changed their personalities (Joys 2011, pp. 411, 412, citing Beatty and Anthony 1965, pp. 4–7). Van Amburgh had been accused of declawing (*Gentleman's Magazine* 1873), and the practice had continued in the United States. In 2002, Cole Bros were exhibiting declawed tiger acts under the 22-year-old trainer, Clayton Rosaire (Rowe 2002). The theme of violence and domination as the subject of the human-cat performance persisted for many decades in the United States, and press and periodical accounts of the antecedent training differed markedly from the accounts of Hagenbeck and Bostock (even though, as criticism became more frequent in the early twentieth century, American trainers began to deny cruel techniques). Clubs, knives, firebrands and thrashings are mentioned by Samuel Hopkins Adams as if commonly used on the big cats: 'Only when they are stubborn or show fight do they suffer . . . "Do not punish until you have to; then punish hard" is the training maxim' (Adams 1900). Harvey Sutherland wrote in *Ainslee's Magazine* that 'Training wild beasts . . . calls for the hard heart that will not flinch from inflicting pain on the refractory being that refuses to be mastered'. Java tigers were impossible to train because they were always crouching, ready to spring (according to William Snyder)—a pair on importation had to be 'punched with red-hot irons' to transfer them into their new cage. Of elephants, they can be trained to adopt certain positions which offer relief from the repeated use of choking chains: 'It is a common thing to see elephants' ears torn into fringes and their thick hide lacerated till the blood comes with the hook the trainers carry. One of the sad features of this life is that the elephant is ungrateful for his education, and watches every chance to kill his dear teacher' (Sutherland 1902). Also dealing with elephants, Lewis later advised that to subdue them, they could be whipped between the eyes, prodded in the flesh between the ear and eye, or a chain slowly tightened between 2 ft that weakened the animal and caused bodily signs of defeat such as bulging eyes, loose bowels and lying down on its side, then followed gentle words and food from the handler (Lewis 1955, p. 40, cited by Tait 2012, p. 93).

In 1908, Maurice Brown Kirby claimed from personal observation that taming and training for performance were achieved entirely by physical restriction with mechanical cruelty—a principle of 'hurt them and they will heed', and 'the only use an animal trainer has for the word "kindness" may be found in its employment when he discusses his professional methods with an interviewer'. And 'when you see [a bear] with his nose in rags [from failed ringing] pacing about a cage, it is probable that he refused to be "broke"'. With a new tiger, the trainer never lost his temper. 'He went about his work with the impersonal determination of a public executioner. Mentally, the man was utterly at ease. He felt no more anger toward the tiger than a butcher entertains toward the carcass of an ox he is quartering' (Kirby 1908, pp. 435–446). Joys describes Kirby's article as a 'bitter, cutting satire', although there is nothing satirical in its tone: in spite of the sarcasm in its title, it is, rather, a vivid criticism, full of irony (Joys 1983, p. 42).[16]

Johnson argues that the illusion on which much of the circus depends has been extended to counter public concern about cruel processes in animal performance and that 'Hagenbeck's so-called "soft dressage" provided a convenient alibi for the numerous trainers who had neither the patience nor the wits to adopt his methods' (Johnson 1990, pp. 14, 68). Stokes (2004) would agree: although Hagenbeck's methods were complimented by trainers, they were not universally adopted. At a House of Commons evening meeting in support of the original prohibiting bill, Mrs Wader-Wallis, a musician, said she had seen a cage electrified to make a lion roar (*Times* 1921). But 'Captain' Woodward told *The Era* that the question now for the select committee was no longer whether cruel devices like electricity or hot irons were used, because it had been shown they were not (*Era* 1922). Ernest Bell thought the trainer and his audience 'are in fear all the time, and this fear can only be overcome by instilling a greater fear in the victim, and pain is the ready and only means to do this.' Of the training of large carnivores: 'It is, in fact, the acknowledged difficulty and risk which make the chief attraction. To accomplish this there is simply no means but fear, and fear has to be instilled by pain and pain alone'. The animal's spirit must be broken. 'No wonder that the performers in these cases have to be safeguarded with iron spikes, red-hot bars, metal whips, fireworks, and pistols. No wonder that we read of animals having their teeth broken off and their claws cut in order to deprive them of all natural means of defence' (Bell 1918). When in the 1920s Nig, a black panther, continued to defy Mabel Stark in front of her audience, she gave it a very loud crack over the head with her thick wooden pole, and 'still

[16] Davis (2007) repeats Joys's assessment of Kirby, as well as her criticism of Jack London.

smiling at him, raised the revolver and fired in his face' whereupon the cowed animal 'slunk over to the perch' (*New York Times* 1922, cited by Joys 1983, p. 75). Alfred Court's own account of his training methods reveals lashings and blows administered with heavy stools, resulting on one occasion in the breaking of a tiger's leg (Court 1955, pp. 37, 39, 45, 130, cited by Joys 1983, pp. 162 and 163). 'The stick and the whip are as necessary as the reward of meat, the soft voice and the caresses' (Court 1954, p. 61, cited by Stokes 2004). Absolute trust and mutuality between participants are the keystone of real performance, Stokes says, but in this case are absent: there is only the illusion of trust and of 'nature' (Stokes 2004). A reviewer of *Lions, tigers and me* by Austrian trainer Roman Proske who retired in 1949 noted that Proske claimed humane methods, but used a revolver and whip. In Europe, such élite trainers criticized the use of the protective chair by other trainers whom they called 'furniture movers'. The reviewer, vice president of the New York Zoological Society, said that like other 'animal men', Proske had no inner feeling for animals, which have only been a means of exhibitionism and livelihood (Carlisle 1956).

Training to make an animal learn and comply was suspected also as a procedure to overcome an animal's reluctance to behave unnaturally or to bring it to overcome an element of terror in a trick. In wild animals, an emphasis on taming, or rather, subduing, was followed by simple training; and in domestic and more amenable animals, there was a concentration on the more complex training that was possible. The disagreement over the actual use of fear or kindness in these procedures reflected the alternative training methods available of motivation through punishment or reward. Although there was disagreement between the two sides about which method was actually in use, the profession increasingly claimed that kindness and reward were the only necessary and successful methods. Critics believed that some forms of performance could only result from harsh training. It was, of course, difficult to establish the truth behind training processes because of secrecy and lack of independent observation, in spite of public challenges issued by each side to the other. It was stated at the select committee that contemporary methods were different from the training of 20 years earlier, when most convictions occurred (1771, 1985). This was then and since often attributed to the adoption of Hagenbeck's method by the turn of the century, based on careful taming and teaching and on education rather than fear, reward rather than punishment, teaching 'to do' rather than 'not to do', and on natural actions (Dembeck 1965, p. 312). Bertram Mills said of his Olympia circus: 'Patience and mutual understanding between trainer and animals, kindness and reward for good work are the methods of successful training' (Recorder 1921). The profession argued that cruelty was usually only imagined, but when it did occur it was the responsibility of 'black sheep' who were often foreigners: cruelty was a feature of individuals, not the training, and could be caused by a human lapse of temper or even a national or cultural characteristic (*794*, *1060/9*, 1950).

The argument over the severity of training methods extended, of course, to domesticated and 'docile' animals as well as ones taken from the wild or with wild temperaments. In an interview of 1922 for a feature on the 'Wonderful art of animal education’, ‘Captain’ Woodward said that methods were much more refined than 30 or 40 years earlier (*Era* 1922). An observer later wrote of his act: ‘The seals were delightful and if cruelty were used there, then the seals took kindly to it and Captain Woodward must have been kindly cruel’ (Pope 1950, p. 350). Woodward confirmed that the proverbial greediness of sea lions had a great deal to do with their training, which took 3–4 h a day: ‘I shouldn’t be able to persuade them to perform by violence, for they are naturally vicious animals.’ This ‘viciousness’ was caused by the fact that although inquisitive, they were exceedingly timid and easily distracted, especially by unexpected noises (Smith 1898). But *The Animal World* quoted an article on the 1923 Performing Animals Bill in the legal paper, *The Justice of the Peace*, in which the writer thought that prohibition was preferable. He recalled seeing a sea lion act in the war in a provincial hippodrome, when military duties later took him behind the scenes. There had seemed to be good trainer-animal relations on the stage, but after a successful performance, the trainer struck the sea lion with a ferocious blow on the head, when the animal scuttled to its holding box. Other artistes said that although the sea lion was docile and cooperative, it was brutally treated in rehearsals—this showed that a terrified animal can indeed perform well (*Animal World* 1923). Elsewhere it was observed that although captive sea lions were not offensive or dangerous, they could bite dirtily and would not respond appropriately to punishment, but rather to a total daily intake given as reward of 20 lb of fresh whiting (Tyrwhitt-Drake 1946, p. 141).[17] However, Gillespie claimed that taming was a short and easy process, for the sea lion was a very confiding creature and, he thought, comparatively fearless (Gillespie 1934, p. 124).

Matilda de Frece (Vesta Tilley) was able to draw on her extensive experience of the music hall to offer reassurance about the absence of cruelty:

> As to the charge of cruelty in the training of animals, in all my experience I have never seen the least sign of it; in fact, I know that it is absolutely impossible to train an animal to perform tricks by treating it cruelly. I have seen a troupe of performing dogs rehearsing on the stage. They were all seated round the stage whilst the trainer was teaching a terrier to turn a back somersault, and, when the dog succeeded and was rewarded with a dainty morsel by his master, it is a fact that the other dogs left their seats and did their utmost to attract the trainer’s attention by attempting to turn somersaults themselves. (De Frece 1934, p. 272)

But some select committee witnesses were uncompromising in their assertions about cruelty, training by fear and the great rarity of kindness in trainers. The cruelty was said to occur during the training rather than the performance, and one who claimed as an artiste to have seen 75 % of animal acts said they were all the product of cruelty (*626/7, 2434ff, 2519, 2815*). These claims were at odds with the explanations of those defending their industry. George Lockhart stated: ‘We rule by

[17] Woodward had reported that a troupe of seven sea lions was consuming 250 lb of fish a day in 1915, when the war had more than doubled the price, through closure of fishing grounds because of minelaying and employment of fishermen in minesweeping. His weekly fish bill was then £24–£25, as opposed to about £7 in peacetime (World’s Fair 1915).

love, not by fear' (*1315*). Positive reinforcement rather than fear was used in training. All good tricks could only be learnt without cruelty, and, for example, cruelty would make a horse vicious and unable to perform. With cruelty, animals would try to escape from the stage, and because no fear was seen in a performance, the training must be kind (*815*, *890*, *1003*, *1282*, *1311*, *2141*, *2164*, *2264*, *3039*). Animals had to be treated like children, and like humans they had individual character and ability (*975ff*, *990*). Cruelty would be foolish because a livelihood was at stake, and trainers were carefully selected (*829*, *988*). There was a difference between cruelty and firmness: for example, ill treatment of elephants would be counterproductive, but they would bluff without authority and the trainer's force of character (*809*, *842ff*, *1373*). Kindness was also more effective than cruelty in the training of wild animals. It was purposeless and dangerous to ill-treat a lion, and James Sanger specified continuous, gentle handling from infancy; otherwise lions would be untrainable and of no use (*964ff*, 1704, *2136*). Mitchell disagreed that one could not train by cruelty, but agreed that kindness was more productive (2306).

In an exchange at the select committee between Sir John Butcher and James Sanger over what was meant by hurting and chastising in training, Sanger claimed that trainers inflicted no more pain on animals than was inflicted on children at school (*915ff*). But Helen Trevelyan later quoted Tom Valentine, 'a famous old circus man', who referred to training procedures involving gelding, starvation, the iron bar, declawing, spiking of elephants' feet, nails in saddles for buckjumping and use of hot steel plates synchronized with music so that animals were conditioned to raise their paws in time with music. Valentine noted: 'Few unauthorized persons are ever permitted to witness the training of animals, and this makes it difficult to discover to what degree their ultimate performance depends on cruelty' (Valentine 1936). Horses were allegedly taught to buck-jump, which always much amused an audience, with spikes under the saddle or under a dummy rider (Bell 1918). Spikes were supposedly fitted to collars and saddles for various purposes. James Burton Cuthbert, a trainer of moose, deer and ostriches, said that fitting nails under saddles for buckjumps was not necessary because use of a cinch was effective (595). Fred Ginnett considered the idea of using nails or pins under the saddles of camels was ridiculous, as they did not perform anyway, and he refused to acknowledge that Hagenbeck had been convicted in Naples for camels injured in this way (832/6). He also rejected arguments that spiked collars turned inwards were found useful by the profession in the quick training of dogs, and offered £100 to anyone who could train a dog with such a collar (*Times* 1913a). In the case of 'equestrian lions', the horse was fitted with a spiked collar to discourage the lion from gripping the horse with its claws, and although Mitchell did not consider the act itself particularly objectionable, he saw some risk to the horse if the collar were displaced (2316). Trevelyan also cited the anonymous, sworn statement given probably to the PADL shortly after the First World War by a former circus hand in which he confirmed he had witnessed severe restrictions of putrid food used as part of the training and performance routine (Trevelyan 1936, p. 95ff). The record of uncompromising and contradictory views of representatives from each side makes it necessary to search for the odd unguarded statement or inconsistency as a means of attempting to establish an accurate picture. But it was usually the case that well-rehearsed arguments were pitted against emotive claims and that a lack of hard evidence one way or the other remained a product of jealously guarded professional secrecy.

The role and use of the whip was frequently discussed at the select committee. It was alleged that dog trainers would use whips with spikes disguised by ribbons. Charles Haverly, the actor-manager, told the select committee that 15–20 years ago he had seen aggressive use of a trident with whip, iron-tipped whips, close confinement and starving in order to subdue animals for training (1530, *548ff*). In order to constrain lions and bears safely in a fixed position so that the trainer could whip or beat them without fear of retaliation, a windlass could be used from below that drew down to the floor a rope attached to the animal's head or neck. This was referred to as Bedale's method (*596*). Fear of the whip on stage was said to result from its ferocious use in training (42ff), but Tyrwhitt-Drake did not use it to establish authority and believed that ill treatment with it would be very dangerous (357f). Some claimed that only cracks of the whip were used, not lashes, and that it was a symbol of authority only or for defence against a savage animal (*1144*).

It was often suspected that elephants and lions were continuing to be regular victims of goads, spikes and hot irons. Frank Bostock told *The World's Fair* that men would be ready with shooting irons and hot spikes in case of sudden emergencies (*World's Fair* 1911d). Pistols, red-hot irons and pointed steel rods were admitted, but explained as a precaution (*437*, *496*, *540-1*). James Sanger and Fred Ginnett said that no hot irons were used, being kept only as props and for show and for effect with tame 'bouncing lions' (*839ff*, 637). For George Lockhart, there was no need for hot irons or spikes, because the animals were easy to train, and E.H. Bostock said that although hot irons were formerly in use, it had been for show only in small menageries. If they were actually used, the animals would be all scars. He had welcomed the 1911 Act because it stopped any possible or apparent use and prevented the profession undeservedly falling into disrepute (*1479ff*, *3250*). Because of their size and thick skin, reasonable use of a goad to control elephants has become customary, but its use in the circus and music hall was said sometimes to have been excessive. James Sanger explained that only light and not heavy striking of elephants was effective, but Charles Haverly accused the Lockharts of cruel use of goads producing sores on elephants 20 years earlier; and elephants at the London Hippodrome were seen to be goaded behind the ears and in terror when made to plunge down a chute into a tank of water (*912*, *556*, 511). A witness alleged that an elephant had gone on the rampage when hot irons were used to make them perform this act on the chute but George Lockhart thought the allegation about the goad used for the chute was incredible because the elephants would have delivered devastating kicks to handlers. Fred Trussell said that Busch's elephants from Berlin performed the chute act ready-trained and that they were motivated only by reward and were attracted to the water (*2649*, *1380*, *2906ff*).

Dogs appeared especially frequently on the music-hall stage. Their popularity was assured because they were animals familiar and attractive to the audience. The trainer also favoured them because they were intelligent, amenable, plentiful and cheap, and as performing animals, mongrels were considered more pliable and resilient than thoroughbreds. It was often stated that dogs enjoyed performing, and there was no need for coercion or cruelty (137), but as with some other animals, it was often difficult to tell from a dog's ingratiating behaviour whether it was fearful or exuberant. Critics of animal performances believed that gross cruelties were perpetrated on dogs because their distress could be disguised and that because mongrels were plentiful, they were expendable. The trainer would reply that, as with any other animal, even in the absence of respect or affection, the time invested in training would make ill treatment or waste of such a profitable asset illogical. But the unnamed former circus hand cited by Trevelyan had included in a sworn statement an account of methods using tight, strangulating straps for back-somersault training and dazing with blows to learn the front-leg balance (Trevelyan 1936, p. 98). Reuben Castang, speaking of the front-leg trick, asserted that a real trainer would not need to hang up and then beat a dog (Fig. 4.5). The trick was easy and natural to train in a small dog, just as it was easy and natural with elephants, which could not be hung up. He felt that no credit was given for animal intelligence, and there was a variety in animals as in people. He did know of cruelty, but not in his circles: his policy was to select suitable trainers who specialized in different animals (430).

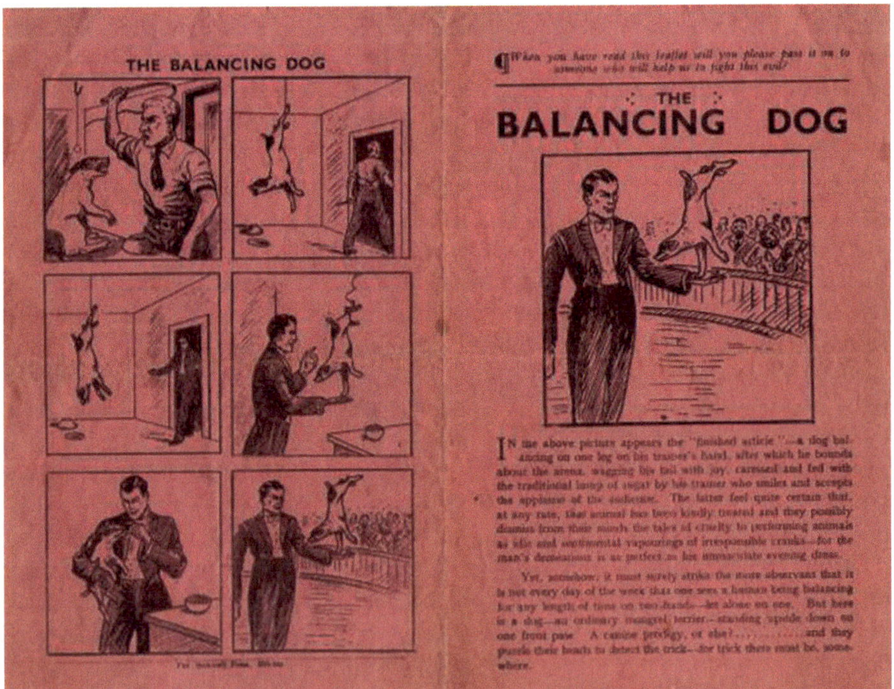


**Fig. 4.5** Part of a pamphlet circulated in the 1930s by the Performing Animals' Defence League, showing the alleged methods of a trainer and his assistant. © National Fairground Archive, University of Sheffield Library

Ernest Bell had written in 1918 of a music-hall turn: ‘We have seen a little dog ride a bicycle across the stage and the audience applauded his cleverness. They did not know that his feet were tied to the pedals and the machine was drawn across the stage by a thin wire.’ And he continued:

> For a dog to walk on his hind legs may not look at all painful, but the training, in which he is held in the position by cords and hit under the chin by a piece of wood if he attempts to come down on his forelegs, is distinctly cruel. Somersaults, which do not hurt when performed, have to be taught by forcibly pulling the dog over at first, with the free use of a whip if he should be slow or sulky. A writer on the subject has said: ‘Have you ever seen a dog taken by the hind-legs, thrown in the air, and alight on its forefeet, balancing on its master’s hand? If you only knew the hours of torture that “friend of man” has gone through to attain that proficiency you would boycott every dog act forever’. (Bell 1918)

Similar training and exhibition of such ‘acrobatic dogs’ were described by Frank Kremnitz of Frankfort in May 1913, when the audience was unaware of the cruelty of the morning’s rehearsal before its approbation of the evening’s triumphant performance (Bensusan 1913, pp. 121 and 122). Carl Hertz claimed that such animal performance depended on manipulating natural play. ‘If rubber balls had not been invented there would have been no performing dogs. Don, the ‘drunken’ dog, was trained by rolling a ball from side to side in front of him. The dog, of course, followed every movement of the ball. Afterwards, the ball was dispensed with, the trainer’s hands indicating the necessary movements. In a similar way dogs are taught to turn somersaults. A ball is thrown, and the dog springs up at it. The next throw is directed behind the dog, which, in consequence, flings itself backwards. Once the idea is grasped, subsequent training is easy’ (EHG 1924). Other allegations of cruelty to performing dogs considered by the select committee included that of a stage carpenter, Sidney Davis, about the weights he had observed on the necks of the Austrian Lipinski’s ‘drunken’ dogs. In this pre-war act (before the time when such as Lipinski would have been boycotted by the industry as ‘enemy aliens’), the dogs entered the stage from the left and proceeded to disappear into a ‘public house’ in the middle, when they would have heavy weights fitted around their necks, hidden by comical costumes of policemen, firemen, messenger boys and so on. These weights made the dogs stagger around and were enough to make it very difficult for them to stand, but men with whips in the wings ensured that they did so (1257/90). It was also alleged that dogs under duress were made to appear happy by having their tails wired to their collars in an upright position. George Lockhart disbelieved this witness’s assertion: he thought the idea absurd because the ears would be a giveaway, and the dog would not in any case be able to perform. Harry Rochez envisaged them doing pirouettes (*1500*, 1359).

A long-established animal act that earned more than usual attention, suspicion and criticism was the ‘Monkey Orchestra’. Harry Rochez patented a version of the act and presented the select committee with persuasive evidence that no cruelty was involved in his own performance. Fred Trussell supported him, describing him as a ‘very kind man’ (*3115*). It is possible that his imitators had made the act controversial, and in one case, there was a successful conviction when cockerels had been used instead. The RSPCA obtained a fine of £5 against the ‘Great Sousa Rooster

Band' at Blackpool Pleasure Beach. Their inspector found 11 cockerels tied up in coloured clothing which they were continually pecking, particularly under the wings and down the legs. It was a very hot day and the showman caused unnecessary suffering—it was advertised as a continuous performance. 'The attendant introduced the birds to the audience, and described that they were highly educated, and trained entirely by kindness.' The birds were attached to musical instruments and were manipulated from the rear of a stage. Only their heads were exposed and they were in pain, continually flicking their heads, rolling their eyes and opening their beaks. Afterwards they were too exhausted and cramped to walk, and their legs and wings were chafed, with raw and bleeding sores (*Animal World* 1921c). Lily Grahame, a pianist and chorus mistress with the stage name of Swonnell, complained of cruelty to exhausted wired-up 'monkey orchestra' musicians at the London Hippodrome between 1900 and 1905. She had an altercation with Walter de Frece about the wiring method used before and after 1910, when a new method was patented, the question being whether wires were attached to restrained monkeys or to the instruments (*2560ff*, *2691ff*). Harry Rochez insisted that in his monkey orchestra at the London Hippodrome, a tiny chord, not a wire, was tied to the monkey conductor, as a simple prompt to the animal. His monkeys were loosely strapped to chairs around their loins. The instruments were self-playing and operated mechanically by staff. The monkeys would naturally grab at instruments when they were moved by the wires. The performance was not intended to be a musical one, but was based on the comedy of the monkeys' boisterous behaviour. The RSPCA had seen the act and had raised no objection, and at a presentation before the Royal Family, the members of the orchestra had been seen close up and displayed no signs of exhaustion or distress. The act was patented in 1910, and there had been one imitator before then, but Rochez had always only used the same method. None of the orchestra monkeys were trained, except the conductor, and one of these had worked for 9 years before being stuffed (1345–1357, 1380, 1395, 1400). Mitchell expressed reservations about monkey orchestra acts because he saw the possibility of cruelty in constrained animals having to sit fixed for a period of time in rows of chairs, and he reported that he had seen one or two very undesirable performances (2274).

Apart from the controversial use of common devices such as whips, cords and restraints, critics found fault with specially devised mechanical apparatus. For example, a possible danger to the horse was found in revolving tables on the stage. Tyrwhitt-Drake opposed revolving tables because he feared the animal might come off, and injury to horses was easy; and the select committee was told of a fatality to horse and rider at the Coliseum. Arthur Aitchley, the horse trainer, thought the revolving stage safe because it had a rubber surface, and the horse would wear rubber shoes. Fred Ginnett differentiated two types of revolving stage. One type was mechanically self-activating and the horse had to follow its movement, and at the Coliseum 5–7 years before, this more dangerous type had been used mistakenly because the horse had also been required to gallop the wrong way round. The second, safer type revolved only as a result of the movement of the horse and stopped when the horse stopped (322, *2566*, *2218*, *2244*, 673).

A further matter discussed at the select committee concerned the use made of animals by conjurors. A well-known magician had admitted that rabbits were handled very roughly when 'produced', tossed into the air or 'disappeared', and they could be very closely confined for some tricks, as were doves and pigeons. He knew of a conjuror doing the vanishing caged canary trick who when he broke the leg of a bird in the course of this trick would set the leg with a matchstick and use the bird for another performance. One well-known conjuror would place a dove on his revolver, fire several shots and the bird would remain perched on the weapon. 'The trick was obviously one of those which should have been ruled out by the managers of the halls' (*Animal World* 1921b). The magician Carl Hertz got his canary 'Connie' into the select committee concealed under his coat, for a demonstration of the kind that had previously satisfied the RSPCA—the collapsible disappearing cage and bird trick (*Animal World* 1921c). It was on this occasion that the select committee recognized and discussed the problem of asking a performer like Carl Hertz to reveal professional secrets (*1642*). *The Encore* suggested that there was a possible problem in the use of mechanical contrivances for tricks, while Arthur Valentine, the amateur conjuror, believed that sales of the collapsible bird cage for amateur use should be stopped (1633) (*Encore* 1922).

## 4.5 Training for Quick Results and Performance to Timetable

Added to discussion about the true nature of the acts for which animals were trained, and the level of cruelty or kindness used in the training process, was the debate about the particular pressures on that process in the entertainment industry caused by the need for quick, economic and profitable results and by the demands of performing to timetable. Mitchell did not doubt Hagenbeck's own patience and kindness in training, 'but with regard to public performances, when the animals have to go through their tricks at stated times, I continue to doubt; in all those that I have seen the trainer displays some kind of whip, and the animals seem to have a very full appreciation that it is more than a symbol of authority' (Hagenbeck 1912, pp. v–xi). Thomas Hardy (1913) had written to *The Times* that some trainers preferred shortcuts through cruelty, and the select committee was told that cruelty might occur when a trainer was impatient by nature. The retired actor-manager Charles Haverly described violent 'shortcuts to efficiency' that were used to subdue lions and tigers, and with dogs, the lack of time and patience would be exacerbated if the trick to be learnt was an extraordinary one. Jessie Hales did not like the cat performances she had seen and found her own domestic cat unresponsive to kind training after over a year (*68*, *599*, *664*, 2106/12). However, James Sanger said that no cruelty was needed for good results, just a light assertion of authority. The horse trainer Arthur Aitchley said the use of cruelty to a dog would actually lengthen the time of training, and it would be better to change the trick if it was taking too long.

It was necessary to be patient, or the animal would not cooperate (*809, 2253ff*). But in a report from Boston, it had been reported that a dog trainer boasted he never spent more than 3 weeks training any set of animals for exhibition; this was achieved by starvation and the whip (Bensusan 1913, p. 126). The 'training to timetable' problem, and its alleged need for motivation by hunger, had been highlighted in two letters to *The Times* in 1913 (Mason 1913; Wain 1913).

According to Bell, the performer must do what cannot be achieved at home over time with a pet. It must be sensational, and economically and quickly achieved—'money is the main object' (Bell 1918). He wrote:

> The training of performing animals is a trade. No reputable tradesman has any objection to the general public knowing the methods which he employs. But, with a very few exceptions, in this particular business it is extremely hard to discover what really takes place. In the majority of cases, and for very good reasons, this work is carried on behind the locked doors of a stable or a room of some sort or another. As the man who trains animals to perform generally lets them out on hire as soon as they are 'broken', it follows that his main desire is to train them as quickly as possible. Rapidity, therefore, and not humanity, is his chief aim. Fear has necessarily to be used as an incentive. Say a dog is to be taught to turn back somersaults. A slow method is to put harness on the dog with leads fastened to it, and practically to jerk the dog over backward time after time. This method, we are told, takes eight to twelve weeks. On each occasion the dog becomes horribly giddy, as he has to learn to do from ten to twenty back somersaults in quick succession. This method is too slow for some trainers, and the wretched dog is 'assisted' by a blow under the jaw with a piece of wood or the butt-end of a whip, until he learns that if he turns of his own accord he escapes the blow. (*Literary Digest* 1920)

*The Animal World* (1921a) published an article reprinted from *The Times*:

> An intelligent trainer can work marvels with young animals by patient kindness. And yet we believe that cruelty lies behind nearly all the public performances of domestic and wild animals—cruelty so great that it is doubtful if most of them should be tolerated. Let their advocate ask himself why a whip is the wand of authority, wielded in public so tenderly and yet so anxiously obeyed. There are two reasons why the memory of pain and the presence of its instrument seem a necessary concomitant of public exhibitions. The first is that many tricks are not developments but violations of natural instincts, learned only when instinctive reluctance was overcome by a stronger stimulus. We admit readily that some performances are without this defect; the tricks of sea-lions, for example. But the second reason almost always exists. A public performance must take place at the minute; when the scene is set and the curtain rises, the 'turn' must be ready. But animals are more wayward and unpunctual than schoolboys—attentive, obedient an hour ago or to-morrow, but often not at the appointed moment. And so pain and the fear of pain are used. There are few of our public amusements which require more careful and more expert criticism than those in which living animals play their part. (*Times* 1920)

At the select committee hearings, Charles Butler explained that cruelty he had witnessed resulted from a requirement to perform quickly and to order. Another witness said that trainers could not afford the time to avoid cruel methods (*453*, 1151/7). Mitchell saw cruelty as necessary for performances to time in 'moody' wild animals, and there was no way out of the problem. There might be no cruelty in an all-day exhibition as in a menagerie, but the timetable factor was a 'very grave difficulty', and in timed performances, at least a music-hall ban was necessary to avoid cruelty absolutely. Even lions simply constraining themselves for a fixed time on pedestals depended on gross cruelty (2239, 2241–2243, 2266, 2288f). Attempts were made by the profession to provide reassurance about the timetabling issue, often by referring to animals' willingness and enjoyment of their performance. Bird trainer Alfred Roelgin said that his birds were always ready to perform, able to participate in the same show three times a day and do the same tricks within half a minute (2362). Frank Stephenson contradicted E.J. Holland's claim of a 1-min 'get-ready' system before an act, claiming it was actually an hour (*Era* 1921b). George Lockhart declared he would not force a trick, and it was claimed by another that fear was not necessary to meet the time and place constraints of the professional (*1421*, 153). Gillespie confirmed that inducement by reward applied on stage when performing to a timetable, but performance of a complicated sequence to schedule did sometimes result in problems, as when an animal omitted part of the routine. Woodward's sea lions were not punished for this and it was looked on rather as a 'whim' and an 'illuminating side-light ... on animal mind and temperament' (Gillespie 1934, pp. 128 and 132). Mitchell had conceded that sea lion performances to timetable were the least objectionable. He told the committee:

> From what I have seen of the performances of sea lions—I have not seen their training—they seem to me to be the least open to criticism of all the performances of any kind of animals that I have ever watched upon the stage. (2271)

But in the majority of cases, he believed that unpleasantness must have lain behind performances, especially because of the demands of the timetable, when the whip was 'more a memory and a warning than a mere symbol of authority' (Mitchell 1929, pp. 244 and 245). In his memoirs he also recalled that Hagenbeck 'disliked most "animal turns" at music halls, where at a given moment the animals had to appear, and, willing or reluctant, be pushed through their "show"' (Mitchell 1937, p. 125). With John Galsworthy, Lord Lambourne (a vice-president of the RSPCA), Winifred Portland, A. Maude Royden, Constance Shaftesbury, George Bernard Shaw, Wilfred H. Sugden and Sybil Thorndike, Mitchell sent a letter in 1927 to *The Times* from the RSPCA's Jermyn Street address, citing a comment of 4 December 1921 in the same newspaper:

> There are experts on this matter who believe that, in spite of regulations, cruelty is inevitable when animals act to a timetable, and all professional performing animals have to do this. (*Times* 1927)

Yeats-Brown wrote to *The Spectator*: 'A circus performance depends for its effect on a precision and punctuality not found in nature. To force such a high standard of mental discipline on an animal that it will perform a trick accurately night after night is cruel because it is unnatural' (Yeats-Brown 1928). To help achieve completeness, precision and punctuality of actions without investing in patient and time-consuming training, the keeper might give the animal unpleasant guidance in the arena itself, in front of the audience. In recent times, the circus vet David Taylor found that a commonplace in chimpanzee training and exhibition at disreputable establishments was discreet manipulation of the sensitive ears with the trainer's thumbnail, filed to a long, sharp point—'a display of brutal sleight-of-hand carried on in full view of the public' (Taylor 1976, quoted by Johnson 1990, pp. 112 and 113).

## 4.6 Black Sheep

Alongside disputes over the possible use of cruel training methods and devices were frequent allegations of the cruel characteristics of named individuals. However, studying the behaviour of the animal towards its trainer might reveal little. Mitchell suggested that although animals could show affection to their trainer, as in the case of Carl Hagenbeck, they were forgiving and not vindictive, and therefore cruelty might not be betrayed by their subsequent behaviour. He gave the example of the bear of the itinerant Basque showman who was imprisoned for cruelty: the Zoological Society of London looked after the bear during the showman's imprisonment, and when it was returned, it greeted him with great affection (Mitchell 1929, p. 244). But Bertram Mills claimed that any cruelty practised in private must reveal itself in public performances (*Era* 1921f). Of course, all witnesses expressed disapproval of the idea of cruelty in a trainer. James Sanger regarded cruel trainers as failures who should be 'wiped out', and he would sack such staff (*885*, *1011*). Some witnesses regarded all trainers as cruel, even Hagenbeck and 'Captain' Woodward, although the RSPCA's Charles Wilson had found a unique kindness in the training of sea lions (*694*, *253*, *398/406*). Indeed, most of the information from witnesses to the select committee consisted of unsubstantiated allegations that were met with simple denials from the other side. At one moment, the American lion tamer, John Cooper, was accused of ferocious beatings 30 years back, and then a later witness praised him (*551ff*, *841*). Lily Grahame reported a catalogue of abuses witnessed during her time as a pianist and chorus mistress at the London Hippodrome in the early 1900s: these were subsequently refuted by other witnesses. She alleged that Salamonski, previously convicted for horse cruelty, returned disguised as a groom to assist his wife. Because Grahame was reprimanded by management after she reported his further cruelty to the RSPCA, she said she had felt inhibited in reporting further cases. But she gave details to the select committee of Cottrell Powell's cruelty to his high-jumping dogs, and of Renz to his horse, equipped with a double bit and 'something inserted under the tail'. She spoke of severely confined elephants, the use of hot irons to make them ascend an incline before diving into water, and of equestrian cats and dogs that were so cruelly treated that the then manager brought the curtain down during the act whereupon the showman, probably Leonidas, threatened to kill him. A Mexican hand-balancing dog was beaten, Morris's bear was apparently beaten twice daily for 2 months and Hettie Chattel, the principal lady in 'Tally Ho!', refused to play when she found that a live rabbit was thrown to the hounds at every performance, so thenceforward the rabbit was killed immediately before it was thrown to them (*2557ff*). Some of the information considered by the select committee concerned animals used only as props by conjurors. Mrs Bradshaw gave evidence that in 1909, the conjuror Leo Curtis used cockerels and pigeons with inflamed eyes and damaged feathers and that they were thrown about and injured after each act and then closely confined all day (*177*). The conjuror who attracted most attention from the select committee was Carl Hertz. Charles Haverly claimed that canaries in a cage that collapsed and disappeared up his sleeve were killed, and that he had seen blood dripping from Hertz's clothing. Since he earned £120 per week, he could write off birds at 6d each (*565ff*, *585*). One witness claimed that Hertz's assistant had told her husband that birds died in every show, and another had seen, many years ago, dead canaries from the night before flattened out on his dressing table (*2825*, 508, 534). Hertz himself claimed that killing birds would be too costly, as their price was £1–£3 rather than 6d each. He could not take 50–100 canaries with him on tour; some birds had performed for 5 or 6 years, and the bird he used in his demonstration before the select committee he had employed for 12 months (*1548ff*).

Cats as well as dogs featured in the hearings. Mrs Bradshaw spoke of Sibbon's cats being thrown from the stage and mistreated after their unnatural performances on trapezes and high wires (*179*). Duncan's Collie Dogs were said to curry favour by wagging their tails at their very cruel owner (*699*). But George Harrop and George Lockhart defended Duncan against allegations made by the PADL, saying that he was humane, and worshiped by his animals (*1173*, *1215*, *1524*). Jack Russell Vokes, owner of Don the Drunken Dog, a fox terrier, complained that he had been targeted for cruelty in the Press, but that in spite of letters of complaint, inspections by the RSPCA had led not to prosecutions but only to appreciation (*2029*, *2037*). Vokes had left his dog outside the House to show the select committee, as well as the sticks he had been accused of beating it with, and as an example of his affection he reported that he had paid £8 to have Don's father stuffed (*2044ff*). Haverly referred to an extreme case of close confinement and the lack of exercise of Peterson's dogs when not performing, and to the notoriety of Edson for beating his dog, 'Doc' (*558*). Zertho the clown and Lipinski toured with 40 dogs, and Albert Hill, the theatre electrician, reported that in 1913 and 1914, he had seen that some dogs had their mouths tied with string to keep them silent on stage, that they were whipped and that they were closely confined as quarantine dogs. Lipinski and his assistant would throw a dog to the floor or lash it for a mistake, as during the riding of a tricycle, when the dog's feet had been tied to the pedals and handlebars (1082/4, 1237). Sidney Davis, the stage carpenter, talking of Leoni Clark's dogs, monkeys, cats, rabbits and others of his tight-rope walking animals, said he had seen the beating of a monkey so that it fell as limp as a piece of rag: training would take place usually before 9 am and a watch kept for any inquisitive strangers. He also complained of the insanitary conditions in which animals were kept by Zertho (1244–1246).

Bears, apes and elephants, rather than big cats, were given as examples of wild animals allegedly persecuted by cruel showmen on the British stage and in the circus in the first quarter of the twentieth century. For the sake of entertainment, the big cats were not versatile enough to warrant treatment much beyond making them either submissive or reactive in a simple show of mock ferocity, and although this was itself thought to involve cruelty, the cruelty of developmental training rather than simple conditioning was of more interest to the select committee. Examples of alleged cruelty were again rather dated by the time they were offered to it. The unpredictable ferocity of bears was supposed to explain the extreme measures allegedly taken to subdue them. Charles Butler spoke of witnessing Spessardy's rehearsals with his Hagenbeck-trained bears in the 1900s. The bears were kept continuously in close confinement, except during performances and in rehearsals, when chains would be fixed each side of their collars and placed through rings on the stage, whereupon the bears would be drawn downwards under control to be hit across the snout with a log of wood. Spessardy had told Butler that through this daily routine over the course of up to 6 months, they would come to know their master (*417f*). Charles Haverly reported that before Permane's conviction at about the same time through the RSPCA, his bears, too, were excessively confined, so that they could not stand up or move from side to side except when released for a performance (*559f*). But as a trainer Reuben Carstang doubted the evidence given by Butler because it was absurd to imagine hitting a bear's nose, its most sensitive part: it would make the bear crazy and kill it through fear (428f). Furthermore, bears acquired ready-trained from Hagenbeck would not be expected to need such treatment.

On a par with the protracted dispute over the nature of Carl Hertz's act was the controversy centred on the alleged mistreatment of the famous ape, Consul, of whom, as it soon became apparent to the committee, there seemed to be a number of versions (840ff, 1025ff). James Sanger was astonished to hear that Consul was supposed, on apparently reliable evidence, to have been thrashed, but E.H. Bostock recounted in possible explanation an occasion when his brother had to beat off an ill-tempered Consul in legitimate self-defence (*821*, *2562*, *2771*, *3176ff*). Kenworthy confronted George Lockhart as a specialist elephant trainer with a number of convictions for cruelty to elephants involving individuals or cases Lockhart claimed he did not know: Darge at Penge in 1900 for wounding an elephant with a lance, Mlle Valsois at Nottingham in 1912 for prodding and beating an elephant and Charles and Gustav Hagenbeck, Leonia Havaria and William Schreida in West London in 1908 for terrifying elephants (*1353*, *1356*, *1358*, *1362f*). But Lockhart had heard of the conviction in 1917 of Alfred Rossi for compelling an elephant to perform when ill, and thought the punishment of £5 and five guineas costs inadequate (*1363*). Lockhart angrily refuted Haverly's allegation that when performing with his father he had surreptitiously goaded elephants in the trunk so as to cause new bleeding and sores, in addition to old ones (*1375f*). Sidney Davis saw cruelty in De Gracie's 'football' elephants, when they were made to balance on their forelegs by use of a harpoon disguised as a crop that caused bleeding and beaten with an iron bar (1236f, 1270ff).

John C. Essex, Joseph Woodward's brother-in-law and cousin, was convicted for his presentation of the 'Great Sousa Rooster Orchestra', described by George Lockhart as probably only an unprofessional sideshow (*1420*). It was noted that his difficulty in finding regular employment after the war was exacerbated by the fact that shortly before the select committee hearings, he had been prosecuted for this cruelty to cockerels dressed in tight bandsmen's uniforms that chafed and exhausted the birds as they apparently performed on musical instruments, which were in reality mechanically operated from the back of the stage. O'Grady suggested that the case was deliberately brought to coincide with the introduction of the bill into the Commons (*322*). John Essex's daughter, Ann (who until the age of eight lived with Joseph Woodward and his wife, Annie, in Southwick, Sussex, where Woodward's sea lions were trained), reported that her father subsequently looked after the animals in Carmo's Circus, before his appointment as first commissionaire at 'Dreamland' in Margate.[18] Another individual who particularly attracted the attention of the select committee was Chief Raven Cody (self-styled brother of the showman and aviator, Colonel Cody). He, too, was dismissed as 'not a proper trainer', this time by Everard Calthrop. According to a local journalist, John Edward Bird, he had beaten an old pony which had failed to cooperate in a performance at a theatre in Warwick on 3 March 1922. There was a strong and sustained reaction from the audience and then an argument with the manager who also failed to get him to leave the stage. The matter was described as the talk of the town for a week, and someone said that if the pony was 38 years old, it ought to be in bed instead of on the stage (914ff, 947, 1838ff).

## 4.7 Health, Veterinary Opinion and Psychological Suffering

The poor survival rate of animals brought from their native habitats in the nineteenth century was exacerbated by ailments characteristic of particular species. The journey to the United States was a long one, since most animals went via brokers in London, and the big cats often developed a kind of consumption, eventually wasting away and dying. Van Amburgh would donate their bodies to university zoological departments, natural history museums or academies of science (*Christian Advocate and Journal* 1863, cited by Joys 2011, p. 113). Tigers, lions and leopards were more delicate than people imagined, often died of tuberculosis and suffered from all sorts of ailments: they had to be looked after 'as carefully as a frail woman' (Moffett 1894). Elephants could also suffer from special problems to their feet, caused by lack of the natural exercise that wore down the sole in the wild. Because 'the unfortunate beast is compelled to stand year in, year out, the horn grows thicker and thicker, and whatever the feet stand upon is squeezed into the sole and works its way into the quick . . . With the advent of the modern wild animal physician, the greatest step in advance has been the appreciation that abundant fresh air, light, cleanliness, and opportunity for exercise are as necessary to the wild patient as they are to man' (Rolker 1906–1907).

Mitchell believed that menagerie animals 'did well', but his view was that music-hall animals did not lead a natural and healthy life (2259). Charles Butler had seen dogs on the stage that were thin and kept short of food, but not apparently untended or mangy because such dogs would be replaced by substitutes, as performers travelled with spare animals (*509–11*). George Lockhart argued that performing animals were in better shape than dogs or horses in the street, while

[18] Personal communication.

Fred Trussell claimed that the average performing animal got the best treatment possible, like a racehorse compared with an ordinary horse, and E.H. Bostock believed that his menagerie animals looked in better shape than those in zoos (*1505*, *3125*, *3221*). As to specific ailments, Reuben Castang, disagreeing with Mitchell's opposition to chimpanzees in the music hall (as expressed in his *Childhood of animals*), and regarding his noticing that the teeth of some of the most advertized performing chimpanzees had been extracted or broken off, assured the select committee that loss or removal was only undertaken because of *pyorrhea* (jaw inflammation), to which chimpanzees were susceptible (444). Although it is difficult to imagine that life on the circus and music-hall circuit was an appropriately natural experience for his sea lions, there is evidence that Joseph Woodward was affectionate towards them and, at least by the standards of the time, most concerned for their welfare. As a practical health precaution, each morning the animals' teeth were cleaned with carbolic tooth powder and their mouths washed out with a solution of permanganate of potash; and in July and August they were taken for an annual 'holiday' by the sea at Shoreham (*World's Fair* 1915).

It was sometimes claimed that life as a performing animal had enabled a restoration of health. A letter from Dan Cleary in *The Era* described how his training of a sickly horse had this effect; and through experience, he said that he knew there was less cruelty in training than in jumping or racing preparation (*Era* 1921d). And *The World's Fair* cited the current edition of *The Empire News* containing clown-trainer Whimsical Walker's account of how a war horse, Verdun, was saved from the tannery and made fit as a performing animal: 'If I thought that any act of cruelty could take place in this show, I'd walk out of it, even if I never had another engagement' (*World's Fair* 1921b). Proof of good health might be found in the longevity of performing animals. However, although he acknowledged that wild animals in travelling circuses and menageries surprisingly did at least as well as zoo animals, Mitchell said it was not possible to be sure about animals' comparative longevity in the wild and in captivity, because of factors such as age uncertainty, possible old or weak condition enabling original capture and the artificial safety provided by captivity (2224, 2251). But Tyrwhitt-Drake believed that animals such as horses lived longer on the stage because they did not work as hard as other horses in harness; and because performing animals became very valuable, and consequently were very well cared for (341ff). In a letter to *The Times*, he later wrote of the longevity of his hyena, which had just died at 27 (*Times* 1937).

The assertion was often made that high value guaranteed good care (e.g. *Encore* 1921b). The select committee was told by Arthur Aitchley that dogs were well cared for by trainers so that they could perform and earn money (*2176–81*); and Frederick Trussell denied that the chimpanzee Consul was ill-treated or injured by his owner, if only because he was a valuable asset (*3085*). He was insured for £20,000, and his death at Berlin from bronchitis cost the insuring companies that amount. He cost almost nothing to keep, and his performances often brought in £2,200 a day. The 'boxing kangaroo' rose in value from about £10 to £5,000 in his heyday at the Westminster Aquarium. 'The prices paid for wild animals as such, apart from any accomplishments which they may have acquired, is much less now than formerly, the wild-beast trade being in a state of depression, for no particular reason except that the public prefers the excitement of seeing animals perform tricks in circuses to watching them in more or less natural conditions' (*Spectator* 1904). A manager would cut an act if the animals were in poor condition and therefore ineffective: performers consequently took care of their animals (*3108*). But select committee member John Swan told Trussell of a conviction in 1917 of a proprietor of a troupe of musical elephants . . .

> . . . for travelling an elephant and compelling it to go through a performance when sick and in an unfit state. It appeared that the animal fell down twice on its way to the show, and although dosed with three bottles of whisky and goaded by an attendant with a short spike, it was unable to do its tricks and subsequently died.

Trussell attributed this to cruelty off the stage and 'a poorer affair' outside the music-hall profession (*3111*). He believed that in the profession a performer would risk his contract rather than his reputation and livelihood by forcing acts by ill animals in front of the audience and manager (*3114*). *Haney's* had emphasized this principle long ago: 'It is a strict rule that no animal who is the least sick is ever allowed to perform or be trained' (Anon. 1869, p. 199).

The opinions of veterinary surgeons were often cited by one side or the other to give legitimacy to their claims, sometimes in court. *The World's Fair* reported the failure of a case brought by the RSPCA in 1911 against two lion tamers of Bostock's Zoological Show at the Festival of Empire at Crystal Palace, where Harold Hilliard was then manager of the menagerie. The lion had been goaded with whips to make it appear fierce, and Arthur Case, veterinary surgeon, said the lion was terrified and in pain. However, the defendants convinced the court that the lion's behaviour was controlled only by light flicks and by cues, and Harold Hilliard said he would instantly suspend a trainer who thrashed an animal (*World's Fair* 1911b). Bertram Mills appointed three vets at his International Circus at Olympia 'to take absolute control of the animals', but an RSPCA inspector told him that opponents of animal performances 'do not care about Vets' that he had appointed, saying that they were paid to give favourable reports regardless of the condition of animals, the opponents expecting therefore that the RSPCA inspected independently rather than rely on such vets' certificates (2150ff). At the time when prohibition appeared a possibility, *The Era* reproduced the views of a veterinary surgeon whose statements appeared rather suspiciously to echo entirely those currently being made by the profession in their defence. Frank Somers, a veterinary surgeon from Leeds of 35 years' experience, 'a lover and student of animals, who has closely observed their training and performances', told *The Yorkshire Evening Post* that electricity and hot irons were not used, only kindness; otherwise the animal would not perform:

> It is entirely incorrect to say that training is nearly always accompanied by cruelty, and that the animals suffer great hardships while in the travelling cages. If those things were so, it would be a sad reflection upon the police, the RSPCA, and the veterinary surgeons of this country.

He described animals on the stage and travelling on the road as the ...

> ... sleekest, finest-looking, and healthiest lot in Britain, owing to the constant care and good food ... I have always been struck in my innumerable visits to menageries, circuses, and variety theatres by the kindness and patience exhibited by trainers.

There was 'no justification for faddist legislation ... which would needlessly rob thousands of honest people of their livelihood', and there were only occasional problems of cruelty from 'vicious foreigners' (*Era* 1921c). But 50 years later, although he believed animals in ACP-approved circuses were healthy and contented, David Taylor recounted in *Zoo Vet* that extreme cruelty in handling, training and exhibition could still be found in a few 'smaller, tatty circuses and menageries if you can penetrate the closed, suspicious, obstructive world behind the scenes' (Taylor 1976, quoted by Johnson 1990, p. 112).

A recognition also began to emerge of the probability of psychological suffering, resulting not just from capture and confinement. In 1922 the American *Literary Digest* cited an article in *The London Nation and the Athenaeum*:

> The psychological rack of satisfying their trainers up to a compulsory pitch beyond the orbit of their natural intelligence is on its own plane as severe as the physical martyrdom endured to achieve it ... There are people with heads even softer than their hearts who maintain an analogy between the tricksiness of their own pet creatures and the vulgar grotesqueries of performing animals. They forget that the latter are professionals, machines for extracting money for mountebanks; that they must not on any account fail; that speed and efficiency must go together, and force and terror are the only agents that will make them. A performing animal that is not drilled down to the final obedience of mechanical response is a contradiction in terms. One marvels at the magnanimity of these miserable beasts—that they so seldom turn and rend their persecutors—until one sees it is not that, but the last abasement of craven terror. It takes some terrorism to make an elephant groan like a man in anguish, a bear shut its eyes and cover up its face, and a tiger whimper. (*Literary Digest* 1922)

In the debate on proposed amendments to Kenworthy's original prohibiting bill, Bowyer protested:

> [I was] at Olympia within the last 18 months, and had seen little dogs, terrified, jumping from the top of that great building on to a trapeze 30, 40, or 50 ft below, and had seen their palpitating fear as, first of all, they were taken up to that dizzy eminence, and then as they were awaiting the word of command at which they were to jump. Some were blindfolded and wrapped up in sacks. Anyone who saw that performance would have no doubt that sooner or later these performances either lead to cruelty in the training or in the exhibition. Animals, after all, exhibit the great qualities of courage, loyalty, faithfulness and trust, and they are not given to us to make our bread and butter out of for business reasons in showing them in performances.[19]

A letter to the *New York Times* from Sydney Coleman of the ASPCA declared that although much of the cruelty was in mental suffering, it was not recognized in a court of law under current penal laws (Coleman 1931). It was nevertheless often claimed that the possibility of psychological cruelty was limited in animals because

---

[19] *Parliamentary Debates* (*Commons*) 1921, vol. 142, column 1429.

of the level of their intelligence, their lack of self awareness and their inability to anticipate, imagine or reflect (e.g. Gillespie 1934, p. 40). A correspondent of *The Times* believed that psychological suffering resulted from anticipation, memory, and the conscious association of moral significance to experiences; but an animal's awareness was 'moment to moment' in nature, so the caged bird suffered less than the caged man. The correspondent argued that if it was thought acceptable to put animals to work, work for entertainment was also acceptable because all work for humans was contrary to animal instincts. In the meantime, cruelty in training for entertainment work would be counterproductive: the objectors were being anthropomorphic (*Times* 1913b). Thomas Hardy immediately replied that he had seen dog performances at country fairs . . .

> . . . where the wretched animals so trembled with terror when they failed to execute the feat required of them that they could scarcely stand, and remained with eyes of misery fixed upon their master, paralysed at the knowledge of what was in store for them behind the scenes, whence their shrieks could afterwards be heard through the canvas. (Hardy 1913)

Ernest Bell reported a case in which psychological suffering had been accepted as the result of cruelty:

> The Performing Animals' Defence Committee has recently carried through a successful prosecution, in which the defendant was a foreigner. The victim in this case was a little dog, whom the man took in his arms and threw up into the air to a height of two or three feet. He then caught him, so that he stood upright on his forefeet on the man's hand. He then turned him over and made him stand with one forepaw on his thumb. The dog was then set to climb a ladder, which was shaken and which he could climb only with the assistance of his jaws. The man, who was at the top of the ladder, then placed the dog on his own head. From the fact that the hair at the back of the animal's neck remained erect the witness judged that he was in terror . . . The case was an important one, as the conviction was made, not so much with reference to any flagrant act of bodily brutality on the stage, as on account of the mental suffering of the dog, as shown by his dejected and terrified appearance. (Bell 1918)

It appeared to be recognized that certain animals in captivity could suffer a form of psychological stress because of the particular characteristics of their species and because of the nature of some interspecies relationships. It was on this basis that George Dalmere of Dalmere's Stable Circus unsuccessfully issued Kenworthy as a critic with a challenge that he (Dalmere) could train for a performance, without cruelty and within one week, 'nervous and incompatible' animals (cats, rats, rabbits and doves, along the lines of the 'Happy Family'): if he failed, he would give £50 to charity, but if not Kenworthy should donate £50 to the Music Hall Artistes' Benevolent Fund. Dalmere proposed that two rehearsals of 10 min each take place daily in the morning and afternoon in front of representatives from the RSPCA and Music Hall Artistes' Benevolent Association. No whips or sticks would be used, and the key of the training room would be held by a third party (*Encore* 1921a, b).

Gillespie qualified his support for animal performances by excepting dancing bears and all cats, especially the domestic cat because of its mind and temperament: being 'solitary and unsocial in its habits and hunting methods, [it] does not readily determine its actions by outside suggestions or stimulus' (Gillespie 1934, pp. 133 and 134). While the probable mental cruelty of an intimidating environment or close confinement gradually began to be better acknowledged, a difficulty in recognizing mental suffering was that the behaviour of the animal could be misleading. Dogs might appear ingratiating when frightened, just as apes often grin when terrified. Fred Trussell said he had never seen any expression of despair and gloom on the face of a performing animal (*3088*). Charles Rodgers explained that for practical purposes, the NCDL had to make a distinction between physical cruelty and the psychological cruelty of training and performance because the psychological cruelty was not so obvious and so the likelihood of success in a costly prosecution was small (*161–2*). The actor Athole Stewart reported to the select committee that his sheepdog had suffered great mental stress from the intimidating environment of the theatre when under a unique arrangement it had accompanied him on stage over a 10-week engagement. If such stress occurred in a well-treated domestic animal, he believed it must be much worse for others put on the stage, and so he expressed support for legislation (*2284/8*). Trevelyan thought dogs were reluctant to perform in front of strangers and in the noisy surroundings of a music hall, so that coercion through fear was needed (Trevelyan 1936, pp. 86 and 87). Frederick Schwartz, a retired merchant in Darmstadt, had long ago used fear to turn his dog into a 'music critic'. He beat it with a rod to train it to react whenever a vocal or instrumental note was out of tune. The dog became a whipping boy in the role of a critic who 'bestowed no praise, but only blame' by barking and growling. In extreme cases, 'his hair stood on end, his eyes became bloodshot, and music-stands, music-stools, and instruments, were strewed in confusion about the room' (Anon. 1869, pp. 94 and 95).

## References

### *(Articles in periodicals are anonymous or untitled unless otherwise shown)*

Adams SH (1900) The training of lions, tigers, and other great cats. McClure's Magazine 15(5): 387–398

Adams SH (ed) (1904) Notes from a trainer's book. McClure's Magazine 24(2):188–197

Altick RD (1978) The shows of London. Belknap, Cambridge

The Animal World (1921a) Performing animals. January, p 10

The Animal World (1921b) Conjuring tricks with animals. April, pp 42–43

The Animal World (1921c) Performing animals: a conviction. July, pp 73–74

The Animal World (1923) The performing animals bill. May, pp 52–53

Anon. (1869) Haney's art of training animals. A practical guide for amateur or professional trainers. Giving full instructions for breaking, taming and teaching all kinds of animals including an improved method of horse breaking, management of farm animals, training of sporting dogs; serpent charming, care and tuition of talking, singing and performing birds; and detailed instructions for teaching all circus tricks, and many other wonderful feats. Jesse Haney, New York

Barry I (1927) The circus at Olympia. Letter The Spectator 1 January, p 12

Bayne W (1921) Performing animals. Letter The Spectator 27 August, p 268

Baumann C, Stevens LA (1975) Tiger, tiger: my 25 years with the big cats. Playboy Press, Chicago

Beatty C, Anthony E (1933) The big cage. Century, New York

Beatty C, Anthony E (1965) Facing the big cats; my world of lions and tigers. Doubleday, New York

Bell E (1918) The performing animal. Contemp Rev 113:93–100

Bensusan SL (1913) The torture of trained animals. In: Trist S (ed) The under dog. A series of papers by various authors on the wrongs suffered by animals at the hand of man. The Animals' Guardian Office, London, pp 109–130

Blathwayt R (1893a) Wild animals—1. How they are captured, transported, trained, and sold. McClure's Magazine 1:26–33

Blathwayt R (1893b) Wild beasts. How they are transported and trained. McClure's Magazine 1:126–135

Bostock FC, Velvin E (ed) (1903) The training of wild animals. Watchmaker, Seaside, reprinted 2009

Busby R (1976) British music hall. An illustrated who's who from 1850 to the present day. Paul Elek, London

Carlisle DT (1956) And in this ring, ladies and gentlemen...! The New York Times 8 April, p BR3

Cather JL (1931) Wild beasts at Olympia. Letter The Spectator 10 January, p 49

Charles J (1952) Naissance du music-hall. Arthème Fayard, Paris

Chindahl GL (1959) A history of the circus in America. Caxton, Caldwell

Christian Advocate and Journal (1863) Menageries and zoological gardens. The Evening Post 10 December, p 395

The Circus Report (1977) Shocking report. 11 April, p 16

Coleman SH (1931) Training wild animals. Letter The New York Times 24 May, p E2

Cooper CR (1923) Under the big top. Little, Brown, Boston

Cooper CR (1924) Lions 'n' tigers 'n' everything. Little, Brown, Boston

Cooper CR (1925) Circus day. Farrar and Rinehart, New York

Court A (1954) Wild circus animals. Burke, London

Court A (1955) My life with the big cats. Simon and Schuster, New York

Davis CB (2007) Cultural evolution and performance genres: memetics in theatre history and performance studies. Theat J 59:595–614

De Frece M (1934) Recollections of Vesta Tilley. Hutchinson, London

Dembeck H (1965, German edition: 1961) Animals and men, trans. Winston R, Winston C. Nelson, London

Dembeck H (1970, German edition: 1966) Willingly to school. How animals are taught, trans. Johnson C. Harrap, London

Diolé P (1974) The errant ark. Putnam, New York

Dolman F (1899) Four-footed actors. About some well-known animals that appear on the London and provincial stage. The English Illustrated Magazine 21(192):521–526

Donald D (2007) Picturing animals in Britain, 1750–1850. Yale University Press, New Haven and London

Dvinsky E (1956) Durov and his performing animals, trans. Griffith P (1960). Foreign Languages Publishing House, Moscow

EHG (1924) Spoofs and the spoofed. The Illustrated London News 20 September, p 550

The Encore (1921a) 30 May

The Encore (1921b) 9 June, p 3

The Encore (1922) 8 June, p 3

The Era (1921a) 2 March, p 16

The Era (1921b) 9 March, p 15

The Era (1921c) 16 March, p 16

The Era (1921d) 23 March, p 18

The Era (1921e) 4 May, p 12
The Era (1921f) 17 August, p 15
The Era (1922) 22 March
Estcourt-Oswald A (1931) Letter The Spectator 7 February, p 185
Ex-Lion-King (1872) Lions and lion-taming. Every Saturday 17 February, pp 173–176
Farm Animals Welfare Council. http://www.fawc.org.uk/freedoms.htm
The Gentleman's Magazine (1873) Three months with a Lion King. March, pp 254–261
Gillespie TH (1934) Is it cruel? A study of the condition of captive and performing animals. Herbert Jenkins, London
Hagenbeck C (1912) Beasts and men: being Carl Hagenbeck's experiences for half a century among wild animals, trans. Elliot HSR, Thacker AG. Longmans Green, London
Hardy T (1913) Letter The Times 19 December, p 9c
Hediger H (1950) Wild animals in captivity. Butterworth's Scientific Publications, London
Hediger H (1955) Studies of the psychology and behaviour of captive animals in zoos and circuses, trans. Sircom G. Butterworth's Scientific Publications, London
Hippisley Coxe AD (1948) Mr Nicolson and the circus. Letter The Spectator 16 January, p 75
Hippisley Coxe AD (1953) At the circus. The Spectator 2 January, p 11
Hornaday WT (1922) The minds and manners of wild animals. A book of personal observations. Charles Scribner's Sons, New York
Horne RH (1838) The life of Van Amburgh: the brute tamer, with anecdotes of his extraordinary pupils. By Ephraim Watts, a Citizen of New York.... Robert Tyas, London
Jennison G (1937) Animals for show and pleasure in ancient Rome. Manchester University Press, Manchester, Reprinted 2005, University of Pennsylvania Press, PA
Johnson W (1990) The rose-tinted menagerie. Heretic Books, London
Joys JC (1983) The wild animal trainer in America. Pruett, Boulder
Joys JC (2011) The wild things. Dissertation, Bowling Green State University
Kiley-Worthington M (1989) The training of circus animals. In: Animal training. A review and commentary on current practice, proceedings of a symposium organized by the Universities Federation for Animal Welfare, 26th–27th September, pp 65–81
Kiley-Worthington M (1990) Animals in circuses and zoos. Chiron's world? Little Eco-Farms Publishing, Basildon
Kirby MB (1908) The gentle art of training wild beasts. Everybody's Magazine 19(4):435–446
Lewis GW (1955) Elephant tramp, as told to Bryon Fish. Peter Davies, London
The Literary Digest (1920) Cruelty charged in training trick animals for stage and movie. 25 September, pp 102–110
The Literary Digest (1922) The "breaking" of performing animals. 6 May, pp 80–82
Mallabar WL (1902) Musical seals and acrobatic sea-lions. How seals and sea-lions (their first cousins) play musical instruments, cricket, and other games, and amuse the public by their strange stage performances. The Royal Magazine 9:191–194
Mason MH (1913) Letter The Times 20 December
Matthews M (2011) The Hippisley family of Chewton Mendip and Bristol, England. http://www.boddyparts.co.uk/hippisley.htm. Accessed 1 Sep 2014
Mitchell PC (1929) Centenary history of the Zoological Society of London. Zoological Society of London, London
Mitchell PC (1937) My fill of days. Faber and Faber, London
Moffett C (1894) Wild beasts in captivity. How they are watered and fed; the special dangers in handling them. McClure's Magazine 3:71–88
Neirick M (2012) When pigs could fly and bears could dance: a history of the Soviet circus. University of Wisconsin Press, Madison
The New York Mirror (1833) Menagerie. January 26, p 239
The New York Times (1894) Big elephant Tip dead: killed with poison after long hours of suffering... 12 May, pp 1 and 5
The New York Times (1922) 26 March, p 20

The New York Times (1964) Soviet bear never naps—well, once in a while. 25 January, p 8
Nicolson H (1948) Marginal comment. The Spectator 2 January, p 11
Nicolson H (1950) Marginal comment. The Spectator 17 March, p 336
Northern Star (1983) Performing cats part of circus enthusiasm. 16 December, p 26
Ogilby W (1829) The menageries. Quadrupeds, described and drawn from living subjects, vol 1. Charles Knight, London
The Pacific Rural Press (1902) Training wild animals. 63(26), 28 June, p 429
Pope WJM (1950) The melodies linger on. The story of music hall. W H Allen, London
Priddy A (1930) The way of the circus, with man and animal. Platform World, Chicago
Proske R (1956) Lions, tigers and me. Henry Holt, New York
Purdy RL, Millgate M (eds) (1984) The collected letters of Thomas Hardy, vol 4, 1909–1913. Clarendon Press, Oxford
The Recorder (1921) The Magnet 1 October, p 3
Regan T (2004) Empty cages. Facing the challenge of animal rights. Rowman and Littlefield, Lanham
Rolker AW (1905) Wild animal trapping. McClure's Magazine 24:431–442
Rolker AW (1906–1907) The wild animal physician and his patients. McClure's Magazine 28: 367–378
Rowe C (2002) Banning the big top: Greenburgh bans live animal circus acts. The New York Times 16 June, p WE1
Salt HS (1894) Animals' rights considered in relation to social progress. Macmillan, New York
Selborne A (1839) The sporting looker-on and reflector. Alfred Selborne's journal. New Sporting Magazine 16(93):72–4
Smith W (1898) Animal training extraordinary. Captain Woodward's musical pets. Chums, July, pp 734–735
The Spectator (1904) The cash value of animals. 27 February, pp 327–328
Stokes J (2004) Lion griefs. New Theat Q 20(2):138–154
Sutherland H (1902) Training wild animals. Ainslee's Magazine 8(6):518–525
Tait P (2012) Wild and dangerous performances: animals, emotions, circus. Palgrave Macmillan, Basingstoke
Taylor D (1976) Zoo Vet: the world of a wildlife vet. Allen and Unwin, London
Taylor P (1925) Training wild animals for circus and stage not cruel. Billboard 30 June, p 62
The Times (1913a) 16 December
The Times (1913b) Performing animals. The psychology of pain in man and beast. 17 December
The Times (1920) Performing animals. 29 December, p 9
The Times (1921) 27 May
The Times (1927) 29 December, p 6
The Times (1937) 13 February, p 13
Todd SW (1918) A school for sea-lions. St Nicholas February, pp 367–368
Trevelyan H (1936) Laugh, clown, laugh! C W Daniel, London
Tyrwhitt-Drake HG (1936) Beasts and circuses. Arrowsmith, Bristol
Tyrwhitt-Drake HG (1946) The English circus and fairground. Methuen, London
United Kingdom Parliament (various dates) Parliamentary Debates (Commons). HMSO, London
United Kingdom Parliament (1921) Report from the Select Committee on Performing Animals, together with the Proceedings of the Committee and Minutes of Evidence. HMSO, London
United Kingdom Parliament (1922) Report from the Select Committee on Performing Animals, together with the Proceedings of the Committee and Minutes of Evidence. HMSO, London
Valentine T (1936) From jungle to circus—some memories. Privately published
Velvin E (1910–1911) Critical moments with wild animals. McClure's Magazine 36:377–385
Velvin E (1911) Critical moments with wild animals. Second paper. McClure's Magazine 37: 155–163
Velvin E (1914) From jungle to zoo. Stanley Paul, London
Wain L (1913) Letter The Times 22 December

Webster J (1994) Animal welfare: a cool eye towards Eden. Blackwell Scientific, Oxford
Wilmut R (1989) Kindly leave the stage! The story of variety 1919–1960. Methuen, London
Wilson DAH (1999) Encouragements and constraints in the development of experimental animal behaviour studies in Great Britain since the late nineteenth century. Dissertation, University of Leicester
The World's Fair (1911a) 18 March, pp 10–11
The World's Fair (1911b) 22 July, p 12
The World's Fair (1911c) Wild animals break their prison bars. Fierce struggles in menageries are described by Frank Bostock. 30 September, p 6
The World's Fair (1911d) 7 October, p 8
The World's Fair (1915) 6 February
The World's Fair (1921a) 5 March, front page
The World's Fair (1921b) 18 June, p 11
The World's Fair (1922) 23 December, p 19
The World's Fair (1923) 24 March, p 11
Yeats-Brown F (1928) Horrors of the circus. Letter The Spectator 15 December, pp 913–914
Yeats-Brown F (1929) Woes of the caged. Letter The Spectator 21 December, pp 939–940
Yeats-Brown F (1931) Wild beasts at Olympia. The Spectator 3 January, p 13
Zoological Society of London (1928) Hints on the transport of animals. Proc Zool Soc, pp 631–652

# Chapter 5
# Unfinished Business

**Abstract** The Performing Animals (Regulation) Act of 1925 legitimised animal performance in Britain by means of a regulatory system of registration and inspection. It was accepted by trade interests as innocuous, but critics in the Royal Society for the Prevention of Cruelty to Animals, Performing Animals' Defence League and Parliament attempted to amend or replace it. In particular, arrangements for inspection were felt to be inadequate because of the trade's ability to conduct training in secrecy, behind the scenes and with the advantage of advanced warning of inspections: these problems had been identified during the select committee hearings. The limited domestic inspections were regarded as ineffective, resulting in a continuing shortage of prosecutions. In addition, the training abroad of animals then imported for exhibition was beyond the reach of British regulations. In spite of their efforts and a succession of bills, the critics were unsuccessful in strengthening the 1925 legislation. This chapter examines the problems associated with licensing and inspection arrangements and discusses the effect of the law on animal performance in an international context.

## 5.1 Secrecy, Inspection and Licensing Proposals

After Kenworthy's original prohibiting bill of 1921 had been subject to standing committee amendments enabling the licensing of training and exhibition, he had stated: 'Rather than admit a system of licensing and inspection, which we feel would only stereotype the evils we are trying to do away with, we would withdraw the whole Bill'. In this same debate, Kenworthy reported that the Home Office itself was unenthusiastic about licensing and that there would be variations in the attitudes of licensing magistrates, so that some would be approached by applicants wherever they lived. Licensing had also failed in the traffic in worn-out horses and for child acrobats, he said. Butcher recommended that licences should only be awarded at Quarter Sessions and after proper investigation of applicants' suitability.



D.A.H. Wilson, *The Welfare of Performing Animals*, Animal Welfare 15,
DOI 10.1007/978-3-662-45834-1_5

However, Colonel Penry Williams believed that holding a licence would provide some immunity from any accusations of cruelty.[1]

The secrecy that traditionally surrounded much animal training, and the corresponding lack of usable evidence, increasingly became a source of suspicion for critics. The reasons for such secrecy on the part of professionals might be that there was indeed something in the methods used that would justify criticism and should therefore be hidden, that there were 'trade secrets' to keep from competitors or that their work was not seen as the business of the outside world and its 'busybodies'. The select committee heard of many instances of difficulty in getting sound evidence of training methods. Charles William Rodgers of the NCDL was one of many critics who complained of the secrecy of training and the lack of access to training sites. He said that no real training was done in front of strangers, and accordingly he had been unable to review dog training at Price and King, Hackford Road, Brixton (*3, 110ff*). L. O. Liesching, the amateur conjuror, confirmed that performers needed to keep the secrets of their methods (125), and Mrs Albert Bradshaw complained that ferreting out cruelty was 'one of the most difficult matters in the world to do' (*259*). Charles Haverly mentioned the difficulty of getting second witnesses: 'You cannot smuggle an officer behind a stage, you know' (*605*), and the theatre electrician, Albert Hill, reported that no-one above stage manager would know what was going on and that trainers were very wary of strangers, including the police (1107f).

Charles Wilson of the RSPCA commented that the Society had never been able to obtain evidence of cruelty off the stage during training in private and spoke of the impossibility of getting behind the scenes. Although he had almost constantly paid visits of inspection at music halls and training quarters for 20 years, this had been fruitless in stopping any cruelty because his visits were not unannounced. Warnings would be given of the approach of a stranger or inspector. Because of agitation against animal performances in 1914, trainers had made a countermove by 'inviting' inspections by posting a notice at their training establishment; and because of the difficulty of inspection, he did not believe licensing would work: all convictions had been at public performances or because of the condition of the animals, but it was the training quarters that needed inspection and that had proved impossible. He had made a report to his manager at the RSPCA about the futility of inspection after he had worked on nothing else in London for 7 months between 1 December 1913 and 30 June 1914 (*276, 281*). To illustrate such difficulties, Jack Thompson, the composer-musician, recounted his experience of the presence of the RSPCA at a theatre when a chimpanzee was clearly being beaten in a dressing room behind a locked door. The cruelty was heard but could not be seen, and so he could not be called as a witness. On another occasion, the RSPCA was summoned to investigate the mistreatment of Zertho's dog, but the dog was removed before the visit (1478ff, 1515ff).

[1] *Parliamentary Debates* (*Commons*) 1921, vol. 142, columns 1414, 1415, 1421 and 1427.

Claims were made that managements, trainers and exhibitors employed advanced warning systems and lookouts so that the efforts of the RSPCA to inspect conditions had been frustrated. Charles Wilson told the committee of the impossibility of effective inspections at Price and Kings because the staff there were vigilant and immediate warnings of a visit would be given, indicating that there was something to hide in what was going on (*329ff*). Colonel Burn referred to this problem when he presided over a House of Commons evening meeting in support of the bill to prohibit animal performances and when evidence was heard from witnesses about training, including Wilson (*Times* 1921). The committee heard from Jessie Hales that everything was in order by the time the RSPCA arrived for an inspection, including the removal of animals, and that the RSPCA suspected that their arrival was planned for (2036, 2077, 2081ff). Charles Haverly quoted a phrase he heard at the London Hippodrome: 'Go slow tonight; Society's men in the house' (*562*). Margaret Martin said that a fireman had told her that lookouts for the RSPCA were a regular rule at the Hippodrome (proprietors, Moss and Stoll) and that this might have been a management instruction coming perhaps via Frank Parker, the stage manager (550, 568–569). Much of the profession as a whole was sensitive about the prospect of formal inspection by outside agencies. An editorial in *The Era* said the idea of unannounced inspections would never be possible, no matter how many acts were passed, but it did not say why (*Era* 1922). *The Encore* described the satisfactory report of the select committee, saying its members had been impressed by the trainers' wish to eliminate the possibility of cruelty, but it was wary of the proposed committee of control as a possible field for busybodies in its powers of 'prohibiting, restricting, suspending and modifying any performance or the training of animals' where cruelty was suspected (Bailey 1922). Charles Wilson had claimed that some acts could not be done without cruelty in training; therefore, this should stand as evidence when inspections were not possible, and if there were some cruelty in a public performance, there must be more in the training that came before (*343ff*), and Albert Hill had witnessed cruelty in the training rather than the performance, in the morning before the manager arrived (1093). Carl Hertz, the conjuror and member of the executive committee of the VAF, exemplified the performer's wish for privacy when he stated that Haverly could not have been at the side of the stage to see his 'bloody' act because Sir Walter de Frece had agreed in the contract not to allow anyone on stage during tricks (*1547*).

When the rodeo arrived in Britain, it generated a special focus of attention by critics, and the question of inspection was raised once more. In the House of Lords, Lord Banbury called for RSPCA inspection of the proposed music-hall presentation of the rodeo, and the Home Secretary warned the promoters against denying inspectors access behind the scenes, or the public would draw their own conclusions: the Government was watching closely (*World's Fair* 1924b). To reinforce its argument that there was nothing to hide, the profession welcomed the idea of informal and invited 'observation' and often claimed that open access was possible or existed anyway. Henri Rey, owner of the training establishment at Brixton, contradicted Charles Wilson about any difficulty of access to his premises. Wilson had visited in April 1914 with 'Captain' Woodward—Rey had not known he was an

RSPCA man and had carried on with his training as usual. There was open access for the RSPCA, when training would continue; indeed, he insisted on RSPCA inspections of both his premises and travelling arrangements (1691f, 1729). James Sanger enabled staff of *The Era* to witness a whole range of (kind) training and see accommodation at his headquarters at Burstow Farm in Horley, Surrey, where he advocated fresh air for the animals. Animals were like children, he said, and cruelty would be self-defeating as well as morally wrong (*Era* 1921a). He gave supporting letters to the select committee from Henry Gurney, a vice president of the RSPCA and a neighbour, who made regular visits under an open access policy (*793–794*, *850*). 'Captain' Woodward invited and recommended inspection, supervision and cooperation with the RSPCA (but perhaps on a voluntary basis rather than in law), and he encouraged professionals to write to the RSPCA stating their willingness to be inspected at any time and to be advised, to allay any public concern created by extremists (*World's Fair* 1921; *Era* 9 March 1921b).

In an article in *The Encore*, Harry Rochez, who claimed that for 30 years he had trained a variety of animals, all inspected by the RSPCA, reported:

> I am now training a little dog for a lady (which will be produced shortly) in the presence of a few people so that should anyone say that this dog has not been well treated I should call them as witnesses.

Following a critical article about animal performances, he unsuccessfully invited the editor of *The Daily Chronicle* and the RSPCA to witness this training. The RSPCA declined because there was no guarantee that something unacceptable might not happen on an unobserved day. Rochez thought there was general ignorance about the training of performing animals and that a range of theatre employees would act against any cruelty during rehearsals. He added:

> Most of the performing animals for many years past have been trained at Messrs Price & King, Veterinary Surgeons, of Brixton, and I do not know of any trainer who would object to anyone interested in the matter being present at their training.

There was no wish for training secrecy (1365ff). Meanwhile, horse racing was less spontaneous and more arduous and dangerous than stage tricks—were the supporters of the bill also against the Grand National (Rochez 1921)? Among others who attempted to reassure the select committee were George Henry Harrop, who said there was free access to anyone at any time at the Blackpool Tower Company (*1263*); Richard Henry Gillespie, who confirmed open access to the RSPCA during rehearsals and the adequacy of inspection (*1948ff*); Arthur Aitchley, who would welcome observation of his training by anyone (*2202*); and Garrard Tyrwhitt-Drake, who offered the RSPCA free access without 'preparation' of conditions and open access to his circus or menagerie for inspections (377ff, 418). Soon after, Lord Lonsdale, president of the circus at Olympia, stated to the press that he was instrumental in inducing the promoters both to cease the public displays of steer roping at the rodeo and to allow the RSPCA full access (*World's Fair* 1924a). Bertram Mills, referring to the International Circus at Olympia, had told the committee that the RSPCA had passes for all hours and that training there was open to the audience (2150, 2160). A decade later, on the occasion of a public meeting of the PADL at Kensington Town Hall against wild animal performances at circuses, Bertram Mills wrote saying that season tickets were available for those genuinely interested in seeing where the animals lived and were trained; he also enclosed a shilling subscription to the League, the sum which members of the audience were invited to pay to help with its funds (*Times* 1931).[2]

Once the objective of prohibition had become unlikely in the short term, some critics then supported proposals for mandatory inspections under the law and without notice, these to be carried out by the full range of responsible agencies, but in the end, powers of inspection were less stringent. Many believed in any case that inspection was a very poor substitute for prohibition. Charles Wilson was pessimistic. He explained that the RSPCA's visiting, *incognito*, of animal performances in the UK was currently carried out regularly and automatically, in each district where events were advertised. The inspector would see the performance and, immediately after it, go to the stage door and ask the exhibitor permission to see his animals before making a report. Having undertaken such duties himself in 1906 and 1914 at a time when a total ban had already been called for, Wilson kept records and compiled a report that showed the inadequacy of such approaches to exhibitors. Because of insurmountable inspection problems, he could not see how new legislation could both punish the cruel and leave alone the 'kind' trainers. At least the impossibility of effective inspection demonstrated that the 1911 Act was insufficient and there was a need for some new form of legislation (*284ff*, *303*, *348*). Charles Haverly spoke of the impracticality of an inspection-based law because training was carried out abroad or in private accommodation. It was therefore not possible to trace it, and in any case, effective inspection in Britain would need an inspector at each theatre, on stage and in dressing rooms (*628*).

Apart from the problems with arrangements for inspection, those of a general licensing system were also discussed in the select committee hearings. Some on both sides opposed the licensing idea, thinking it either an unnecessary intrusion or ineffectual in comparison with prohibition. Henry Gurney had suggested in a letter to James Sanger in 1913 the licensing of training establishments. Sanger said he had no fixed opinions about the idea, but it got no further because the propaganda against animal performances collapsed (*1079ff*). R. H. Gillespie pointed out that licensing would be difficult to organise for non-English acts (*1811*). Furthermore, if licensing were adopted, different magistrates' benches might express different views of applications, and whereas a liquor licence would last 12 months, one for performing animals might have to be applied for on a weekly basis (*2020f*). And Everard Calthrop, speaking of the idea of a process of censorship of animal performances, as in plays, wondered whether a strong presence of the RSPCA on a committee of censorship might lead to prejudice (1991). The professional journals held differing views over the licensing idea. *The Encore* reported that 'Captain' Woodward had asked trainers for their views on the licensing proposals of the bill

---

[2] Toole Stott heard that a member of Mills's staff had found his way on to the League's committee and was able to warn of demonstrations at Olympia (Toole Stott 1971, p. 62).

and wondered whether it could stigmatise them. *The Encore* accepted the licensing proposal of the new bill because it represented a defeat of those who had been pressing for prohibition and because it would require accuracy in opponents' charges of cruelty in terms of time, place and so on (*Encore* 1921). *The Performer* welcomed the registration and stiffer punishment agreed in the new proposals (*Performer* 1923). But *The Era* published a letter from Frank Stephenson advising a campaign against the prospect of restrictive licensing (*Era* 1921b).

Others saw advantages in a licensing system. In 1913 *The Times* had reported that the RSPCA had met with the profession and agreed on a form of RSPCA licence as a guarantee against cruelty. Foreign trainers would need a foreign animal welfare society licence. The idea was that logs would be kept with details and evidence of training methods, animals, equipment, inspections and engagements, and inspections should be allowed at any time and without warning (*Times* 1913). This early agreement appears to have been no more than an informal arrangement, and the 'licence', if used, had no practical status or effect. Many professionals soon after favoured formal licensing, if only as a fallback from the spectre of prohibition. George Lockhart supported the licensing of training establishments to prevent cruelty and the idea that if lost because of cruelty, a licence should be lost forever (*1334*, *1471*). Fred Trussell and E. H. Bostock both supported the licensing idea (*3137*, *3163*, *3189*). Everard Calthrop suggested that an animal's behaviour after a performance was the evidence of whether a trainer deserved a licence, and such assessments should also apply to foreign acts. There should be few imported wild animal acts, only after an examination and licence had confirmed that no coercion had been used in the training, but for this purpose it would be difficult, though not impossible, to find unbiased experts (1897, 1857, 1899). Finally, the bellicose Albert Voyce agreed with registration as 'the best way out of it' and handed the select committee a written suggestion about registration or licensing proposals (2494f).

*The Animal World* quoted an article in the legal periodical, *The Justice of the Peace*, on the bill that passed its second reading on 23 March 1923. The article complained that it was proposed everyone had the right to be registered unless prohibited:

> We should have thought that a public application to a committee of the local authority, to be supported by evidence of the fitness of the person applying, and of the suitability of his premises and methods, with power for persons or societies, desiring to oppose the application, to attend and be heard, would have been more satisfactory ... It is one thing to make regulations to ensure that no cruelty is used, and quite another thing to enforce such regulations in respect of the necessarily private training of the animals in question. (*Animal World* 1923b)

Following the passing of the new legislation in 1925, Trevelyan Thomson asked the Home Secretary why he had persuaded Exeter Local Authority to reverse the appointment they had made under the Act of an RSPCA inspector, who was also a special constable, as their inspecting officer. The reply was that the appointment as a special constable had been made specifically to enable inspection under the Act, and this was not a purpose of special constable appointments nor a procedure intended by the Act.[3] The Act has remained a disappointment to the present day to those who had hoped for effective inspection as an alternative to total prohibition. Such disappointment was the basis of Helen Trevelyan's critical publication. She objected to the Act's restriction of visits by authorised people to training centres at 'reasonable times' and the lack of provision for going on stage or behind the scenes during a performance, making it seem that there must be something suspicious to conceal. For her, the Act allowed cruelty to remain hidden: it required proof of cruelty but prevented the means of getting that proof because of loopholes affecting access for proper inspection. She illustrated the problem by referring to the late Captain H. E. Hobbs, who pointed out that dogs showed affection to the man on stage, as opposed to their trainer behind the scenes (Trevelyan 1936, p. 103ff). But by 1930, Lord Jessel was offering reassurance that according to the Home Office, London County Council inspectors were required to visit every place of entertainment at which performing animals were exhibited: 'It is their practice to secure a seat from which the best view of the performance can be obtained, and animals are inspected in their housing and training quarters. On an average two inspections are made weekly by each officer'. Earl Russell reminded him that it was behind the scenes at the moment when animals were about to do tricks that cruelty could occur, as a consequence of the 'training to timetable' requirement, but that inspection at this time was not permitted by existing legislation.[4]

Wilmut describes an unexpected weakness in the system of RSPCA inspection, involving the trainer of Duncan's Royal Collies:

> Duncan always took good care of his dogs, but of course the RSPCA kept a tight watch on all animal acts. Duncan had a friend who worked for the RSPCA, and whenever his friend wanted an afternoon off, Duncan would phone the RSPCA with a complaint against himself. The friend would be sent off to investigate, would go nowhere near Duncan, and would return to report that there was no cause for complaint. (Wilmut 1989, p. 115)

## 5.2 The Effect of the Law

The historic lack of successful prosecutions, in spite of a widespread suspicion among the interested public that cruelty existed in various stages of the performing animal process, led to calls for better means to identify and punish cruelty. When cruelty was discovered and led to a conviction, the rarity of such an event was used by some to suggest that the problem was the tip of an iceberg and by others connected with the profession to confirm that the cause for concern was minimal and certainly no greater than in other forms of animal cruelty. Both sides agreed that punishment by courts should be made more severe, but the profession was reluctant to accept that new systems of inspection or regulation should be introduced in order to provide an easier mechanism for prosecution. The RSPCA managed to secure a

[3] *Parliamentary Debates* (*Commons*) 1926, vol. 199, columns 1941 and 1942.

[4] *Parliamentary Debates* (*Lords*) 1930, vol. 76, columns 910 and 913.

modest succession of minor convictions, accompanied by equally modest fines, and there was an impression that its interest was often ignored or resented. For example, in mid-1919 a conviction for exhibiting a lame performing donkey resulted in a 20-shilling (£1) fine. The following year saw the conviction of a circus proprietor for cruelty to a bear performing with a hyena which attacked it. The RSPCA inspector had advised separation and the removal of the injured bear from performance but had been ignored. A fine of £10 and £1.2s (£1.10) costs ensued, then a failed appeal together with more costs (*Animal World* 1919, 1920).

At the select committee, George Lockhart said that any 'cowed' aspect in a performing animal was sufficient to reveal cruelty, and the existing law only needed strengthening. He thought that if cruelty existed, it warranted a greater penalty of the order of 6 months' imprisonment and the confiscation of animals (*1321–1322*, *1365*, *1368*, *1472*, *1492*). Carl Hertz considered the current law sufficient (*1582*), but R. H. Gillespie declared that he would agree with greater penalties and a tightening of the law (*1939*). Everard Calthrop favoured the existing legislation and was noncommittal about adding a licence requirement, but he thought that penalties should be substantially raised because they had not been a deterrent (1987). Fred Trussell also proposed stiffer penalties as a real deterrent (*3113*), and Arthur Aitchley was in favour of harsher sentences of 6 months' or even 10 years' imprisonment, as well as exclusion from the profession (*2190f*). Sidney Davis on the other hand hoped for the elimination of animal performances: this was the only remedy, because the professionals always managed to pay the fines (1342f).

Charles Haverly stated that any Act that introduced licensing could only be effective if the manager concerned, such as Stoll or Gillespie, were included as an accessory under a conviction: indeed, that would stop animal performances at once (*629*). It was in the meantime noted that in one or two countries wild animal performances were prohibited (*1111*), and Arthur Valentine, the amateur conjuror, offered the select committee a list of certain tricks germane to his specialism that ought to be eliminated (1671). Some years later, T. H. Gillespie returned to the matter of legislative control of trainers and owners when he advised that it would be appropriate in the case of chimpanzees, whose behaviour on reaching sexual maturity could be objectionable (Gillespie 1934, p. 143). Some witnesses in sympathy with the profession advocated a policy of 'no second chance' for those convicted. It had become apparent that a performer might continue with the same act even after a conviction for it (*23*). Everard Calthrop hoped that the law would prevent the possibility of the cruelty that had resulted in over 100 convictions in the previous 30 years (1844): he believed there should be no second chance of exhibiting animals for those convicted (1860). But Bertram Mills was satisfied with the current law and believed that unless there was set up a central, impartial board of scrutiny, a conviction of a trainer or performer should not mean a life ban, because there were differing regional prejudices affecting the profession (2187–2188).

The select committee learned of the difficulty in bringing successful prosecutions. Charles Rodgers of the NCDL recounted the problem of making a charge of cruelty stick. Complaints were investigated, but for every single prosecution, there were hundreds of cases that were thought to be cruel but would not stand much chance of a successful issue (*33*). In any case, his was not a prosecuting league, and only extreme cases were taken up (*63f*). Charles Wilson reminded the committee that actual convictions resulted from what was before the audience or in the public sphere and could not apply to training, which was unobservable: it was not possible to get behind the scenes unannounced. That was why there had been only 100 prosecutions in 29 years, and many of these did not even involve the stage (*2843, 2875, 2889*). Robinson (1913), as an inspector with the RSPCA, had received reports of cruelty to elephants by a foreign woman trainer and managed to infiltrate the rehearsal at the music hall with the help of a veterinary surgeon disguised as one of the artistes. With the further help of other real artistes who were afraid of dismissal, he secured an arrest and conviction. Lily Grahame did not continue reporting her experiences to the RSPCA because she regarded it as impotent (*2675*), and Sydney Arthur confirmed that because critical artistes and stage hands would not usually have other corroborating witnesses to incidences of cruel training, only to performances, they were reluctant to complain (*780*). Furthermore, some complaints were exaggerated. Sir Algernon West, chairman of the Theatres Committee of London County Council, withdrew and apologised for a critical statement he had made about the alleged cruelty of the use of a chute in the 'plunging elephants' act, after an enquiry showed that he had been misled (*607ff*).

In place of prosecution, the RSPCA frequently settled for warnings. In this way, Charles Wilson managed to persuade Madame Caselli to withdraw a trapeze act with dogs (*2886*). He said that the RSPCA gave about 20 warnings to every one conviction, and they were usually effective (*2894*). Non-specific warnings were also given by the VAF, perhaps mainly to reassure the public. In 1914 a bill was posted at Olympia, endorsed by the RSPCA, announcing the VAF's policy to detect and expel any member found guilty of cruelty (*2881, 2891*). In an interchange between James Seddon and Charles Haverly, it was mentioned that during the 30 years before the select committee was formed, there had been no convictions for cruelty in training and few for mistreatment in accommodation, transport or performance—only about three per annum, in spite of a number of societies and individuals actively looking for evidence of cruelty, and only the more extreme cases were able to be identified at the performance stage (*631ff*). At another time, O'Grady suggested that it was itinerant showmen at fairs who received the bulk of convictions (1968). In an apparent attempt to justify the legislative status quo, Walter de Frece pointed out to Fred Trussell that only 29 out of 100 convictions had been for music-hall cruelty in the course of 29 years (*3064*). But Everard Calthrop complained: 'I cannot understand how the RSPCA occasionally fade out of the vista of a case altogether. I cannot understand why they do not act strongly when they have a good case before them' (1993). The explanation was, of course, that restricted access meant the RSPCA was usually unable to acquire conclusive proof through effective inspection and witnessing. The profession, on the other hand, would say that a lack of convictions was proof that there was no cause for concern.

Kenworthy referred James Sanger to a list of 102 convictions since 1892 related to performing animals, including cruelty while travelling, as in the case of Sylvester of Fossett's Circus in June 1917, regarding an unfit camel (*1046ff*). In 1900 there had been a conviction by Thames magistrates for the wounding of an elephant with a lance at the London Hippodrome, resulting in imprisonment with hard labour (*2938*). In 1902 Mlle Valsois was convicted for prodding and beating an elephant (*2943*), and in 1904, Buffalo Bill was prosecuted for using nails under the saddles of buck-jumping horses (598). In 1907, Branie and Hassan were convicted of horse cruelty at the London Hippodrome and each fined 40 shillings (£2) plus costs (*2950ff*). The following year at West London Police Court, one of the Hagenbeck family was fined five guineas (£5.25) plus £30 costs for permitting his trainer, Schrieda, to terrify elephants that had been made to descend a chute into water, and Schrieda himself was fined £5 plus five guineas (£5.25) costs (*275*, *2947*). Charles William Rodgers had seen an act by Les Marcellis several times up to 1914. This involved cruelly terrifying a dog by making it climb a moving ladder and sit on a man's shoulders. He managed to get the exhibitors convicted at Tower Bridge Police Court in July 1914, whereupon the dog was given to the NCDL. Rodgers described the difficulty of realising a prosecution:

> I may add, to show the difficulty one has to contend with in these matters, that I applied to one Court for a summons on those facts, and it was refused, the magistrate observing: 'Why should not a dog crawl up a ladder?' (*4*)

In the case of a Greek performer, Leonidas Arniotis, in 1914, 'The League employed two experienced detectives for over 6 weeks to try to detect this man in training his dogs, but they failed to catch him training'. Subsequently, a League veterinary surgeon saw the act with dogs that were unfit. Arniotis later attempted to produce in defence at the court hearing only his fitter dogs. He was fined £10 plus costs and had two previous convictions, the first case having been in 1905, for cruelty to a cat at the Lyceum Theatre, when he was fined £2 plus £2.4s (£2.20) costs (*5*, *275*). In 1919 Rodgers also prosecuted 'Dandy George' for abandoning a performing dog in a locked box at a railway station for over 24 h, for which he was fined £10 plus costs (*6f*). Charles Wilson gave the select committee an account of further recent convictions. In 1916, James Westwray, a travelling showman, was fined £2 for jabbing a lioness in the face to make it ferocious. Alfreda Rossi was fined £5 with five guineas (£5.25) costs in 1917 for forcing an elephant, which later died, to perform with pneumonia. The illusionist, Thomas Moore, known as The Great Como, was fined £5 with three shillings (£0.15) costs in 1919 for throwing a pigeon against a wall in his act and dazing it. Following a previous warning, John Bostock was fined £10 with one guinea (£1.05) costs in 1920 for allowing a hyena to attack a bear during a performance. In 1921, Frank Felix was fined £1 with five shillings costs for cruelly terrifying an acrobatic monkey. And in June 1921 John C. Essex was fined £5 with four guineas (£4.20) costs for the suffering caused to the 11 cockerels that appeared in tightly fitting uniforms in his 'Great Sousa Rooster Orchestra' (*272ff*).

R. H. Gillespie said that as a manager and an inveterate reader of newspapers, he saw some convictions reported in *The Stage*, *The Era*, *The Performer* and *The Encore* (*1917*). *The World's Fair* was perhaps the most thorough in reporting all such cases, as in the issue of 29 July 1911, referring to a show rider who was reported as fined for twisting and biting the ear of a bucking horse, described by its owners as wild, dangerous and untamed, when saddling it as part of the 'Wild Australia' act. This treatment, which drew blood, was excused as 'common practice in Australia'. The use of a flank strap was alleged, too, to make the horse buck. The RSPCA brought this action at Penge, and each defendant was fined 40 shillings (£2) (*World's Fair* 1911). In the mid-1920s *The World's Fair* reported a number of cases brought against American operators of the touring rodeo. A performer was heavily fined £20 plus five guineas (£5.25) costs as a result of an act at the Poplar Hippodrome. The RSPCA brought this action against Tex Millar for ill treatment of a steer that was unfit and had slipped on matting, after which it was allegedly kicked and its tail pulled when it took time to stand up (*World's Fair* 1924c). Soon afterwards, *The World's Fair* recounted the RSPCA's successful prosecution, resulting in a fine of £2 plus costs, of Fred Ginnett's daughter, Poppy, for performing a horse in an unfit state after an inspection and an earlier warning (*World's Fair* 1925). Sir Edward Marshall Hall, Recorder of Guildford, exclaimed: 'Sometimes I wish these ill-treated animals could turn into lions and tigers and take care of themselves'. But *The Animal World* quoted *The South Wales Echo*: 'The only lions and tigers who can take care of the persecuted animals are the Magistrates and Recorders of Quarter Sessions, and frequently they pay too great heed to the whining excuses of the brutes who have shown no mercy to the animals, but who expect it for themselves' (*Animal World* 1923a).

The 1925 Act itself did not later result in an increase in convictions, and the same anxieties were expressed when a conviction was well publicised. For example, Peter Freeman asked the Home Secretary to consider legislation prohibiting animal performances in the light of a recent conviction of an exhibitor for cruelty, the reply given being that the conviction showed that existing legislation worked in punishing such cruelty.[5] In the session of 1947–1948, Barnett Janner asked the Secretary of State for the Home Department for the number of police prosecutions to date under the 1925 Act. He was told that the figures were not available, but it had been found that there had been none, at least for 1941, 1943 and 1945. Janner referred to public anxiety about performing animals and asked Chuter Ede to keep a close eye on any offences committed.[6] Since 1925, there had been frequent Government assurances to parliamentary questioners that the Act was sufficient, but these were often given alongside answers to accompanying questions that the import of performing animals was not separately controlled.[7] For many critics, the

[5] *Parliamentary Debates* (*Commons*) 1929–1930, vol. 236, column 453.

[6] *Parliamentary Debates* (*Commons*) 1947–1948, vol. 452, columns 1540 and 1541.

[7] For example, *Parliamentary Debates* (*Commons*) 1953–1954, vol. 523, column 143; 1955–1956, vol. 548, column 283.

foreign dimension of the problem remained a reason to continue to call for prohibition.

Malcolm Clay, secretary of the ACP, has commented that the practical difficulties of the 1925 Act were that it licensed the trainer and not the animal, did not lay down any standards of animal welfare and did not impose any qualification on the applicants for licences.[8] Mike Radford considered the Act inadequate and the power of inspection in urgent need of reform. While the Protection of Animals Acts could enable prosecution and removal of animals for cruelty, poor condition of premises or animals alone is not enough for a local authority to act, and the power of entry only applies when animals are present. The place of entertainment itself has not required licensing for animal performance. Indeed, the 1925 Act might be regarded as enabling legislation: when a local authority refused a circus an entertainment licence because it considered the concept of performing animals to be morally unacceptable, the court held that, in the absence of an express statutory power, the authority was not empowered to prohibit acts of which it disapproved, being the very types of performance which Parliament required to be regulated by it under the Act (Gerry Cottle's Circus Ltd v City of Edinburgh District Council 1990). The Act is 'rudimentary in the extreme' concerning registration. A local authority cannot refuse or revoke or make conditions or set standards—that would require a court order on evidence of cruelty. Registration is guaranteed on payment. There is a need only to give particulars of animals involved and the general nature of the performances for which they are to be trained, and trainers in their applications for licences do not have to describe any apparatus (Mitchell 1975). 'The degree of regulation under the Act is so inadequate that its contribution to animal protection is negligible' (Radford 2001, pp. 292, 302–303, 372–373).

In the United States, circus animals were only given federal protection by the Animal Welfare Act of 1970, which was then supplemented by state laws (Joys 1983, p. 267). It was to include amendments after 1970, expanding standards of care to animals in circuses, carnivals and exhibitions and broadening the transportation regulations for nonfood animals (Beers 2006, p. 178). An announcement of a joint resolution for the 1990s was made in the *New York Times* by 104 American animal protection organisations. They resolved to secure greater enforcement of laws and regulations which provide protection to animals used for exhibition or work purposes. They also resolved to work together to secure the enactment of laws to prohibit, inter alia, abusive training practices, practices that are dangerous to the animals and the capture of animals in the wild to be used for exhibition or work purposes; to limit the breeding of captive, wild animals and to prohibit their cruel disposition; and to require that the behavioural and physical needs of exhibition and work animals be considered (*New York Times* 1991). In 1999:

> PETA dug up dirt on Ringling Bros and was a formidable adversary. It had circulated reports by the U.S. Department of Agriculture, for example, describing horrible conditions at the circus's Center for Elephant Conservation in Polk City, Fla. On Feb. 9, 1999, USDA

[8] Personal communication, 24 March 2005.

> inspectors found two tightly chained baby elephants with lesions and scars on their legs, evidently caused by constant friction with their restraints. The manner in which they were chained limited the offspring to 'some side to side swaying,' according to the USDA report. But that wasn't all. Asked about it, the elephant handlers told the inspectors that baby elephants were 'routinely' chained to forcibly separate them 'from their mothers,' which they called 'an industry standard'. The handlers tried to block the inspectors from taking pictures and angrily asked them why they were making such 'a big deal' about it. (Stein 2001)

However, when the Clyde Beatty–Cole Bros Circus was also under scrutiny in 1999, Patricia Derby, director of the Performing Animal Welfare Society, said that USDA (United States Department of Agriculture) had investigated it previously for cruelty but was 'absolutely gutless when it comes to enforcing the Animal Welfare Act' (Miller 1999). Before 1970, cases had to be brought locally as a result of actions by such as the Humane Society of the United States. In this way, for example, Clyde Beatty was arrested at Pittsburgh following a charge of cruelty against his animals with the whip and pistol (he initially denied the charge but subsequently pleaded guilty and was fined) (*New York Times* 1936). Regan points out that subsequent inspections under the Animal Welfare Act have been inadequate because of bureaucratic delays and inadequate training (Regan 2004, p. 132). A former USDA inspector has also indicated that inspectors were encouraged not to file violations (Rowe 2002).

## 5.3 Attempts at New Legislation[9]

In the 1930s, during the aftermath of the controversy and of the resulting legislation in Britain, the level of demand for animal acts was discussed in the House of Lords.[10] Lord Jessel reported that although Bertram Mills had dropped animal performances in 1930, public demand induced him to reintroduce them the following year.[11] On the other hand, Lord Chesham referred to a Home Office report stating that in 1929 there were 206 British trainers, but by 1933 the figure for trainers in England was 113, and Lord Danesfort claimed that the number of trainers of scheduled animals had gone down considerably, reflecting a decline in demand.[12] The 1925 Act has not been superseded, but unsuccessful bills of amendment were soon forthcoming after its passage. The first was the Performing Animals (Regulation) Amendment Bill of 1930. It was presented in the House of Lords by Danesfort and then brought to the House of Commons, and proposed no exhibition or training for performance of any ape, big cat, or hyena, with exemptions applying

[9] Some passages in the early part of this section are taken from Wilson (2008).

[10] Reports in the press continued to indicate that audiences were losing favour with animal performances (e.g. *Animal World* 1926).

[11] *Parliamentary Debates (Lords)* 1933, vol. 87, column 730.

[12] *Parliamentary Debates (Lords)* 1933, vol. 87, column 737.

to the Zoological Society of London and any other organisation which had as its principal object the exhibition of animals for educational or scientific purposes. During the debate on the second reading, Earl Russell referred disapprovingly to a circular from the SGGB describing the bill as a gross breach of faith. He supported the bill and advised that if necessary another select committee should be appointed.[13]

On 24 April 1930, Joseph Woodward wrote to Lord Auckland about the progress of Danesfort's bill, as 'a voluntary worker on behalf of many colleagues' opposing it (MacMichael Papers). He referred to Auckland's interview in *The Evening News* of 9 April when he had said that trainers for circuses and music halls could not give the necessary time or patience or bear the expenses if training only by kindness. Colleagues had asked Woodward to request Auckland to qualify this observation, because young semi-domesticated, menagerie-bred lions and so on were used: they were trained while they grew up, and so there was plenty of time. Chimpanzees, which the bill also sought to prohibit, were the easiest to train because of their fondness for mimicry: Woodward invited Auckland to the Holborn Empire Theatre to see their performance and quarters for himself. This letter was forwarded to Edmund MacMichael, director of PADL, who wrote himself to *The World's Fair* on 20 May 1930 about comments made by Woodward in the previous week's issue that he (MacMichael) was like a solicitor working as directed for a salary. He said that if he were not an employee, he would still denounce the cruelty. As to Woodward's further comments, because he and others had been denied unannounced access behind the scenes and because training mostly took place abroad, he could not be convinced there was no cruelty. The PADL's leaflet 'Doing men out of jobs' dealt with the matter of compensation resulting from the prohibition he continued to campaign for. He added: 'I have never made a personal enemy of any trainer or proprietor' because 'in my opinion the community which allows cruelty is responsible for it' and should end it and give compensation. He would help to secure this. *The World's Fair* replied to MacMichael saying his letter was too long and forwarded it to Woodward. MacMichael then suggested it should be printed in instalments and said the paper also had a responsibility in the whole matter and should itself advocate compensation. MacMichael's further letters to E. G. Fairholme, Chief Secretary of the RSPCA, from 27 October 1930 refer to a joint committee of the RSPCA and PADL that decided in favour of the inclusion of bears in the schedule of the bill, urging continued cooperation. Bears were the only carnivore often to appear at music halls. The PADL had 'put out of action' in 1929 Jim the Bear and Keutsky's Bears, but in August a bear had killed its trainer at Hastings: joint action was advised concerning this bear, now in Bertram Mills Circus (MacMichael Papers).

A similar Lords bill, now including bears, and devised largely by the RSPCA of which Danesfort was vice-chairman (as well as being a leading figure in the NCDL), was presented by him in 1933, to prevent cruelty to the specified animals

[13] *Parliamentary Debates (Lords)* 1930, vol. 76, column 915.

(usually trained abroad) and danger from them to their trainers and exhibitors. It was later defeated by only one vote. He cited an extract from *The Times* of 3 April 1933 by its Berlin correspondent, noting that the authorities there (perhaps, in retrospect, for more than one reason) had banned bear performances led by gypsies or other itinerants, because of the plight of the animals. Danesfort hoped the bill would strengthen the 1925 Act, which had been emasculated through compromise, and he called in support the original evidence given by Mitchell to the select committee and to recent cases of human injury. Danesfort referred to a mistaken statement from the Home Office that there had been no cases made out for cruelty to performing animals since the Act of 1925, when there had been recent convictions (not for training but for exhibition and management) for the death of a foreigner's lion as a result of a 'wall of death' display and for a case of close confinement of lions. But there was a lack of convictions under the Act because training was done abroad or behind the scenes, out of the reach of the RSPCA inspectors, who had no legal power of inspection. Furthermore, Danesfort did not think the police and local authorities adequately exercised their powers to carry out those inspections permitted when a performance was not taking place. Meanwhile, there had been a slight increase in the training and exhibition of lions, tigers and bears since 1930, and there had been many recent cases of fatal injuries sustained by trainers who were usually foreign. On behalf of the Home Office, Lord Lucan replied that since the 1925 Act no further evidence of cruelty had been brought to the notice of the Department, including the two lion cases, perhaps explained by a meeting of chief constables in 1928 and a related systematic inspection of exhibitions in the boroughs when no concerns arose. Police inspection methods and results were checked in 1929.[14]

During this second reading of the 1933 bill, Lord Carnock attempted to downplay arguments about unnatural behaviour by claiming that he had seen chimpanzees enjoying a whisky and soda.[15] In the debate on the second reading, reference was made to Lord Lonsdale's belief in Bertram Mills's humane standards, and Lord Carnock complained that people spoke from their hearts rather than from their heads.[16] Lord Mount Temple charged the PADL, just after its Kensington meeting 2 years previously, with having attempted to blackmail Sir Walter de Frece and other opponents of the earlier bill by threatening posters and leaflets critical of them.[17] In support of the bill, Lord Tenterden said that it was entirely against the nature of animals like bears, tigers and lions to perform tricks, and the idea that they could be trained by kindness as opposed to starvation and torture defied belief: the

---

[14] *Parliamentary Debates (Lords)* 1932–1933, vol. 87, columns 723f and 741f.

[15] *Parliamentary Debates (Lords)* 1932–1933, vol. 87, column 740f.

[16] He regretted too that it had been made illegal to have a real dog in a Punch and Judy show: 'One of the traditional amusements of this country has been ruined and has had to give place to cinemas from America'.

[17] *Parliamentary Debates (Lords)* 1933, vol. 87, columns 733 and 734.

itinerant circus was a barbaric relic and trainers deserved it if they were mauled.[18] The Earl of Kinnoull quoted Lord Auckland from 1930:

> Trainers cannot afford to keep them the necessary length of time to train them by kindness. It generally takes anything from one to two years to train a lion by kindness, and the trainer has neither the patience to do this nor the money to support the animal during that time.[19]

Kinnoull then went on to point out that chimpanzees at their tea party at London Zoo did not have to perform to a timetable.[20] On the other hand, it was clearly necessary for performing animals in the circus and music hall to be trained so that they would perform on cue. This challenging need for reliability differentiated their position from that of the animal casually performing irregular tricks in a domestic setting.

It was still held by representatives of the profession that cruelty was inefficient and unproductive and directly antagonistic to obtaining good results and that it would be difficult, if not impossible, to train animals to perform by anything savouring of cruelty. It was necessary to win confidence and affection through patience, the whole essence of animal training being inducement by reward, leading to habit formation and a possible love of showing off and applause. Punishment could only teach an animal not to do something (Tyrwhitt-Drake 1936, p. 135; 1946, pp. 130 and 138; Gillespie 1934, pp. 108, 111 and 125). In spite of such reassurances, protesters continued unsuccessfully to challenge the profession to arrange controlled demonstrations of training by kindness,[21] just as trainers had earlier made the equivalent unanswered challenge. Meanwhile Lord Jessel felt that the 1933 bill should not be passed without the investigation of another select committee.[22] Toole Stott claimed that Bertram Mills then helped to defeat the bill by asking him to hire a Fleet Street photographer to accompany him to London Zoo in the guise of Australian tourists, so that photographs could be taken of animals performing tricks: these were then distributed to Members (Toole Stott 1971, p. 63).

The possibility of a return visit of the rodeo arose in 1934, when the National Sporting Club and Tex Austin planned to ask permission for 120 cowboys to visit Britain. Lord O'Hagan opposed the idea because of the experience of 1924, when in steer roping, steer-wrestling and 'bulldogging', animals' necks, legs and horns were broken—'one steer went bellowing across the arena with one horn hanging down and flopping about'. In steer-riding, he said, the steers were supposed to be wild animals but were not, and a cinch was used to infuriate them with pain, to make

[18] *Parliamentary Debates* (*Lords*) 1933, vol. 87, columns 739 and 740.

[19] *Parliamentary Debates* (*Lords*) 1933, vol. 87, column 734. At a public meeting of the PADL at Kensington Town Hall, to protest at wild animal performances at circuses, a letter had been read from Lord Auckland, who had owned and trained lions in the United States. He stated that training by kindness took 1–3 years but that most trainers needed quicker results for revenue, and therefore cruelty was used (*Times* 1931).

[20] *Parliamentary Debates* (*Lords*) 1933, vol. 87, column 736.

[21] For example, the challenge of Naomi Jacobs referred to in Trevelyan 1936, p. 106.

[22] *Parliamentary Debates* (*Lords*) 1933, vol. 87, column 719ff.

them buck. Some court cases had resulted, and the public's indignation at the Wembley exhibition resulted in the omission of steer roping. O'Hagan cited the views of an American who regarded rodeos as 'a purely artificial sport devised to satisfy jaded appetites for spectacle' and 'not typical of any authentic frontier life'. He commented also on the effects on the minds of children: there would be 'a deleterious effect ... on the young people of to-day'.[23] Two days after O'Hagan made these comments, the second reading of the Protection of Animals Bill was debated in the House of Commons. It had been introduced to prevent the lassoing, wrestling and riding of beasts exhibited as unmanageable at a rodeo, which was beyond the reach of the Performing Animals (Regulation) Act of 1925. Sir Robert Gower said that the whole press had been unanimous in condemning incidents of 1924 and that even *The Sporting Life* described them as 'a direct challenge to the British idea of sport and of fair play for the dumb animals and the under dog .... It is not sport'. Thomas Groves was concerned that the clauses of the bill should not prevent the exhibition of boxing kangaroos or the riding of lions in the circus. Alfred Denville claimed that the entertainment industry regarded rodeos as degrading (no doubt also as a source of foreign competition). William O'Donovan believed that 'one of the most compelling reasons for passing such a Bill into law is the need of maintaining the eminent and supreme dignity of man ... we cannot tolerate cruelty for private gain'.[24]

Meanwhile, the campaigning had not diminished, prompting Gillespie to complain: 'I hate the uninformed fanaticism which attacks [performing animal acts] with violence and intemperance proportionate to its lack of knowledge' (Gillespie 1934, p. 146).[25] Helen Trevelyan, the actress and novelist, had given up her acting career to work on behalf of animals and devoted much of her energies to the performing animal issue. Founder of the Fellowship of Faithful Friends, she published *Heaven's Rage* in 1934, a work containing a foreword by Frank Swinnerton and attempting an unsentimental demonstration that animals can only be trained to perform to a timetable by means of fear and that their unnatural tricks were part of a 'broken-spirited existence of perpetual rehearsals, performances and travelling in cramped quarters'. Trevelyan intended this book as a riposte to Gillespie's *Is It Cruel*?, and Swinnerton said in his foreword: 'If once we could be made to understand how and why the animals do their tricks, only the most callous among us would remain indifferent'. Two years later Trevelyan published *Laugh, Clown, Laugh*!, with a foreword by Canon H. R. L. (Dick) Sheppard and a preface by E. Keith Robinson, secretary of ODFL. Here she criticised the negative effect on animal welfare development of the fanatics, the sentimentalists and the

[23] *Parliamentary Debates* (*Lords*) 1934, vol. 91, columns 531–534.

[24] *Parliamentary Debates* (*Commons*) 1934, vol. 288, columns 670, 671, 676, 678, 679 and 681.

[25] Tyrwhitt-Drake also complained of inexperienced busybodies and cranks opposing zoos and menageries with constant agitation and misrepresentation (Tyrwhitt-Drake 1936, p. 74). And in the following decade, he expressed resentment at criticism from people with no practical experience, grouping together all 'antis' (those against everything from performing animals to dancing) as killjoys (Tyrwhitt-Drake 1946, p. 130).

squeamish. She advocated the use of sound argument and reliable evidence and also practical measures, such as Battersea Dogs' Home's stipulation that dogs taken must not perform in public and that a sworn statement must be signed to that effect. She had the courage to seek first-hand evidence for herself and described her attempt to arrange a visit to a leading circus proprietor's training school under a false name, when soon after her arrival her book *Heaven's Rage* was flung about the room during an argument about cover-ups and lies (Trevelyan 1936, pp. 16, 91, 104 and 105).

Another bill was presented in the Commons by Daniel Frankel, Labour MP for Mile End, in 1937, to prohibit the use for performance of dogs, seals, monkeys, apes and baboons or any other animals imported into Britain of a greater age than 6 months. The bill provided for rights of inspection of training premises under the 1925 Act by representatives of recognised humane societies (the RSPCA, ODFL or others approved as having similar objectives by the Secretary of State) who could also be complainants, and they should be allowed to attend performances and exhibitions of performing animals for the purpose of inspection (if necessary and without interference on or near the stage), with or without previous notice, when apparatus and materials could also be inspected. Frankel's measure would also make the particulars of registration and court orders under the 1925 Act and this proposed Act more readily available for public inspection. The aims of the bill were to ensure that performing animals were born in Britain and to improve the machinery of legislation. Penalties could for the first time include imprisonment; confiscation and care of animals by a humane society or, if necessary, destruction; and the possibility of permanent disqualification. Finally, each animal would be identified individually at registration. The same exemptions applied as those included in the 1929–1930 bill.

Before the outbreak of war, an unsuccessful bill much like Danesfort's of 1933 was presented in the Lords by Charnwood in 1938. He had written to *The Times* to protest at the confinement of large, active beasts, even if they continued to appear healthy (*Times* 1937a). In a 'good' menagerie, two to four lions had a space of 24 by 9 or 10 ft, sometimes divided into two by a partition, never having a wider range of movement. Feeling repugnant about this did not result from 'feeble sentimentality', and the question needed public attention. John C. Lockwood, Conservative MP for Hackney Central between 1931 and 1935, replied that these feelings were shared widely. In the last Parliament, the Home Secretary had told him that existing legislation was sufficient, but mental cruelty needed to be considered, too (*Times* 1937b). A little later, Sir Robert Gower, one described as opposing all wild animal captivity except in certain, special near-ideal conditions, wrote that the dangers of the natural state were better than imprisonment:

> Happiness is perfect harmony between function and environment, and a truly civilized community will not take enjoyment in seeing wild animals leading empty, unnatural lives (Gower 1937).

Soon after the end of the war, and with the constraints of rationing, Mrs Leah Manning unsuccessfully urged the Minister of Food to reject as far as possible all future applications for food licences from performing animal showmen.[26] In 1946, the Secretary of State for the Home Department, Chuter Ede, in a written answer to Peter Freeman, revealed the number of trainers, trainer-exhibitors and exhibitors of performing animals registered by local authorities in England, Scotland and Wales at the time. There were four trainers in Leeds and one each in Essex, Lancashire and Northamptonshire; 213 trainer-exhibitors in the London County Council area, 88 in other counties and 82 in the county boroughs and City of London; and 235 exhibitors in the London County Council area, 30 in other counties and 124 in the county boroughs and City of London. This gave a total of seven trainers, 383 trainer-exhibitors and 389 exhibitors.[27]

A 35-min meeting of the Departmental Committee of Enquiry into Cruelty to Wild Animals was held at the Home Office in Room 240 on 16 January 1950, to examine the subject of performing animals. The committee was chaired by Mr Scott Henderson KC, with Miss Frances Pitt, Professor Peter Medawar and others, and was attended for this hearing by Edmund MacMichael as director of the PADL and Lionel Dole (PADL treasurer). As well as his verbal statement, MacMichael submitted a typed memorandum for the meeting on the history of the performing animals issue and the PADL's activities, which had become concentrated on Bertram Mills Circus (MacMichael Papers). The first circus campaigned against, in the same month that he became identified with the PADL, in April as secretary, was Carmo's at Ilford in spring 1929, when the circus claimed Lord Lonsdale's and Bertram Mills's support for it. MacMichael said he was expelled from the RSPCA by its council because he tabled a motion that Lonsdale be required to sever his connection with Bertram Mills Circus, when he misled prominent guests about cruelty at Mills's annual luncheons costing around £4,000, making the PADL's legislative aims very difficult. MacMichael said he was wrongly arrested three times for distributing leaflets to guests. He said animal performance was illegal, because training was necessarily cruel and it therefore transgressed the 1911 Protection of Animals Act. The 1925 Act was 'innocuous to traffickers in performing animals'. The public remained unaware of the Act and therefore of the difference between the tricks of pets and performing animals. To date, there had been no police prosecutions for cruelty under the Act. MacMichael appeared before the committee in order, firstly, to produce evidence of cruelty to a performing British wild mammal, the episode of the 'Channel-towed Seal' (the PADL had unsuccessfully asked for an enquiry the previous October 1949)[28]; secondly, to offer irrefutable proof of cruelty in the secret training of performing animals; and thirdly, to ask the committee to promote his Performing Animals (Prevention of Cruelty) Bill. In notes probably made for a report-back to the League's advisory committee after the meeting, MacMichael noted that a Royal Commission or select

---

[26] *Parliamentary Debates (Commons)* 1946, vol. 423, columns 137 and 138.

[27] *Parliamentary Debates (Commons)* 1946, vol. 419, columns 312–316.

[28] The nature of the so-called Channel-towed Seal or the related episode cannot be discovered from MacMichael's papers.

committee was needed rather than this Committee of Enquiry.[29] The 'irrefutable proof of cruelty' MacMichael had referred to consisted of *Behind the Scenes at the Circus* by A. J. D. Mole. This was a voluntary statement to MacMichael made on 5 March 1951 about what Mole had seen as a groom at Billy Smart's circus from 1949 when after a few months he resigned. He referred to habitually cruel horse training by the spiteful Vinicky (a Czech whose real name was probably Toni Schmar). Obvious injuries were not noticed by the circus audience. Mole describes 'lungeing' training, when the horse was hoisted by a strap to the bridle and whipped from behind. Vinicky allegedly had a dual personality, different when in or out of the ring. Clumsy influenza injections were administered by Ronald Smart, and his wife made bloody tramlines with spurs that Mole had to disguise. The only exercise the horses had was in the ring, he claimed. Elephants were wounded behind the ears with goads, and he recalled that monkeys made to ride on horseback were terrified, and he spoke of the injured noses of highland steers whipped and forced to fall. However, the audience was deceived by the apparent affection of animals shown to cruel trainers, although there had been objections from women cleaners at Belle Vue after the training. Mole left the circus secretly because a 'buck nigger' had been threatened with a beating-up by the Smart family when he handed in his notice.

In 1954, Peter Freeman unsuccessfully asked the then Secretary of State for the Home Department, Major Gwilym Lloyd-George, if he would receive a deputation from animal welfare societies to be given information about performing animals, especially those trained abroad.[30] But it was not until 1956 that the training of performing animals was once more fully discussed, this time in the House of Lords when David Kenworthy, Lord Strabolgi, followed in the footsteps of his father and attempted, unsuccessfully, to secure the commitment of the Government in supporting new legislation.[31] He was backed by the Earl of Listowel, who called for an enquiry. Strabolgi referred to the problem of the cruel methods of foreign trainers such as those of Alfred Court, described in his *Wild Circus Animals*, published in Britain in 1954. Court had appeared at Blackpool in 1939. Meanwhile, a training camp in Denmark that supplied performances for London was described as an animals' Belsen. Strabolgi referred to an account given in the House of Lords in 1930 by Lord Auckland concerning personal friends' eyewitness account of whipping of lions after a performance when they had returned to their seats to look for lost car keys and to a *Daily Mail* report of 24 December 1955 that described how 2,000 children at a circus in Manchester saw a tussle with a terrified Russian bear who refused to perform a cycling trick and tried to escape, as the majority of the

---

[29] He also records that in his verbal statement he had said that he regarded wars as retribution upon nations for their ill treatment of God's animals, and now that the atomic bomb was involved, he alone was most likely to be able to guarantee freedom from the threat it posed because of his work to correct the human–animal relationship.

[30] *Parliamentary Debates* (*Commons*) 1954, vol. 535, columns 1096 and 1097.

[31] *Parliamentary Debates* (*Lords*) 1956, vol. 195, columns 668–717.

audience cried 'Shame!' That Christmas an inspector of the RSPCA had witnessed seven lionesses and one lion performing with the utmost reluctance. One, a lioness, he said, was obviously terrified. The first act of this animal on entering the cage and taking its place was to pass a motion, and it then made frantic attempts to squeeze a way between the bars of the cage and had to be threatened with the trainer's stick and struck once to force it to perform the trick. Strabolgi considered the sight of a magnificent animal such as this being in terror of a human being as pitiful. On the completion of the 'act', the animals made a mad rush to get out through the exit. If animals were treated like that in full view of the audience, he shuddered to think of what they had to undergo when they were out of sight.

> Then I have seen, also in a London circus, an elephant sitting in a jeep wearing an ostrich feather hat. It seems to me an odd way to amuse people by making a noble animal like this look ridiculous. One wonders what some future historian will think of the kind of civilisation which finds amusement in making bears ride motor-cycles, or bicycles 10 feet high—difficult enough for a human being—or the sight of a bear standing on a high stool wearing a frill round its neck and a dunce's hat on its head. In all these degrading acts noble animals are made to look foolish ... Quite apart from the effect on the animals, I submit that these circus performances are bad for children and for their education. The present trend in education is to teach the child to appreciate and understand an animal in its surroundings, and this aim is vitiated by circus performances where an animal is made to do unnatural and degrading acts.

Listowel thought that this 'denaturing' removed what was 'probably the most attractive quality about a wild animal—namely, the quality of being itself'. He was concerned that the 1925 Act did not extend inspection to places where the public were not admitted, such as film studios, and he wondered whether council inspectors were properly qualified for their work. The Bishop of Norwich helpfully observed:

> Some of your Lordships, like myself, are old enough to remember performing pigs. With what joy they set about doing their counting and their playing of cards! I do not know whether they were trained by cruelty but I hardly think so. I think it is possible they preferred that kind of life to being turned into sausages.

He wanted to encourage the training of more domestic animals, 'for their joy and ours', although he opposed the subjection of wild animals to unnatural performances.

Lord Ailwyn was particularly concerned with the cruelty of capture, incarceration and transport, as described in *Catching Wild Beasts Alive* by Joseph Delmont (1931), to be trained later in 'some unnatural trick or absurd antic in the ring, to pander to the crude appetites and emotionalism of a thrill-seeking crowd'. He recalled his travels to Malaya:

> A few years ago, when travelling in Malaya, I was taken to see three six-months-old tigers recently caught, each incarcerated in a box measuring some six feet by four feet. The month was February. They were not due to be moved until April. In those boxes they were destined to live and eventually to travel to England. About mid-May at earliest they might, if they survived, be released from their appallingly cramped quarters and be placed in a cage for the rest of their lives. And these are God's creatures, my Lords. In the event—for I followed up this case—one of these unfortunate animals died on the voyage, a second died on arrival

> in this country, and the third survived. So much for the capture and transport of jungle beasts ... Less than 200 years ago, people used to go and gape at the lunatics at Bedlam Hospital, as entertainment—I seem to remember a Hogarth picture depicting the scene. We have grown out of, and progressed beyond, that squalid form of inhuman entertainment. Has not the time come for us to acknowledge the rights of the animal kingdom and cease to exploit it for man's amusement? Above all, I plead for the wild animals in captivity, and particularly for those sentenced, through no fault of their own, to imprisonment in the cages and boxes of a travelling circus, condemned to exist in darkness or artificial light, deprived of fresh air, sunshine and exercise—relegated, miserably, to a living death.

An animal trainer had told the Earl of Huntingdon that in order to train satisfactorily a savage animal, such as a tiger or an elephant, its spirit must be broken completely through fear, pain, hunger and thirst. He did not think conditions had changed very much since the days of *Michael, Brother of Jerry*, and he referred to the fact that the PADL's £1,000 challenge (raised from £500 by 1953) had never been taken up, and winning it would have given the industry great propaganda.

Lord Dorchester told of the recent case at Washington where a gorilla and a full-grown panther fought for more than 2 h—it ended when the panther tore off the right arm of the gorilla, and the gorilla was shot. In Evesham, an Indian lion tamer, who later died, was mauled in front of an audience containing many children. At Bertram Mills Circus in 1955, the keeper of five tigers was clawed by what he called his tamest tiger, showing that these animals were not tamed by kindness. Lord Dowding found it strange that if the RSPCA should be given privileged access to inspect slaughterhouses, this right of access should be denied for performing animals, and he added:

> As a nation we are not animal lovers: we are sentimentalists about our animals. It is true that we cherish our domestic pets and that we have qualms about the condition of old and worn-out horses; but where the interests of animals run counter to our sports, our amusements or our pockets, the animals receive scant consideration at our hands ... Religious training is largely responsible ... Cruelties are perpetrated on animals without official protest from the various religious authorities ... Animals have not souls, is the teaching.

Lord Gifford, a director of a company selling circus seats, asserted that every animal trained in the three largest circuses was either born in captivity or imported as a cub and that the PADL's charges were unfounded.[32] The Joint Parliamentary Undersecretary of State for the Home Office, Lord Mancroft, thought there had been long-term, hysterical exaggeration of the issue, and he had himself watched a considerable number of circus acts in training. He felt an enquiry would throw up nothing new. Since the 1925 Act came into force, 1,191 persons, including 488 aliens, had registered under it, and no court cancellation of registration had ever been reported, suggesting the value of the Act in preventive power rather than in punishments awarded. Before the debate, he had approached the chief constables

[32] The PADL's pamphlets were clearly becoming noticed and referred to in Parliament. Richard Pilkington, during a Commons debate on 16 February, asked the Secretary of State for the Home Department whether he had read *How to Break an Elephant's Heart* and *The Truth About the Circus*, which had been sent to him (*Parliamentary Debates* (*Commons*) 1956, vol. 548, column 283).

of the areas in which the permanent training headquarters of the various dozen circuses in the country were situated, and they said there was very little indeed to complain about, the animals seeming well cared for and the training carried out humanely: their opinion was that the Act was being scrupulously observed. Of trainers, 'I was very much struck by the sense of responsibility which they showed. Clearly, it seemed to me, they are men who realise how much of a burden of humanity rests upon their shoulders'.

The final Performing Animals Bill to be debated in Parliament during the life of the PADL under MacMichael was introduced to the Lords in 1965 by Lord Somers. Its purpose was to respond to the inadequate inspection arrangements of the 1925 Act, and many familiar arguments against animal performance were repeated during the debate for the second reading.[33] Somers said he had been snowed under with letters and telegrams of support, and the only objection had come from the chairman of the ACP. Somers's bill was being promoted by CAPS,[34] of which Somers was then president, and the RSPCA and supported further by the World Federation for the Protection of Animals, the National Equine and Smaller Animals Defence League, the Catholic Study Circle for Animal Welfare, the Crusade Against all Cruelty to Animals and the World League Against Vivisection and for the Protection of Animals. The PADL had its own proposals for legislation, and this bill did not meet its requirements.

Somers thought that the public were very gullible, and one trainer had been quoted as saying, 'God made fools for showmen to deceive'. However, he offered new evidence from trainers 'behind the scenes': two of them at least were so absolutely revolted by what they saw that they tendered their resignations—they had no reason to hold anything against their former employers since they were not dismissed. According to the first trainer, after a baboon had torn a trainer's trousers during the show, it was brought to the back of the Big Top and held by three men while the fourth took steel cutters with blades of 3 in. and handles of 18 in. and cut the fangs that had done the damage flush with the other teeth. This was done without any injection to ease the pain and without any vet's advice. The baboon was said to be holding its jaw for approximately 10 days. This was only reported to the secretary of CAPS because there was no other witness. The trainer also referred to lions in very small cages and witnessed the 'breaking' of two lion cubs which had heavy stools thrown at them to keep them back, bamboo canes rammed down their throats and blank cartridges fired at their noses and under their tails (the latter to make them move more quickly): these left powder-burn scars. There was a three-cage trailer for transporting the lions. At one end was a lioness who was on heat, in the middle a very young cub and at the other end a fully grown lion:

> On opening them up in the morning, we found that the lion had ripped down the shutter separating the sections of cage and killed the cub. I told the trainer, and he came along and

[33] *Parliamentary Debates* (*Lords*) 1965, vol. 263, columns 405–484.

[34] This society was founded by retired teacher Irene Heaton in 1957 and, like the PADL, aimed at abolition rather than reform.

for approximately fifteen minutes fired blank cartridges at the lion, and then beat him with anything he could find handy—iron forks, rakes, etcetera.

He had also seen dogs in small pens or boxes for days, let out only to do their act at night. The second trainer saw an electrical instrument for training elephants, causing such distress he found it necessary to leave the building during working hours. The instrument was a series of metal rings connected to each front foot and joined by a cable to a mains electrical supply. The elephants were given a verbal command which, if disobeyed, resulted in a lever being pulled when 'several hundred volts' were circulated through the animals causing them to raise their feet in anguish. Their screams were 'more than a hardened criminal could stand'. Baroness Summerskill said she had been informed that the only way trainers could compel elephants to stand on their heads was to apply an electric shock onto the elephants' hind legs, combined with goads to its tail and to its most sensitive parts:

One is surprised that these miserable beasts do not turn and rend their persecutors—until one realises that the explanation lies, of course, in broken spirit. It is the daily and persistent domineering of the mind of the animal which causes the final submission.

The Earl of Huntingdon thought that to assume a wild animal 'which instinctively loathes, hates and fears [fire], is going to jump through a hoop just because you offer him a bit of meat, is folly'. The Bishop of St. Albans thought that animals were degraded in direct proportion to their skill in human accomplishments. The Earl of Haddington[35] added:

Surely no dog was ever intended by nature to climb up a ladder and then jump off a platform at the top into a blanket. Good heavens! It reminds me rather of a man making his first parachute jump, but the parachutist knows that if he pulls the rip cord he will come safely to the ground. A dog can have no such idea of that . . . To dress animals up and make them do these tricks and for human beings to sit there and laugh at them is a most degrading spectacle. Surely, if humanity is never going to rise higher than that it is a very poor outlook for the world.

Lord Burden quoted lines from Ralph Hodgson:

'Twould ring the bells of Heaven
The wildest peal for years,
If Parson lost his senses
And people came to theirs,
And he and they together
Knelt down with angry prayers
For tamed and shabby tigers
And dancing dogs and bears,
And wretched, blind pit ponies
And little hunted hares.

However, Lord Silkin believed that the majority of people did not regard animal performances as degrading spectacles, but that:

[35] His sister and wife were life members of the PADL.

> these sights of animals are instructive, interesting, entertaining and that there is no cruelty inflicted on the animals. In those circumstances, it would be wrong for us at the present time to attempt to pass legislation which would prevent the exhibition of animals for entertainment purposes.

Lord Belhaven and Stenton said that great patience, bravery and skill were needed to train a circus animal, and patient men were not very often cruel. Viscount Massereene and Ferrard asserted that caged animals would not feel the same emotions as incarcerated people and that freedom in the wild was:

> a precarious existence which is liable to end in sudden terrifying death or a slow prolonged death from starvation ... I maintain that if [a] lion had the power to reason and could choose, he would far prefer the well fed security of the circus.

The Earl of Gosford questioned the accuracy of Somers's witness statements, and Lord Stonham classed himself as the first contributor to the debate who had gone behind the scenes to see circus training—he also found the personal statements forwarded by CAPS dated, unspecific and without an explanation of why alleged cruelty was not reported to the authorities. He referred to the departmental committee of 1950 chaired by John Scott Henderson, which produced a report that was reprinted in 1958: the PADL had not produced convincing evidence to the committee, which was satisfied with existing legislation.[36] Since 1925, there had been 1,395 registrations, now with about a dozen new ones annually. Chief constables confirmed continuing regular inspections and the scarcity of complaints. Every animal act which had performed in a tented circus in Britain since the war was trained here, he said. A polar bear trainer told him that if you hurt an animal, you frighten him and he does not know what you want him to do:

> I am a slow trainer. First, you have to make contact. You cannot chase him, or he will get you. He takes his food on a long pole. Week by week the pole gets shorter, until there is no barrier between you. If an animal will not feed out of your hand in six weeks he is no good for training and you send him to the zoo.

This trainer added that he would accept the £1,000 challenge without taking the money. Stonham was prepared to accept the trainers' honesty:

> Knowing that tens of millions of people watch performing animal acts on television, I sought the views of Mr Derek Burrell-Davis, the senior producer in BBC Television Outside Broadcasts, who for the last ten years has been responsible for the majority of circus shows televised by the BBC. Your Lordships will be aware that a television production of this kind involves lengthy preparation, and the whole production team and crew virtually live with the show for days. Thirty or forty technicians are involved every time, and inevitably they see what goes on behind the scenes in every department. This, of course, is also the case on the Continent, and Mr Burrell-Davis has televised circuses in many countries, including Russia. He assured me—and I have his authority to quote him in the actual words he used—that his difficulty has not been in seeing what he wished to see but in resisting invitations from trainers to share their complete intimacy with their charges. He has never witnessed any cruelty.

[36] Janet Fookes MP later described the Scott Henderson Report as a somewhat weak, anaemic and wishy-washy document (*Parliamentary Debates* (*Commons*) 1973, vol. 866, column 829).

Lord Stonham made it clear that there would be no Government facilities to assist the further progress of the bill. Somers felt Stonham saw what he wanted to see and promised that evidence could be examined in detail if a select committee were appointed after a successful vote for a second reading, but the bill then failed by 14 votes. After this failure, correspondence in the press continued as usual. *The Illustrated London News* was criticised for its past annual appreciation of Bertram Mills Circus, part of a 'revolting trade', where open training sessions were really rehearsals. Correspondents advocated encouragement of more local authorities to ban circuses with animal performance, and Irene Heaton of CAPS repeated the long-standing claims that the circus environment for wild animals was unnatural and that in them the animals' spirit must be broken by instruments of domination later not seen in the ring by the audience, by fear of punishment and by being chained up (Gray 1966, Mitchell 1966, Heaton 1966).

The subject of circus animals arose again in the House of Commons debate on the Tourist Trade Facility Bill in 1967.[37] Bertram Mills Circus would not have a home, because its lease of Olympia had run out and there was no building in London capable of economically accommodating it. Sir Harmar Nicholls proposed that London's Royal Parks be available for use for enclosed, chargeable events such as circuses, but members felt many foreign as well as domestic visitors now disapproved of circuses using animals, and the arguments about capture, training and exhibition were made once more and at length. Bob Brown stated that one of the worst things that can happen to a child is genuinely to believe that kindness is employed when an animal is trained. Reginald Prentice, Minister of Building and Public Works, was opposed to animals' performing difficult tricks with the strong possibility of cruelty and felt that if the bill succeeded, there would be much external criticism, but the bill went no further. This bill was described in *The Illustrated London News* as deceptively named—'it should have been called a Circus Facility Bill for Maxwell Joseph's new Bertram Mills Circus' (Mitchell 1967).

## References

### *(Articles in periodicals are anonymous or untitled unless otherwise shown)*

The Animal World (1919) Performing animals, July, pp 77–78
The Animal World (1920) Performing animals: a conviction, November, p 122
The Animal World (1923a) A new bill, March, front page
The Animal World (1923b) The Performing Animals Bill, May, pp 52–53
The Animal World (1926) The "Daily Mail" on performing animals, March, p 29

[37] *Parliamentary Debates (Commons)* 1967, vol. 742, columns 856–901.

Bailey AL (1922) Report on performing animals. The Encore 8 June, p 3
Beers DL (2006) For the prevention of cruelty. The history and legacy of animal rights activism in the United States. Swallow Press, Athens
Delmont J (1931) Catching wild beasts alive. Hutchinson, London
The Encore (1921) Performing animal cranks routed, 5 May, p 3
The Era (1921a) 2 March, p 16
The Era (1921b) 9 March, p 15
The Era (1922) 22 February, p 11
Gillespie TH (1934) Is it cruel? A study of the condition of captive and performing animals. Herbert Jenkins, London
Gower R (1937) Letter The Times 26 August, p 8
Gray I (1966) Performing animals. The Illustrated London News 19 February, p 3
Heaton I (1966) The case against circus animals. The Illustrated London News 7 May, p 44
Joys JC (1983) The wild animal trainer in America. Pruett, Boulder, CO
MacMichael Papers (uncatalogued and unpublished material relating to the director of the Performing Animals' Defence League between 1929 and 1968, kindly lent to the present author by Edmund MacMichael's daughter, Mrs Esther Denham)
Miller EK (1999) Animal rights groups protest circus visit. The New York Times 1 August, p LI3
Mitchell K (1966) The Illustrated London News 19 March, p 3
Mitchell K (1967) Hyde Park peril. The Illustrated London News 11 February, p 34
Mitchell K (1975) Circus animals. The Illustrated London News 22 February 1975, p 4
The New York Times (1936) Clyde Beatty fined as 'cruel' to lions, 27 May 1936, p 25
The New York Times (1991) Advertisement, 29 January, p D22
The Performer (1923) 1 August, p 7
Radford M (2001) Animal welfare law in Britain. Regulation and responsibility. Oxford University Press, Oxford
Regan T (2004) Empty cages. Facing the challenge of animal rights. Rowman and Littlefield, Lanham, MD
Robinson (Chief Inspector—Northumberland and Durham Society for the Protection of Animals) (1913) How a foreign trainer was brought to book. In: Trist S (ed) The under dog. A series of papers by various authors on the wrongs suffered by animals at the hand of man. The Animals' Guardian Office, London, pp 133–140
Rochez H (1921) Are trainers of performing animals getting fair play? The Encore 24 March, p 9
Rowe C (2002) Banning the big top: Greenburgh bans live animal circus acts. The New York Times 16 June, p WE1
Stein J (2001) The greatest vendetta on Earth. Why would the head of Ringling Bros-Barnum & Bailey hire a former top CIA honcho to torment a hapless freelance writer for eight years? Salon Media Group http://www.salon.com/2001/08/30/circus_2/, 31 August. Accessed 19 February 2014
The Times (1913) 19 December, p 5
The Times (1921) 27 May
The Times (1931) 3 January, p 12
The Times (1937a) 19 August, p 8
The Times (1937b) 23 August, p 15
Toole Stott R (1971) Circus and allied arts: a world bibliography, 1500–1970. Harpur, Derby
Trevelyan H (1934) Heaven's rage. C W Daniel, London
Trevelyan H (1936) Laugh, clown, laugh! C W Daniel, London
Tyrwhitt-Drake HG (1936) Beasts and circuses. Arrowsmith, Bristol
Tyrwhitt-Drake HG (1946) The English circus and fairground. Methuen, London
United Kingdom Parliament (various dates) Parliamentary Debates (Commons). HMSO, London
United Kingdom Parliament (various dates) Parliamentary Debates (Lords). HMSO, London
United Kingdom Parliament (1921) Report from the Select Committee on Performing Animals, together with the Proceedings of the Committee and Minutes of Evidence. HMSO, London

United Kingdom Parliament (1922) Report from the Select Committee on Performing Animals, together with the Proceedings of the Committee and Minutes of Evidence. HMSO, London
Wilmut R (1989) Kindly leave the stage! The story of Variety 1919–1960. Methuen, London
Wilson DAH (2008) Politics, press and the performing animals controversy in early twentieth-century Britain. Anthrozoös 21(4):317–337
The World's Fair (1911) 29 July, p 4
The World's Fair (1921) 12 February, p 8
The World's Fair (1924a) 19 July, p 10
The World's Fair (1924b) 19 July, p 15
The World's Fair (1924c) 25 October, p 18
The World's Fair (1925) 8 August, front page

# Chapter 6
# Agitation After the Act

**Abstract** Dissatisfaction with the Performing Animals (Regulation) Act of 1925 led to the adoption of confrontational tactics by the Performing Animals' Defence League under the leadership of Edmund MacMichael. After a disagreement, MacMichael had replaced its founding chairman, Ernest Bell, to become director of the League; and then relations with the Royal Society for the Prevention of Cruelty to Animals soured as he entered into bitter arguments with its chairman, Sir Robert Gower. At the same time, MacMichael waged campaigns against circus figures Bertram Mills, Lord Lonsdale and Billy Smart, and the uncompromising behaviour resulting from his personal beliefs led to conflict with a range of national authorities. Meanwhile, problems in the use of trained animals for the international film industry and for advertising attracted new levels of attention. From the middle of the twentieth century, scientific contributions to the practice or assessment of training animals for commercial purposes became increasingly evident, as in the case of the Brelands' work in the United States; and in more recent debates on animal welfare problems, scientists have been called on by lobbying organisations to support their arguments.

## 6.1 Edmund MacMichael and the Performing Animals' Defence League

The Performing Animals' Defence League was the most active organisation opposing animal performance from its foundation in 1914 until 1968, when the Captive Animals' Protection Society (founded in 1957) gradually began to take over the campaign. Attempts to amend legislation began almost as soon as the 1925 Act had been passed, and within a few years the work of the PADL intensified under its new director, the highly principled, uncompromising and eccentric Captain Edmund MacMichael. During the 1920s the music hall continued with its post-war transformation into variety theatre (considered to be more respectable), and lost more ground to other forms of popular entertainment such as the cinema and radio, as well as live jazz and revue, 'something between music hall and the spectacular



D.A.H. Wilson, *The Welfare of Performing Animals*, Animal Welfare 15,
DOI 10.1007/978-3-662-45834-1_6

musical comedies of the period' (Wilmut 1989, p. 17). The attention of the PADL therefore became more focussed on the circus and increasingly also on the use of animals in films. MacMichael wrote to *The Spectator* in 1931:

> With the decline in performing wild animal turns on the stage we may get an increase on the films, for the owners will still want to make them pay. But letters to the Press, to cinema managers and to the film censor, *fortiter in re suaviter in modo* [vigorous in the matter but pleasant in manner], together with editorial comments, will do an enormous amount towards making cruel films unprofitable. I shall be grateful for notifications by your readers of any films objectionable in this respect (MacMichael 1931b).

The PADL's further interest in the problem of confinement promoted a more general awareness of this aspect of animal welfare. Accordingly, it changed its name to the Performing and Captive Animals' Defence League in the late 1920s, although MacMichael and others would frequently continue to refer to it as the Performing Animals' Defence League. In August 1936 the League's letterhead described it as 'The Humane Society dealing with Cruelty in Circus, Cinema, Music Hall, Zoo, Fun Fair, Rodeo, etc.' Its letterhead in 1967 used both names and had added bullfighting to the list. Long before then, the use of titled personages as patrons, presidents and vice presidents had ceased. Among these, in the late 1920s and early 1930s, had been the Countess of Warwick and Nina, Duchess of Hamilton, an energetic animal welfare campaigner until her death in 1951.

Shortly after the passage of the 1925 Act, a key adversary of opponents of animal performance, Albert Voyce, disappeared from the scene. He had been the professionals' most vigorous salaried champion in the controversy but resigned from the VAF in unfortunate circumstances. He had been a scena artiste with a turn on the stage worth £12, 'and with his well-oiled hair and high starched collars, he exemplified the image of the typical refined "pro"' (Rutherford 1986). Albert Voyce first appeared with Sydney Verno on the halls in the 1890s as 'The Fin-de-Cycle Bicyclists' and 'The Singing Cyclists' (Fig. 6.1). After Verno's death in 1897, Barney Stuart replaced him. By 1905 the duo was working as 'sketch vocalists' instead of a cycling act. They advertised themselves at the Canterbury, Collins and the Royal in April 1906 as 'popular singing comedians'. They were never big stars and were earning £25 a week in 1912. Voyce's great granddaughter, Lauren Donaho, believes that Voyce was a stage name adopted to reflect his specialism and that his real surname was Morgan.[1] Moving from the stage to the office, Voyce then became president of the Music Hall Artistes Railway Association and later chairman of the VAF.

The Executive Committee of the VAF had instructed 'MB' (presumably the National Organiser, Monte Bayly) to send Voyce a letter on 1 September 1927 concerning the apparent misappropriation and fraudulent cashing of a crossed cheque for £10 sent by a member from New York for passing to David Allen and Son, solicitors or agents. Voyce gave false assurances that he had paid the money,

[1] Personal communication, 6 July 2009.

**Fig. 6.1** A caricature of Verno and Voyce by George Cooke. © Victoria and Albert Museum, London

and the VAF was then obliged to honour the payment of £10 itself. The executive committee therefore resolved to ask him for an immediate written report and to attend a meeting on 8 September. (MB added as a postscript: 'My dear Albert Voyce, I trust you will realise that this letter is written with very great reluctance by myself, and I would that it was written to anybody but you'.) At the meeting of 8 September, Albert Voyce, Monte Bayly and Fred Herbert (Secretary) left the room during two and a half hours' discussion of the matter. Eight voted for and four against acceptance of Voyce's explanation, and he was granted 4-week sick leave. At the following executive committee meeting, the minute of this acceptance was not confirmed, but Voyce's further written explanation of 22 September was entered in the minutes instead. It was decided not to pursue the matter, as the cheque could not be traced. It was also agreed because of his continued ill health to extend the period of sick leave until the date of his departure from the VAF on 1 December 1927: Voyce, Bayly and Herbert had all tendered their resignations on 18 August and these had been accepted, to take effect at the same time. (Bayly subsequently took up the post of General Secretary in December on a salary of £8 per week but suffered a heart attack in August 1928 and died in November.) In his convoluted written explanation of 22 September, sent to Fred Herbert, Voyce referred to his having been on the verge of a nervous breakdown since the previous May. The cheque had been mixed up with his mother's papers and cashed and the £10 kept for his collection, and he denied any intention to misappropriate, later reimbursing the VAF with the money (VAF Archives). In this way, Albert Voyce left the VAF and the performing animals controversy: it is interesting to speculate on the nature of the confrontations with Edmund MacMichael that would have resulted had he continued in office into the 1930s.

MacMichael succeeded Ernest Bell as effective controller of the PADL at the beginning of the 1930s and embarked on a vigorous and colourful campaign of agitation and direct action to within 3 years of his death in 1971 (Figs 6.2 and 6.3). Shortly before Bell, its founding chairman, retired from the League, *The Times* reported: 'To mark his retirement from the editorship of *The Animals' Friend*, a magazine which he founded 30 years ago, Mr Ernest Bell was on Saturday presented with a book bound in non-animal "leather", with pages of vegetable parchment' (*Times* 1929). Bell switched his attention to other causes; for example, he served as treasurer in the Pit Ponies' Protection Society and in the National

**Fig. 6.2** Edmund MacMichael in 1919, aged 30. MacMichael Papers

**Fig. 6.3** PADL logo in use in the 1960s. MacMichael Papers

Society for the Abolition of Cruel Sports, created in 1932.[2] Meanwhile, MacMichael's personality immediately began strongly to affect the behaviour of the PADL, as well as external perceptions of it, as it would over the course of four decades.[3] Even by 20 July 1936, he had replied to D.N. Singh about his suggestion of a play based on his recent life, saying he could not help finance it at the moment because of the costs of litigation of 2½ years, but that he remained interested: 'A play based upon my life during the past few years would be a real thriller, but would be voted too impossible for words'.

Edmund MacMichael was born in 1889 and served in the Devonshire Regiment as an intelligence officer and German-speaking spy in the First World War. At its end he was given the job of minding the Kaiser's sister, afterwards becoming a journalist on *The Daily Herald*, a newspaper that would take a strong stand against animal performance. Esther Bright—his father-in-law's sister, the daughter of Ursula and Jacob Bright (radical-liberal industrialist MP for Manchester), a suffragette and theosophist and close friend of Annie Besant—sparked his interest in animal issues and vegetarianism and with her considerable inheritance later helped him financially during his numerous court cases, as against the circus owner Bertram Mills. Esther became a life member of the PADL and was also a member of the Women's Freedom League in the 1920s (Fig. 6.4). MacMichael adopted many of the principles and beliefs of members of his extended family such as Esther Bright. He became a vegetarian, advocated Nature Cure (naturopathy) instead of conventional medicine, opposed the germ theory of disease and helped in the work of the National Anti-Vaccination League. G.N. Gould, a veterinary surgeon, wrote to *The Worthing Gazette* on 18 May 1932, arguing against MacMichael's advocacy of alternative medicine for animals as well as denying the role of vivisection in veterinary training and practice. In a letter of complaint of 9 December 1932 to the British Broadcasting Corporation (BBC) that the anti-vaccination point of view would not be given air time, MacMichael stated: 'There is ample proof of the presence of gross cruelty to animals as well as deaths, maimings, blood poisonings, defective eyesight and according to some authorities cancer as a result of this unscientific and out of date practice [vaccination]'. And in a letter to *The Daily Mirror* of 17 July 1933 about germs and cholera, he stated that serum was 'part of the diseased blood of a tortured animal' and offered to address a public meeting as a leading scientist. On 18 May 1944 he enrolled in the British Health Freedom Society, for recognition of alternative medicine in the National Health Service.

MacMichael also adopted unorthodox religious views, referring to the Aquarian Gospel as revealing Christ's true teaching on the treatment of animals: in AD 325, 'Correctores were appointed who crippled Our Lord's teaching by cutting

[2] Born in 1851, he became chairman of George Bell and Sons Ltd, the publishers, and also president of the Vegetarian Society, chairman of the National Anti-Vivisection Society and treasurer of the National Council for Animals' Welfare. He died in 1933.

[3] The following account of MacMichael's correspondence and activities is based on evidence contained in his surviving papers (referred to here as the MacMichael Papers), which are not catalogued.

**Fig. 6.4** Esther Bright with, *left–right*, her nephew, Pennington Bright, Jiddu Krishnamurti (thought by theosophists to be the reincarnation of 'The Lord') and Edmund MacMichael. Photograph courtesy of Caroline MacMichael Dales and Juliet MacMichael

out those portions which did not tally with the orthodoxy of that day'. In a letter to John H. Brocklesby of 20 July 1937, he claimed that these 'correctores' deleted all God's commandments and teaching about the subhuman creation—11 out of 12 commandments were cut out, leaving only the Golden Rule, for example, the second commandment: 'Ye shall not steal the goods of any; nor gather lands nor riches to yourselves beyond your need or use'.[4] MacMichael wrote to *The Times* on 12 October 1932: 'Granting the existence of a Power higher than man, the

[4] A source of inspiration for MacMichael might have been Ouseley's *The Gospel of the Holy Twelve*, known also as *The Gospel of the Perfect Life* (1904). This rediscovered 'gospel' was probably a creation of Ouseley rather than authentic, but it was attractive to MacMichael because it advocated animal welfare and vegetarianism. MacMichael believed that *correctores* at the Council of Nicaea in AD 325 had edited those divine texts that did not tally with the orthodoxy of the time, including the commandment: 'Ye shall not take away the life of any creature for your pleasure, nor for your profit, nor yet torment it'.

attainment of human happiness is out of the question unless mankind is prepared to grant an ever increasing measure of justice to those weaker than himself'. Accordingly he would write on 8 September 1942 to the Board of Examiners for the Higher Certificate about a request for permission for his daughter to do dissection: 'In view of the Higher Law: "Ye shall not take away the life of any creature for your pleasure, nor for your profit, nor yet torment it", it is no business of mine either to grant or to withhold "permission" – any more than to do so regarding burglary or murder'. On 5 April 1944 he promised support for the ordination of women if they would reform the church's attitude to animal cruelty.

Edmund MacMichael took an interest in government and politics as well. By 1926 he was the assistant secretary of the British Committee on Indian Affairs and supported home rule for India. Perhaps partly as a result of the post-war poverty he had witnessed in Germany when in charge of the Kaiser's sister, he became critical of policies adopted during the growing confrontation with Nazi Germany and then refused to cooperate with measures affecting the British civilian population once war had started: he believed a peace should be negotiated as soon as possible. By October 1939 he had not registered for food rationing and objected to identity cards. He would not participate in or pay subscriptions for fire watching around the PADL office in Buckingham Street through the Adelphi Fire Prevention Committee—it was 'alike unlawful and impossible for me, personally, to take precautions against an avoidable evil'. For the same reason he was opposed to and disobeyed blackout orders, seeing them as unlawful regulations and 'A.R.P'ism'. On 28 March 1940, he was in court over his failure to observe the 'unlawful' blackout. On 3 December 1940, MacMichael wrote to his printer's solicitors about his failure to pay a bill of £1.19.6 for his anti-blackout pamphlets, giving a highly charged account of the state of national justice, blasphemous politics and a disastrous war leading to decimation and annihilation as the reason for delay.

Apart from immediate concerns related to the cause and ending of the war, he argued for a system of social security and a universal guaranteed minimum income, partly to negate vested interests as revealed in the treatment of animals in performance and vivisection. MacMichael wrote to the Christian socialist and joint founder of *The Daily Herald*, George Lansbury, on 6 October 1931 mentioning the idea of forming a Humanitarian Party. He said he was preparing for the next election when he would travel the country asking audiences to ask their candidates to support a policy of national security of minimum income in return for service to the community by those who were employable and to support anti-cruelty legislation. On 24 September 1937, MacMichael wrote to his MP Sir Arnold Wilson, asking him to speak out against recent ineffectual animal legislation, enclosing his own Performing Animals Bill and expounding his general philosophy on war and on security and equality of income. By then, he had been imprisoned eight times, 'without having committed or been accused of any crime'. He wrote to John Stone on 19 March 1940: 'I think I ought to make it clear that I look to no "orthodox" statesman, politician, diplomat, etc., to guarantee us an honourable, lasting and abiding Peace, and that the only man who has earned the right to declare that he, D. g., can do so is ... Yours truly, E. T. MacMichael'. On 23 June 1941,

MacMichael wrote to the press and Prime Minister and spoke in Hyde Park calling for a plebiscite, an 'undeclaration' of war, equality of income (resulting in the ending of vested interests) and decimalisation. In a letter of 10 June 1942 to Miss Loat of the National Anti-Vivisection League, he described his philosophy of 'God, Mankind and the Animals'. He advocated equality of income to remove the vested interests that make cruelty profitable and open to divine retribution on mankind, as in the nation's suffering in the current war. He had been to prison 10 times by then 'to make quite sure that my faith was the right one' (and, as he mentions elsewhere, to 'make my name'). He would not help in the war effort, but one of his aims was to meet Hitler in the name of the people of Britain and with clean hands, to persuade him to introduce his policy into Germany and end the war; if necessary, this would be achieved by a challenge of single combat 'on horse or on foot with sword, lance, battleaxe or mace'.[5] In a letter to W. Adair-Roberts of 30 September 1942, he said he was 'not a member of the Labour party, though in general agreement with their aspirations'. Later, he opposed the idea of the Common Market because he envisaged increased organised cruelty to animals. MacMichael gave an account of many of his political beliefs in issues of the PADL *Newsletter*.

On 8 May 1956, MacMichael wrote to the Soviet ambassador, asking him only to attend the Moscow State Circus at Harringey if it was free of animal performance. This should concern the ambassador, he said, because of the need to avoid war, and attendance at performing animal acts would alienate some influential people. MacMichael quoted a message he had sent to the International Humanitarian Bureau to be embodied in a petition to the disarmament conference held at Geneva in 1932:

> Unless and until our rulers act on the natural assumption that nations have no right to expect, at the hands of that power higher than man, treatment better than that which, openly and secretly, they allow and encourage to be inflicted on the weak, whether human or sub-human, there can be no peace. This League therefore directs attention to the sufferings of animals in all zoos and circuses and in the production of films and to the evil effect, especially upon children, of exhibitions involving captivity and torture.

This statement had not been used, with, as he believed, dire results in the form of the Second World War.[6] MacMichael concluded that individual cruelty was the responsibility of individuals, but organised cruelty was the responsibility of nations, leading to divine retribution, usually in the form of war.

On issues such as these, MacMichael would speak more than once at Hyde Park Corner, but his energies were reserved mainly for the performing animals campaign. In conducting this he would apply his independence and determination, and his belief in putting conscience above the law if he believed the law was at fault. In

[5] MacMichael's eminent correspondent and sometime leader of the Labour Party, George Lansbury, was similarly a pacifist who worked to prevent the Second World War. He managed a (conventional) meeting with Hitler before the outbreak of hostilities. Like MacMichael he had gone to prison for his politically inspired views and actions.

[6] Since then, in 1950, the World Congress for Animal Protection at The Hague attempted to ban animal performances as part of a Bill of Rights for Animals (Joys 1983, p. 267).

the work for the PADL, he was outspoken and unrestrained but claimed he tried to avoid personal enmity and bad feeling. Much of his correspondence for the PADL is laced with religious and classical quotations. Kathleen Hanchett-Stamford, whose husband, Derek, a former RAF pilot, joined the PADL in the late 1950s as a part-time assistant to MacMichael, told *The Times*: 'The league was the animal rights activists of their day ... Captain MacMichael liked being in trouble and going to court, where he would quote Greek and Latin' (Thorley and Brown 2008).[7] He always invited or challenged his correspondents to personal meetings to thrash out a disagreement or a claim made on behalf of the PADL. Mrs Albert Bradshaw acted as an intermediary between MacMichael and B. Coppinger Prichard when they fell out at a Kensington Town Hall meeting. MacMichael wrote to him admitting '[I] ought not to have used the terms I did', and he apologised on 16 April 1931. At the same time, he would often copy his letters to the local press in the hope that it would give publicity to his assertions. This was especially true of his lobbying of organisations, as when he contacted councils to try to persuade them not to let their land to circuses.

MacMichael was imprisoned many times for contempt of court and for refusing to pay rates when he thought they would be used to support circus visits on public land or to pay tax because it could fund vivisection. In 1958 he refused to pay income tax unless he was compensated for what he considered past false imprisonment; this resulted in 3 years of lengthy correspondence about summons and appeal with the Lord Chief Justice. Because of his single-minded determination to resist a perceived injustice at any level, when on 14 September 1933 he was fined in court for jumping traffic lights, he questioned this by letter after the decision, making detailed calculations of time, distance and speed. He felt equally ready not to pay a fine for a parking offence in 1965. His granddaughter, Caroline MacMichael Dales, says that he loved a fight and to pit his wits and considerable intelligence for a cause he believed in. In a case in the early 1960s, he acted as *amicus curiae* in the High Court when a printer was prosecuted for publishing a pamphlet showing some of the cruel instruments used in the training of circus animals. He was eventually told to sit down and shut up—when he would not he was sentenced for contempt of court. 'The difficulty came when he was asked to apologise and of course he could not because he thought he had done nothing wrong'.[8]

The PADL remained vigorous until MacMichael's final days, as shown by miscellaneous evidence in his surviving papers. The League had discussed concerts with 'star artists' in the 1940s and staged a send-and-buy Animals' Fair on 27–28 November 1964 at the Royal Horticultural Hall, when no animal products were

[7] She became an active member herself, and her daughter remembers being taken to Bertram Mills Circus in the 1960s when she 'stood on her seat and booed and booed until they dragged her out'. See also *The Argus* (2008): 'I remember going to the London Palladium and seeing a performing dog, which was appalling. I was sitting in the front row of the circle and stood up and booed until the orchestra stopped'.

[8] Personal communication, 17 June 2008.

allowed. On 18 April 1967 Lady Chichester joined the League, with a letter from her secretary at 'Sheila Chichester's London Woman. Her Guide to Shopping and Fashion. Francis Chichester Limited'. By this time, the south-western branch of the PADL at Torquay was still active, its secretary, Wilfred P.H. Warner, having held his post for around 25 years.[9] On 5 June 1967, MacMichael asked if any of its members would like to attend the Annual Convocation of the Brotherhood of the Essenes at Glastonbury Town Hall and Tor. Full details were available from the Secretary, The House of the Holy Grail. Meanwhile, new links were made with other emerging animal welfare movements: in 1967 the Hunt Saboteurs sometimes targeted circuses, and MacMichael attended a factory farm rally in Trafalgar Square in 1967.

When MacMichael died, he left Derek Hanchett-Stamford in charge of the PADL. Tactics then softened, and campaigning relied on the writing of thousands of letters rather than direct action, concentrating on local authorities who let travelling circuses use their land. By the late 1970s Hanchett-Stamford had effective control over the PADL's assets, and in the early 1990s, he moved its offices to the Gothic mansion, Sid Abbey in Sidmouth. On his death in 2006, the League's assets amounted to nearly £2.5 million in property and shares, and after a court case Mrs Hanchett-Stamford kept the assets from the Treasury so that she could donate them to the Born Free Foundation. She told *The Times*:

> It is sad to be the last member of the League, but the performing circus animals have gone . . . I'm glad that the League's work has been successful and that the money is now going somewhere where it will be useful (Thorley and Brown 2008).[10]

### *6.1.1 Confrontation with Ernest Bell and Realignment*

MacMichael's forthright approach to his work for the PADL resulted in some friction within the organisation just before Ernest Bell retired from it and MacMichael took over. His surviving papers include a 'diary of events', probably compiled as a statement of defence after he was asked on 23 December 1929 at the League's shop in Kensington by Miss Jessey Wade as honorary secretary to resign as salaried secretary. He replied that this request should come from the committee, but Wade then threatened to dismiss him. On 1 January 1930 he received a letter from Bell giving him a month's notice of dismissal. Bell had told a mutual friend that MacMichael could not fit in with influential members of committee, and in the letter to MacMichael, his manner of carrying out the duties of secretary was described as unsatisfactory. MacMichael wrote back refusing to accept the dismissal as it had not been before the executive committee that had appointed him.

---

[9] After the war, the Scottish Branch in Glasgow had been run from London following 'difficulties' involving its honorary secretary and its failure to distribute posters sent up to it.

[10] The performing circus animals had not *quite* gone. . . .

In subsequent exchanges, Bell admitted he had done some very good work, but 'several members of the committee disapprove of your rather high handed and extravagant way of conducting the work without getting their sanction and your retention would mean their resignation, which would be fatal to the Society'. Bell said he was worried that in 6 months there would be no money available to pay for rent or secretary. MacMichael thereupon called a special committee meeting, which was cancelled by Jessey Wade because he had no authority to call one; she also informed members of committee of his notice of dismissal. Bell then wrote to him that he was 'lamentably ignorant of the legal position of a salaried official' and of proper behaviour towards the committee. MacMichael replied that he deprecated the tone of Bell's letter but apologised for calling the special meeting. In his statement of defence, MacMichael continued that Bell as treasurer as well as chairman was responsible for paying accounts for literature, the source of his alleged extravagance. Since he had taken over the secretaryship on 4 April 1929, there had been little increase in the cost of literature, in spite of greatly increased activity, resulting in only one animal turn remaining in West End music halls, no wild animal turns at Olympia (that year), four successful shop campaigns (in London and Brighton and two in Leicester), two van campaigns and numerous poster and literature campaigns in the provinces: 'Our present forward policy has had a really remarkable effect upon public opinion and the question has again come to the fore as a political issue'. The number of subscriptions in 1928 had been 134, but the number in 1929 was 321, and 200 new members had joined since 18 December. He concluded by stating that 'The Animals, at any rate, have no cause for complaint as regards the activity of the League during my secretaryship'. No doubt this episode resulted less from accounting worries than from problematic relations with committee members and MacMichael's impetuousness and uncompromising nature, but there is no record of any discussion of this, and soon MacMichael took over complete control of the PADL as director. A vote of confidence in him was later passed by the committee, but its composition soon changed as MacMichael took over control of the League.

This was clearly a time of great stress and much hard work, and from late 1931 to early 1933, he occasionally suffered nervous exhaustion. On 30 December 1932 he wrote to Lady Cory, puzzled by her complaints, not made directly to him, about the lack of acknowledgment for her donations: he asserted that receipts were always sent. A special meeting of the committee was held on 6 January 1933 to hear various complaints against him, when he was supported. His friend Robert H. Spurrier, now editor of *The Animals' Friend*, published by the National Council for Animals' Welfare, wrote to MacMichael on 10 January 1933 regretting his stress and ill health from over-exertion. Spurrier was critical of Lady Cory, but was restrained in his own correspondence with her. In 1934, the Annual General Meeting of the PADL decided to:

> ... dispense with the services of an Executive Committee, and to empower the Director to appoint an ADVISORY COMMITTEE, leaving him sole discretion as to its composition. This decision was arrived at in order to avoid dissension on vital policies.

That year, after lack of progress in parliament, the PADL began to base its policy on its own unofficial Performing Animals (Prevention of Cruelty) Bill, which was later approved by the Quarterly Conference of Animal Welfare Societies in January 1946. It called for constant supervision of training and provided for compensation for those losing their livelihoods as a result. MacMichael now accepted the impossibility of enacting his bill by orthodox channels. The League's stated objectives by 1944 were to see the passage of a Performing Animals Bill, making illegal public performances by animals involving cruelty in training, with just compensation to the trade; to see passage of a bill against cruelty in the production of any film permitted on the British screen, foreign or not; to prevent the establishment of new zoos; and to oppose the introduction into Britain of rodeos and bullfights (PADL 1944, pp. 4 and 5).[11] In the PADL's *Newsletter* no. 26 of August 1945, MacMichael wrote:

> Private Members' Bills, we hear, will be given no time for some considerable period. Ordinarily this fact should have evoked a storm of protest, but so far as Animal Welfare Bills go it is rather good news. This sounds a dreadful statement, but it is perfectly justified in respect of all the Animal 'Protection' Acts of Parliament passed in the 1930s. There is hardly one that is worth the paper it is written on.

In 1962, Evalyn Westacott set out to advertise the continuing work and unchanged aims of the League under MacMichael, and in order to do this based her book on a reprint of chosen extracts from evidence given before the original select committee, noting the need to do this because of the rarity and dilapidation of the surviving reports of its findings.[12] She noted that a £1,000 challenge to the profession to train under supervision an untrained animal to perform a circus trick on demand and without cruelty remained unclaimed (Westacott 1962, p. 7).

### *6.1.2 The Campaign Against Bertram Mills, Lord Lonsdale, Billy Smart and Others*

Throughout his campaigning career, MacMichael advertised his belief that the 'Hallmark of the performing animal trade is cruelty, deceit, fraud, bribery and corruption', one he formed as a result of his early conflicts with Bertram Mills. One cause of Bell's concern that MacMichael was upsetting influential people on the committee and maybe elsewhere as well might have been the gathering intensity of the campaign he started in late 1929 against Bertram Mills Circus at Olympia.

---

[11] At this time, MacMichael's salary was £500 per annum.

[12] The 1921 report had not been available since 1930 and that for 1922 since 1935 (*Parliamentary Debates* (*Commons*) 1946, vol. 422, column 6). According to MacMichael's statement to the Scott Henderson Committee, 3,080 copies of the 1921 select committee report were printed, 708 of the 1922 report. Both had remained out of print in spite of requests for reprints, although the PADL was permitted to reprint them wholly or in part (MacMichael Papers).

MacMichael's dispute with the Mills family would last for the rest of his life. On 5 November 1929, he wrote to Mills inviting him to withdraw wild animals from the tenth Olympia programme. He explained that although he had no desire to combat individuals, he would if necessary lobby prominent people are invited by Mills to attend the promotional luncheon inaugurating the circus. A month later, a letter had been duly drafted and authorised by his committee, to be sent to Earl Jellicoe, Field Marshal Lord Plumer, Viscount Curzon and others, asking them either to boycott such shows of performing animals, especially wild animals, or to ask the management of the circus for their future exclusion; PADL literature was also to be enclosed. By then, a part of MacMichael's first draft had been cut out:

> I ought, perhaps, to point out that the animals are mostly trained abroad and are generally exhibited by foreigners and that while foreigners have just as much right to live as Britishers, the position is certainly anomalous when foreigners performing objectionable turns shall be allowed to have their living earned for them when so many excellent British artistes, whose turns in no way offend against the higher human instincts, are vainly looking for work.

Signed top copies of the final draft of the letter remain in MacMichael's papers, and it is possible they were not in the end sent, either for reasons connected with the attempt to dismiss MacMichael from the PADL or because Mills did not include wild animals at Olympia that year.[13] *The Kensington News and West London Times* reported that there were only elephants, horses and a dog, but 'Even the omission of the wild beast will not satisfy those who place a ban on all performing animals – but these are obsessed by the one idea that there must be cruelty in the training' (Kensington News 1929). At the luncheon itself, Lonsdale had given an assurance that there was no cruelty in the Olympia circus (Performer 1929).

Wild animals did, however, appear at Mills's simultaneous circus in Manchester, for 4 weeks from 21 December. *The Hull Evening News* (1929) reported on 9 December 1929 that their arrival was delayed by bad weather at sea. On board was a large shipment of 'Captain' Alfred's trained animals from Breslau via Bremen: 70 lions, 30 horses and ponies, two tigers, 5 bears and 13 monkeys. A Mills representative described the animals as extraordinarily good sailors: 'They adapt themselves to conditions'. On arrival they would be taken by train to Manchester City Hall for this, Mills's first show outside London. When they did arrive, *The Manchester Guardian* reported that a cub had been born on the 2-day journey. A second was presumed to have been eaten by the mother 'when strange circumstances make them over-fearful for the offspring's safety'. Only 61 out of the 70 expected arrived, apparently because the original figure had been an exaggeration (*Manchester Guardian* 1929). Wild animals were destined to return to Olympia the following year, and MacMichael claimed that Mills's agent 'specifically induced' MPs to block the House of Lords bill to prohibit certain wild animal acts in 1930. On 18 December 1930, he wrote to the Lord Mayor of London about

[13] Nevertheless, 600,000 people attended in spite of the absence of wild animals (Yeats-Brown 1930).

**Fig. 6.5** Hugh Cecil Lowther, fifth Earl of Lonsdale, by Bassano Ltd, 1928. © National Portrait Gallery, London. Ref. x85187. Reproduced with permission

the wild animal turns, including a tiger on horseback, being billed for Olympia, asking him to boycott the circus unless they were withdrawn and mentioning the large PADL protest meeting in Kensington Town Hall planned for 2 January 1931. On 31 December, Mills approached Lord Auckland about his speaking at this meeting and offered him first-hand evidence through an invitation to see accommodation and training at Olympia, unescorted with a season ticket. He added:

> I have refused to allow Mr MacMichael such facilities unless he pays for admission to the building, as he is a paid servant of a Society and not an independent person and because he has repeatedly made the mis-statement in public that nobody is allowed in buildings or training quarters to see the training...

...whereas local authorities and the police were. Mills wrote to MacMichael at Kensington Town Hall on 2 January 1931 about a letter from him and (absent) ticket inviting him to the meeting which he could not make. He enclosed one shilling so MacMichael could not again say he did not subscribe to PADL funds and also asked that if his letter were read out at the meeting, his other of 31 December to Auckland should be read out in full, too.[14]

On 13 December 1930 MacMichael had written to *The Spectator* that Lord Lonsdale could earn public gratitude by encouraging the trade to omit performing animals. This letter followed one from Lonsdale explaining his controversial position as both president of Bertram Mills Circus at Olympia and a vice president of the RSPCA (Fig. 6.5). He claimed to know the difference between 'kind' tricks and others and had accepted the presidency on condition he could vet ones proposed for the programme, as did Mills, who also hated cruelty. Lonsdale also said that:

> It is little realized that most of the lions and tigers in circuses are taken away from their mothers when but 3 or 4 days old and reared by spaniels or retrievers; that they have never

[14] At the meeting it was decided to demonstrate at Olympia on a secret date. It was also reported that MacMichael had written to the Prime Minister against his attendance at a wild animal circus, and also to the Home Secretary, who replied that MPs should learn first-hand of matters they had to deal with. A resolution was passed urging the government to enable passage of the current Performing Animals (Regulation) Amendment Bill (*Times* 1931).

> seen a jungle or forest; that they have all been bred as household pets as their mothers have been for most of their lives, and they are taught exactly the same as everyone teaches a puppy.

He added that he held no brief for local and travelling menageries, for he disliked them owing to their continual movements and enclosed cages. At the end of this letter, the editor of *The Spectator* announced: 'We appreciate Lord Lonsdale's temperate letter; but would make it clear that it does not move us an inch from our position with regard to the exhibition of wild animal turns in circuses. Such turns should be prohibited by law' (Lowther 1930). MacMichael also said in his reply: 'Lord Lonsdale's plea that it is all right if they have never known freedom must be rejected at the outset. Would he or anyone else care to be forced to remain in the room he was born in for the term of his natural (or, rather, unnatural) life?' (MacMichael 1930). He further pointed out that whilst Lord Lonsdale stated that he objected to travelling menageries, one was attached to this very circus (MacMichael 1931a). At this time, copies of a private letter from Stephen Coleridge and a reply from Lonsdale were also submitted to *The Spectator*. Coleridge had complained of his approval of an unnatural trick, the tiger on horseback, and the inconvenience of his publicly supporting both sides of a serious controversy, because the RSPCA was officially supporting the current bill to prohibit the exhibition of performing animals. Lonsdale responded: 'I did not realize that as a Vice-President I was supporting a Bill to prohibit performing animals in this country. Such was never my intention where such performances did not involve any cruelty'. He said his presidency at the circus was conducive to the interests of the RSPCA, his being a member who went into detail and saw that no cruelty was employed. He declined to vacate either of his positions. Meanwhile, the horse was miserable without the tiger, and 'what can be more interesting to children than to see two animals of different breeds having a friendship not caused by cruelty?' The RSPCA did not understand such acts, and they advertised against them to catch the eye of the public in the hope of gaining funds (Lowther 1931). A council member of the RSPCA then commented that the correspondence was entirely between Coleridge and Lonsdale, that the council had not requested Lord Lonsdale to resign the vice presidency and that membership in the RSPCA did not compel the adoption of any particular policy in regard to matters not yet included within the scope of animal protection acts (Ward 1931).

MacMichael wrote directly to Lonsdale on 26 June 1931 about the 'harm which your story of the tiger and the horse had caused'.[15] He referred to his booklet dealing with Lonsdale, showing his word could not be trusted and doubting his character, which a third party thought libellous and deserving of Lonsdale's

[15] Lonsdale had replied to a letter from Dr Agnes Estcourt-Oswald on 11 March 1931 saying that it was a lion, not a tiger, on horseback in Bertram Mills Circus. MacMichael wrote to Lonsdale from 30 March 1931 about inconsistencies of statements about this subject and appealed to him to publicly correct himself.

attention. He also mentioned the resolution he had put forward on 2 June 1931 for the agenda of the RSPCA Annual General Meeting:

> That in view of the cruelty inseparable from trained wild animal performances and travelling menageries, the Earl of Lonsdale, Vice-President of the Royal Society for the Prevention of Cruelty to Animals, be asked to sever his connection with any such shows in which he is interested.

The Resolution was 'irregularly omitted' from the agenda and when the Chairman later refused to allow him to move it, MacMichael left the meeting in protest because he was not allowed to speak. He advised Lonsdale:

> If you will devote the remainder of your days to undoing some of the harm you have done to the cause of granting justice and fair play to animals in those trades, professions and occupations where they are the victims of men's greed and cruelty, you will die – and you've got to die anyhow – a happy man, respected by everyone whose respect is of any value at all (MacMichael Papers).

When Stephen Coleridge, director and treasurer of the National Anti-Vivisection Society and editor of *The Animals' Defender and Zoophilist*, had written on 26 December 1930 to Lonsdale about the inconsistency of his two positions and asking him to abandon one, Lonsdale declined and replied 'If your Society [the RSPCA] wish to erase my name – well, there it is, and I shall bear no ill will'. MacMichael said that Lonsdale's dual role affected the work of the PADL because some in the RSPCA accepted his accounts of the circus and then criticised opponents of performing animals, as did the president of the West Caernarvonshire and Anglesey Branch in *The North Wales Chronicle* on 12 June 1931.

On 7 December 1931 MacMichael wrote again to the Lord Mayor of London about his opening of the circus at Olympia that would include 'wrestling tigers' and 'educated chimpanzees'. He cited Mitchell's evidence at the select committee against the use of apes. He said that his presence would be an embarrassment to school teachers who had to teach respect for authority whilst promoting fair play concerning animal turns, and he asked him to use his influence to get the Olympia animal turns omitted or to boycott them. In February 1934 when Mills stood for re-election to London County Council, in which he was chairman of the entertainments licensing committee, MacMichael distributed leaflets on his alleged circus cruelty. Mills achieved an injunction against them under the Municipal Elections Act the day before the election, but lost his seat. Then MacMichael entered a claim against him for £30,000, being prepared instead to accept a promise of no more performing animals in his circus. Bertram Mills died in 1938 with the case still undetermined (Fig. 6.6).

The campaign against the Mills family was continued after the war (Fig. 6.7). In 1951 MacMichael had published his *Murder at the Circus. An account and criticism of the trial for murder arising from the death of J. E. Newall at Bertram Mills Circus in Glasgow, September 1950*. He also wrote to the Earl of Huntingdon answering some points raised in a letter from Bertram Mills Circus, sent to influence the House of Lords debate in January 1956. An invitation to inspect training had been given, but MacMichael had proof that Cyril Mills would not

Fig. 6.6 Bertram Wagstaff Mills, by Howard Coster. © National Portrait Gallery, London. Ref. x2029. Reproduced with permission

Fig. 6.7 A large brown bear riding a motor cycle during a performance of the Bertram Mills Circus at Olympia in December 1951. © Illustrated London News Ltd/Mary Evans Picture Library

supply a circus ticket accompanied by a guarantee that there had never been any cruelty at any stage of an exhibited animal's training. In a general promotional pamphlet dated 12 December 1958, MacMichael referred to Edward VII calling Lord Lonsdale the 'biggest liar in Europe' and said that he was given legal advice not to apply for summons against Lonsdale and Mills on 1 December 1932, with proof of the 'cold-blooded torture of two horses in the show', because he would fail to get a conviction. In the pamphlet, MacMichael quoted Tertullian and suggested that C.B. Cochran, the theatrical manager and rodeo promoter, died as retribution for attending a Mills luncheon a couple of weeks previously in December 1950 (Fig. 6.8). Finally, on 25 February 1967 and when he was suffering from a long bout of bronchitis, MacMichael sent a memorandum from his hospital bed, not later acknowledged, to Sir Harmar Nicholls and seven MPs about the promotion of the bill that would enable Bertram Mills Circus to have a pitch in Hyde Park. He referred to 8,000 signatures on a protest form and asked that the bill be withdrawn. He announced that he had been arrested annually from 1932 when he leafleted outside Mills's luncheons with 12 sandwich-board men. He added that Mills, when a mortician, allied himself with Lord Lonsdale to resuscitate the performing

**Fig. 6.8** Cover of PADL pamphlet of 1958 criticising Mills's long-standing Olympia circus luncheons. MacMichael Papers

animals' 'ramp' after the select committee enquiries, which themselves resulted from the PADL's agitation.

In the 1950s, MacMichael began an equally bitter confrontation with Billy Smart's Circus. On 14 September 1955, he sent a telegram to the Commissioner of the Metropolitan Police demanding the cancellation of Billy Smart's Circus parade on the Queen's highway because as a circus with performing animals it was illegal.[16] The next day he asked the Commissioner if he had read the pamphlets he delivered soon after the telegram. If the parade were not cancelled, MacMichael would appear with a banner board. Then on 24 July 1958, the police removed MacMichael from Brighton Council Chamber for causing disorder when protesting at the letting of public land to Smart's Circus. On 7 August 1958, the Brighton

[16] MacMichael always contended that such events inevitably contravened the animal welfare legislation of 1900 and 1911.

*Evening Argus* reported the PADL's leafleting of traders to ban the advertising of Smart's Circus. The majority of shopkeepers allegedly ignored the leaflet. A circus spokesman said such leaflet distributions were quite common, and of this fact, when on 15 August MacMichael asked the Chief Constable of Brighton to make a surprise visit, he said 'E'en the ranks of Tuscany/Could scarce forbear to cheer!' (In the leaflet, a Jack London Club announcement now advised: 'The most effective form is a vocal protest from the auditorium – in other words, Boo it off the stage!')[17] In a statement to the police on 29 June 1959, he alleged that 5 days before he and two ladies carrying banners and leafleting in Leamington Spa against the letting of public land to Billy Smart's Circus were attacked by two men from it; and he complained to Crawley Police on 10 September 1963 that circus hands had spat at sandwich-board men. Smart allegedly resorted to this because he was not confident of legal action against the PADL. The clerk of the council concerned was an accessory because he had allowed the letting of the land.

On 7 April 1964, MacMichael wrote to the Chief Constable of Sheffield about a bear's attack on a horse at a televised showing of Smart's Circus. He noted that on 28 August 1960, a girl had been killed in another special performance for the BBC. He wanted proceedings to be instituted and details of registration, a vet's report, and information on whether the occurrence was on public or private land. On 17 April the police replied that they lacked information and witnesses, but understood that the registration was in order; the circus had been on public land. A *Daily Express* report of 5 November 1966 under the headline 'No chimps' announced that the Royal Albert Hall would not allow animal acts in Billy Smart's Circus Ball, held there for the National Society for Mentally Handicapped Children, but on 9 November MacMichael asked the Royal Albert Hall by telephone and letter still to cancel the contract. The following year he composed a memorandum, on 30 June 1967, 'to whom it may concern'. It stated that *The Portsmouth Evening News* refused to publish his notice of a threat to the local authority of a public denouncement if Billy Smart's Circus was allowed use of Southsea Common. He had responded: 'Failure to publish our notice will render you liable to a charge of misprision of felony (concealment of crime)'. He referred then to Smart's forthcoming trial on a drunk driving charge, when, as he later stated on 31 July 1967, MacMichael intended to be *amicus curiae* to draw attention to other crimes. In MacMichael's papers is a printed page, 'Elephant Hook and Goad', which describes how an active member of the League surreptitiously took this instrument from a German trainer ('in uniform – a man of average height, slightly built, fair hair, square face, thin lips, a German with a very hard, cruel face'): Susie the elephant was at Billy Smart's Circus on 8 July 1967, when the trainer allegedly attacked her ear offstage, making her scream.

[17] In 1945, the PADL began to compile a list of "active members" who would engage in such "inside protests" as the "most effective of all weapons at our disposal", and expenses could be paid (PADL 1945a).

Other circuses were targeted as well, with posters, leaflets and loudspeaker vans. In February 1944 MacMichael was in court again because of an action by Robert Fossett for defamation. On 22 October 1954 he informed the police that he had found 14 Fossetts in the Performing Animals Register, but none was registered to train the polar bears then being featured, and the registrations were probably out of date. The PADL publicised John Rankin MP's raising of Robert Brothers' cramped 300-mile lorry transport of five elephants to Glasgow in the House of Commons on 20 January 1965. On 22 November 1966 MacMichael told *The Bristol Evening Post* that their correspondent R. Walker of the ACP was wrong to say there had been no action for cruelty since before the Second World War. Thomas Fossett was convicted in 1943, after a 3-month case when MacMichael was consulted by the Stipendiary Magistrate at West London Police Court, Sir Gervais Rentoul KC, and performing dogs were produced in court. Two of these died because of epilepsy caused by their training (PADL 1945b). He asked *The Bristol Evening Post* to publish his 'Hard Facts' leaflet as an advertisement—it named all circus owners as liars and swindlers on the cruelty issue. On 28 November 1966 Bournemouth Times Newspapers wrote to MacMichael requesting he omit circus names from the advertisements he wanted to place in four papers. His response was 'NO! NOT ON YOUR LIFE', and whereupon his cheque for payment was returned, MacMichael threatened proceedings, and *The Bournemouth Times* declared that it would not be threatened.

The PADL had also remained active in variety theatres when opportunities arose. A veterinary surgeon, David Wileman, made a vocal protest at the Wood Green Empire on 21 August 1945, during a performing pony turn. At the Finsbury Park Empire, he launched handfuls of leaflets from the gallery on to the heads of the occupants of the stalls. Side shows at fairs suffered protests as when Miss M.P. Tate picketed a performing lion display at Wimbledon Common until the council removed it, but not before Miss Tate had been assaulted by its irate woman owner. The trade was described as equally active, following the PADL around, covering up or tearing down its posters. MacMichael therefore wanted 10-foot posters, and he appealed to wealthy members to fund the purchase of vans with loud speakers, amplifiers and microphones (PADL 1945a).

### 6.1.3 Relations with Sir Robert Gower and the RSPCA

As a campaigning organisation after the First World War, the RSPCA had entered a relatively quiet period in its history (Ryder 2000, p. 146), but there was new vitality in the issue of performing animals when in the late 1920s and in the 1930s, Lord Danesfort, its vice chairman, led moves to strengthen the legislation of 1925. In 1928, a letter to *The Times* from Robert Gower, chairman, Lord Danesfort and Captain E. G. Fairholme, chief secretary of the RSPCA, complained of an increase in numbers of performing animal turns in Britain and highlighted points made about training to a timetable and foreign training. They also complained of the practice of introducing domestic with wild animals in performances and appealed for a public boycott of a form of entertainment that was against public decency (*Times* 1928a). This followed a similar letter a year earlier from John Galsworthy, Lord Lambourne, Mitchell and others (*Times* 1927), and it was followed in a week by yet another in support, of the same kind and from the same RSPCA address, signed by Winifred Portland, Frances Evelyn Warwick, Archibald Fleming and Mitchell (*Times* 1928b).

MacMichael nevertheless became frustrated with the relatively moderate tactics used by the RSPCA in a variety of animal welfare issues. He wrote to A. Amos, a life member of the PADL, on 19 March 1931 about the need to get the RSPCA to spend about £2,000 against stag hunting whilst the bill against it was in the Commons and before the Field Sports Protection Society got 'really active'. He spoke of the RSPCA Council dragging its feet on hunting issues and not reflecting its membership; they would have to live it down 'when we get control'. At that time, many of the more prominent individuals in the animal welfare movement were associated with a number of societies. For example, in 1931, Gower was president and Bell was treasurer of the Pit Ponies' Protection Society, and as at 10 November 1933, Jeffrey Williams was its secretary, and MacMichael, Spurrier and Dr Agnes Estcourt-Oswald on the executive committee. Ernest Bell became a founding Honorary Treasurer of the new National Society for the Abolition of Cruel Sports in 1932, alongside other founders including Margaret Bradish, Henry Salt and Jessey Wade. However, there was frequent evidence of dissension between some of these individuals, a characteristic of animal welfare organisations that has persisted to the present day. In the matter of opposing animal performance, the major disagreement was that between MacMichael and Gower (Fig. 6.9).

MacMichael wrote to Sir Robert Gower on 23 June 1931 about Lonsdale's unacceptable conflict of interests, the validity of constitutional measures to address it within the RSPCA and Gower's unwillingness to affect Lonsdale's position. This was the start of extremely strained relations between MacMichael and Gower. MacMichael had requested an opportunity for Lonsdale to choose between his roles with the RSPCA and with Bertram Mills Circus rather than that Lonsdale should be removed from the RSPCA vice presidency. (Lonsdale had not responded

**Fig. 6.9** Sir Robert Vaughan Gower, by Bassano Ltd, 1928. © National Portrait Gallery, London. Ref. x124147. Reproduced with permission

to MacMichael's request to use his influence to remove wild animal turns from the circus.) He said to Gower:

> To have as a Vice-President a man (especially a man in his position in Society) who is not only interested in performing animal turns, but who is leading astray prominent people inside the Society as well as members of the general public during this particularly stiff battle to end these cruel exhibitions, is a state of affairs which no serious Society can tolerate for a moment longer than is necessary.

He concluded that he would raise the matter at the RSPCA Annual General Meeting. He was subsequently prevented from doing so by Gower.

On 3 July 1931, MacMichael wrote to Miss Warner of the committee of the RSPCA Members Association saying that the AGM had been illegal because of what had happened and that another one was needed. Gower had been in the wrong, but 'the unrestrained emotions of a few Members gave him [Gower] an excuse for some of his actions', and it was very important for members not to forget themselves and so damage the cause. On 14 July 1931 he wrote to Stephen Coleridge, editor of *The Animals' Defender and Zoophilist*, about his article on page 21 of the July issue on the RSPCA AGM and its wrecking. MacMichael said it was not just some members' fault: Gower had acted unconstitutionally and unfairly in attempting to manipulate the election of members of Council and without first allowing discussion of the Report and without MacMichael's resolution having been put on the agenda.

The RSPCA had indeed planned a winter campaign in 1931 against circuses, menageries and trained animal turns. Its Policy and Publications Committee called on 6 November 1931 for a ban on foreign-trained animals and quarantine exemption and on wild animal performances, the creation of an organised Jack London-type campaign, the lobbying of the National Union of Teachers at their Congress through the supply of an invited speaker and a request to clergy again not to take children to circuses. These requests were approved by Council on 19 November. However, the cautious approach of the RSPCA under Gower provoked MacMichael. *The Sussex County Herald* of 16 April 1932 under the heading 'Cruelty to animal entertainers' reported that at the RSPCA Annual Meeting of the Mid-Sussex and Eastbourne Branch, Sir Robert Gower said that exaggeration did harm to a good cause, although the large majority of trained animals suffered. On 25 April 1932, MacMichael wrote to Gower saying that there was no exaggeration and asking who he thought had exaggerated and when, and could he give examples of animal turns without suffering. Gower replied in a letter of 29 April 1932 saying he had been slightly misreported. He would not name names because that would be disruptive. He did not want a 'personal and possibly acrimonious controversy' (presumably with MacMichael). On 2 May 1932 MacMichael wrote to Gower saying he had been quoted in inverted commas, and he had asked the editor of *The Sussex County Herald* for a full record. Gower's anonymous accusations were very serious. MacMichael knew more about the performing animal problem than he, he said, and no one had exaggerated; rather, he had minimised it. MacMichael wanted evidence of exaggeration and non-cruel acts. Gower's statements were 'calculated to do untold harm to the Cause'. Gower replied on 4 May 1932 that he had nothing to add, but had always advocated prohibition and asked to end their bickerings. On 11 May 1932, MacMichael replied that there was therefore no available evidence for his claiming that some acts were not cruel or for exaggeration. He advised a campaign for legislation against wild animal performances as in the bill passed in the House of Lords in 1930, as a first step towards prohibition. 'The Constitution is wittingly and unwittingly being brought into disrespect by persons and bodies in authority who countenance and encourage wild animal performances'.

In what may have been a standard letter to newspaper editors of 29 June 1932, MacMichael gave an account of what had happened at the 1932 RSPCA Annual Meeting that day. He had used the rules to attempt to replace as chairman Gower, who illegitimately moved it out of order. MacMichael thought that the chairman of the Annual Meeting should not be the chairman of the RSPCA Council, because he would be biased towards the Council's views. The Council's actions were supposed to be scrutinised at Annual Meetings, and MacMichael had wanted to do this; also, his motion to stop cruelty in the production of animal films was thwarted by Gower. In a second standard letter of 30 June 1932, he maintained that the 1932 Annual Meeting and its decisions were therefore invalid. Gower had threatened to eject him from it. On the same day he wrote to Lt Colonel Lord Stamfordham, Private Secretary to the King, who was Chief Patron of the RSPCA, and to Sir Godfrey Thomas, Private Secretary to the Prince of Wales, who was President of the RSPCA, asking for the reconvening of the Annual Meeting with the Prince of Wales as chairman. In another letter, he offered to meet Thomas.

In a letter prepared for *The Times* of 3 July 1932, he referred to Stephen Coleridge's mistaken account of the meeting, which he had not attended, and his mistaken assertion that he was antagonistic to Gower for personal reasons. The letter was not published, whereupon MacMichael drafted a letter of correction that he sent to Coleridge in the hope that he would send it to *The Times* to qualify his earlier one. In a letter composed on 3 July 1932, MacMichael explained to Gower why he had refused to shake hands with him after the previous Wednesday's meeting:

> No earthly power is going to make me shake the hand of a man or woman which I know to be dripping, as yours is dripping, with the blood of these helpless victims whom you have betrayed, and this also applies to those whom I know to be conscious followers of yours... treachery I cannot abide... If you want to escape it, there is only one way that I can think of: 1. You must make a public confession of your treachery; this you must do NOW. 2. You should make a private, and where necessary a confidential, confession to me of all the facts of this unsavoury business of cruelty on the Screen... 3. Then you must leave the Movement. 4. Then I shall ask to shake your hand, and shall invite you to return [to the struggle against cruelty]

A 'top copy' of this letter remains in MacMichael's papers, suggesting it might not have been sent.

A motion from the Rev. T.J. Bass of the RSPCA Council was prepared for its meeting on 21 July to expel MacMichael from membership of the RSPCA, because his conduct on 29 June had been prejudicial to the interests of the Society. MacMichael then asked Sir Godfrey Thomas to discover the views of the Prince of Wales, but without success. MacMichael, Sir James Douglas and Alasdair Alpin MacGregor (president of the League for Prohibition of Cruel Sports)[18] sent a letter on 18 July 1932, not later acknowledged, to the Chief Secretary and Council of the RSPCA, describing how new rules had not been followed by Gower on 29 June 1932 regarding the decision on the chairmanship of the Annual Meeting, which was illegal and should be reheld. These points should be considered before the proposal for MacMichael's expulsion. On 22 September 1932, MacMichael wrote to Dr Agnes Estcourt-Oswald, a friend on the RSPCA Council, that the rules were not followed at the Annual Meeting and that 'the RSPCA must be cleansed'.

The following year, on 15 June 1933, MacMichael wrote to the Rev. Basil G. Bourchier, president of ODFL, about the danger to it from the RSPCA and the need to keep Gower and his associates off its council. Then on 27 June, J.M. Coubrough, a member, compiled a circular letter (perhaps drafted with the help of MacMichael) to all other members of the RSPCA, expressing dissatisfaction with the expulsion of MacMichael, MacGregor and Douglas by the Council of the RSPCA sitting in camera and saying he had therefore sent a resolution to the 1933 Annual Meeting demanding an extraordinary general meeting with an impartial chairman to decide on the allegations of the three about the conduct of the society and whether to demand a public enquiry into its affairs and administration. This resolution was not on the agenda, but Coubrough would present it. The Chief Secretary had said it was refused because of the rules. Coubrough said that those expelled should be heard at an extraordinary general meeting, with reference also to MacMichael's charges against the chairman. A letter to MacMichael of 2 July 1933 from 'Michael' declared: 'Gower really is a cesspit of hypocrisy'.

On 5 July 1933 Douglas wrote to the editor of *The Sunday Referee* about its article of 2 July concerning the RSPCA's 'hush hush' meeting to decide on the expulsions. He said a public enquiry was the only way to ensure public confidence. MacMichael approached Frederick Messer (who was re-elected to Parliament in 1935, having lost his seat in 1931) on 10 July 1933, asking him to chair a public meeting held by MacMichael, MacGregor and Douglas at Caxton Hall on 21 July about whether to demand a public enquiry into the administration of the RSPCA. On 17 July, Coubrough offered to be chairman, and the meeting was held successfully.[19] MacMichael wrote to MacGregor and Douglas on 12 July 1933 noting that the International League against the Export of Horses for Butchery had had similar problems with the RSPCA. He had just consulted the editor of the *Referee* about its championing the case, and they discussed an advertisement in the *Referee* and other papers appealing to RSPCA members to write to headquarters asking that its funds

[18] In 1931, MacGregor, the Scottish author and photographer, had applied for summons at Bow Street Magistrates' Court against Gower and Fairholme for alleged assault at the RSPCA annual meeting that year.

[19] By August 1936, J.M. Coubrough was treasurer of the PADL.

should not be used in any legal support of the Council. In February 1934, MacMichael accepted nomination for the committee of the League for Prohibition of Cruel Sports, of whose members most wanted to see the RSPCA trouble cleared up. Then divisions occurred in the League because the secretary was alleged to have had secret correspondence with the British Field Sports Society. MacMichael suggested the secretary should offer to resign, but he refused. The problem became a tussle between the secretary, J.C. Sharp, and MacGregor, who left the presidency after being excluded from the committee. MacMichael was asked to support MacGregor, but hesitated because he thought he had acted against his own advice in the matter.

In spite of pressure from MacMichael and the PADL, the RSPCA continued to deal with the performing animals controversy in its own way, benefiting from its prestigious status and from its representation in Parliament in the form of Gower and Danesfort. As a result of Lord Danesfort's resolution, the RSPCA Council confirmed, through the Performing Animals ad hoc Committee (which first met on 18 January 1934 and became soon after the Performing and Captive Animals ad hoc Committee, then the Performing and Captive Animals Advisory Committee), Society policy 'to prohibit all animal performances which involve cruelty to the animals in the training, confinement or the exhibition' (RSPCA Archives).[20] This committee resolved on 17 July 1935 to lobby educational authorities against children's attendance at animal performances. On 22 July 1937, the Council approved a special subcommittee to assess what information was available about performing animals and what evidence was available to support the Performing Animals (Regulation) Amendment Bill (as promoted by ODFL), which sought the prohibition of the training and exhibition of big cats, hyenas, bears and anthropoid apes. The Special Committee on Performing and Captive Animals reported on 30 September 1937 that the campaign was founded . . .

> not so much upon the direct proof of cruelty ... but upon the basis that the public performance of wild and other animals for monetary gain, and the confinement in Menageries of such animals, is, *ipso facto*, cruelty.

To support the campaign, the need was recognised to get proof at home and from abroad, to estimate the significant expense of such a campaign and to appoint an organiser. The issue of performing animals should be kept separate from that of menageries in order to avoid confusion, but both issues should benefit from a joint campaign. It was said of performing animals:

> We do not possess very much more evidence in 1937 than we did in 1921–22 . . . it would therefore appear necessary to give greatest prominence to the ethical side, at any rate in preliminary propaganda. Apparently the only way to obtain full and definite evidence of cruelty in the training (where it mainly occurs) is to employ investigators and send them to the places where the training actually goes on [abroad].

[20] Meanwhile Danesfort had confirmed to Parliament that he had nothing to do with the PADL (*Parliamentary Debates* (*Lords*) 1933, vol. 87, column 743).

This was understood to be very costly and perhaps too difficult. 'Stratified' appeals were to be made to intellectuals, professional classes, working class movements, educational authorities, the stage and various national organisations; and use made of special meetings and stalls, with special photographs:

> showing zoological and biological aspects of the evil, ... [and] performing animal outfits such as clothes (with their waterproof linings) and the gadgets used by a troupe of performing dogs, monkeys, etc.

Lobbying would include the police, constituency petitions to MPs, playlets with a message, the wireless, posters, lunches with talks, letter stamps, a petition and press publicity. It was estimated that this would mean circulation of at least 300,000 appeals involving about £1,900 in postage and £1,200 in printing, £1,650 for stalls at £10 each and over £10,000 in advertising and general publicity (RSPCA Archives).

Looking back in 1967 at his relationship with the RSPCA, in a letter of 6 April, MacMichael wrote to a Miss Sleigh, '... The RSPCA have never run a straight and honest course with regard to the big organised cruelties – and we have suffered more than any other Society in this respect'. Elsewhere he wrote that the PADL had done 99 % of the work, whilst others had reaped the credit. He observed:

> From the dawn of history until now, and from now until the end of time, there has never been, is not, nor ever will nor can be a performing animal act that has not involved gross brutality, coldblooded cruelty or sadistic torture in the misnamed 'training'.

### 6.1.4 Problems with the Establishment

Throughout his campaigning career, and as a result not just of his work for the PADL but also because of his philosophical beliefs, MacMichael came into continuous conflict with established institutions in the form of central government, local authorities, the police, the courts, radio and television. On 30 January 1936 he sent a petition to King Edward VIII, in which it was said that George V allowed no performing animals in Command Performances. The petition called for prohibition with compensation—a delay would tempt Providence at a time of international tension. The PADL leaflet describing the methods of training a dog to balance on one front leg was enclosed. MacMichael later complained that the petition took 3 months to get from Whitehall to Buckingham Palace and failed. On 22 December 1941, he wrote to the Under Secretary of State at the Home Office regretting the failure of all his four petitions, the first two concerning performing animals and animals in films legislation (favourably commented on in *To-day's Cinema* on 29 July 1937). The third petition had advised the course of action to be taken after Dunkirk (to 'undeclare' war). The letter is a long complaint with ancient historical references and attacks on public figures, especially Herbert Morrison whom he suspected of frustrating his petitions because of resentment at MacMichael's attack on other public figures well fed by Bertram Mills and because

**Fig. 6.10** MacMichael leading a poster parade with sandwich-board men outside the Home Office in Whitehall on 5 July 1943, in connection with the Ministry of Home Security Holidays at Home Scheme, under which local authorities could officially engage shows containing performing animals. From PADL *Newsletter* no. 27, winter 1945. MacMichael Papers

he was unrepentant that retribution in the form of the war had resulted from cruelty to performing and other animals. He also began to circulate propaganda against Arthur Harris's 'mass bombing'. He was in touch with the Peace Pledge Union and Prisoners' Aid Society, writing to prisoners and political internees, and he offered one of these a job as a sandwich-board man for his 'poster parades' against animal performance (Fig. 6.10).

His papers include a transcript of MacMichael v Home Office (1957, No. 2655). In a Notice of Motion, MacMichael as plaintiff (and bearing the costs of the application) and the PADL on 3 December 1957 began an action in the High Court of Justice, Chancery Division, against the Home Office as defendant to restrain it from permitting a performance by one or more performing animals at Wandsworth Prison on or about 4 December 1957. On 23 October 1967 he wrote to the Private Secretary of the Queen, Sir Michael Adeane, objecting to the plan to include a performing animal act in the Royal Command Performance on 13 November 1967:

> To include one in a COMMAND performance is to put the clock back 46 years at the very least . . . and here one must take refuge in the second verse of the British National Anthem – 'Confound their politics, frustrate their knavish tricks' and in Virgil's Aeneid III 628 – 'Haud impune quidem; nec talia passus Ulysses ['But not without punishment; Ulysses did not tolerate such [barbarity] . . .'] – etc.

He ended by proposing a personal visit 2 days later at 3 pm.

In a letter of 10 June 1937 following a dispute with the clerk of the council in Letchworth over an episode involving staff and police at the swimming baths, MacMichael admitted he reacted to the clerk's communication too strongly: 'What, in short, I rashly took to be the adder of injustice and prejudice is, I am relieved to find, but the grass snake of benevolent subtlety'. In 1945 the League successfully persuaded Leicester City Transport to prohibit cards in passenger vehicles advertising circuses (PADL 1945a). But his disagreements with local authorities were later centred on the letting of public land for circuses. In MacMichael v. Salmon and Morrison in 1943, MacMichael took court action against the clerk to London County Council with whose permission Sanger's circus was held on its land and against the Home Secretary for allowing laws against cruelty to be flouted, but the case was only partly heard because of lack of evidence (PADL 1945b). On 30 September 1965, he published an appeal to the ratepayers of Guildford to withhold one shilling from their rates payment because the council had let public land to Billy Smart's Circus in 1963 and to Chipperfield's circus in 1964, and therefore, the council was liable for the cost of his poster parade, nearly £100. On 2 June 1966, he wrote to Godstone Rural District Council against the letting of land to Walter Luckens, someone said to be notorious for cruelty, on 6 June, and complained that League posters had been removed from 20 trees—he wanted any information on the culprits in order to recover costs. Then on 28 September 1966 a public demonstration was held in Horsham Park and a call made for a resolution to Horsham Urban District Council against the letting of public land for circuses, following newspaper advertisements and comment. 'This foul abuse of God's animals [was] for the monetary profit of rogues' (Chipperfield, Fossett, Mills, Robert Brothers and Smart). On 28 July 1967, he asked Bournemouth council to deny land to Billy Smart, but failed, and the most recent letter in MacMichael's papers was sent on 16 March 1968 to a League member, Miss Francis, about the task of denying Hoffman's Circus public land.

Because he was prepared to place his moral conscience above laws that he felt unjust, and then to ignore or challenge them without compromise, it was inevitable that MacMichael came into frequent contact with the police and courts. In 1940 there was a High Court case between MacMichael (plaintiff, self-represented) and the Commissioner of Police of the Metropolis (defendant). Ironically, he regarded his direct action as a form of voluntary police work, because he felt that the proper authorities were not enforcing the inadequate laws that already existed. Much of MacMichael's public activity involved demonstrations, and the police often became involved as a result of their arrangement or consequences. The Commissioner of Police of the Metropolis had told MacMichael on 13 September 1933 that meetings were not allowed outside Labour Exchanges because of past disorders. On 29 April 1954, MacMichael wrote to the Chief Constable of Leicester about his plans for a poster parade along the route to be followed by a circus and that any regulation affecting the distribution of printed matter should be waived because the streets would be used to advertise an illegal show. On 10 November 1955, he complained to Hackney Police about Roberts and his wife who allegedly threatened the manager of Superboardmen Limited. They had a performing animal show at the Hackney Empire. Police had advised the sandwich-board men to keep away because they were aggravating the situation. MacMichael asked why the Roberts were not arrested, because, being based on cruelty, animal performance was illegal.

On 7 January 1960, he requested the Chief Constable of Hertfordshire to charge Marcus La Touche, whose dogs had died in a van blaze, for out-of-date registration, and on the same day he wrote to the Commissioner of the Metropolitan Police urging him to take action on the rodeo film 'Rin-Tin-Tin' on Children's Hour on Independent Television. The Chief Constable of Brighton received a letter on 10 July 1960 about the League's 'sandwich-board parade' against the letting of public land to the 'illegal' Chipperfield's circus. Two policemen wrongly stopped it and names of the sandwich-board men were taken, when the PADL was doing a police job. The Chief Constable replied referring to the provisions of the Prevention and Obstruction (Sandwich Board, Costume and Animal) Order of 1934, but that he had decided not to charge. MacMichael then asked for a summons against himself or a cheque in compensation for lost time on parade. All this was reported in *The Evening Argus*. A leaflet about Chipperfield was soon forthcoming from the League: 'These Scoundrels'.[21] On 12 July 1961, he wrote to Oxted Police, asking them to prohibit a parade of Fossett's elephants (and if not, to have a vet inspect them for injuries) and pointing out the illegality of using public land.

MacMichael wrote on 10 and 28 August 1961 to the Commissioner of the Metropolitan Police about Augusto's (alias 'Captain Cook' or Ellis Cook) Bijou Miniature Circus at Fulham, and of a complaint about the stench in a small unventilated van, used to transport performing dogs from Sunderland and back. A prosecution was needed under the Protection of Animals Act 1911. Augusto's registration was allegedly also out of date. He encouraged Hemel Hempstead Police on 4 December 1961 to charge Mrs Mavis Horsley, a dog trainer, and Count Andre Lazard, an exhibitor, because of cruelty in June and because they were not registered. The case was heard on 14 March 1962, and Horsley was fined. MacMichael said both committed perjury to save Lazard, who had a previous conviction for cruelty; therefore, further proceedings were necessary. On 12 November 1964, a message was sent to Brooke Bond:

> The Performing Animals' Defence League denounces your 'free' circus stunt as mischievous and corrupt. Bribing children to enjoy the fruits of secret cruelty is vicious, depraved and detrimental to character-training and education. Call it off!

MacMichael asked the police to ban it. On 25 March 1966 he informed the Commissioner of the Metropolitan Police that concerning the Brooke Bond chimpanzee at Safeways supermarket, neither Brooke Bond nor Safeways had registration, and he called for a plain-clothes visit of inspection.

[21] In spite of serious ongoing hostilities, MacMichael and Dick Chipperfield discussed the question of cruelty in training for the BBC's 'Round-up' radio programme, a broadcast on 28 July 1961. The BBC Programme Catalogue records that 'No holds are barred in this discussion which generates much heat but no light at all. No conclusions are reached'.

A telegram was sent on 21 June 1967 to Worthing Police, reporting that Hoffman and Timmis of Hoffman's Circus were not registered and that a £50 fine per animal was needed. On 8 July 1967 he asked the Chief Constable of Portsmouth under which Act of Parliament Jean Pyke of the South West Branch of the PADL might have been 'in danger of arrest' for trying to hand a leaflet to the Mayor. Three days later he complained that his question was passed for attention to a subordinate (noting also that circus rogues are the 'veritable scum of the earth'). A reply of 21 July from the Deputy Chief Constable of Hampshire Constabulary mentioned police reorganisation as the reason that the related newspaper report was incorrect and that the police warning had been to do with possible obstruction. On 24 July, MacMichael replied that if he had known the police were on duty at the circus before it left, he would have asked for their intervention on behalf of Susie the elephant, mistreated with a disguised hook on the end of a pole.

Between 1934 and 1942, MacMichael met 27 judges and a host of magistrates, sometimes for affixing strips to circus posters. In 1937 Frederick Messer asked the Home Secretary whether he was aware that MacMichael had been committed to prison in consequence of his activities in defence of animals, and when he was likely to be released, as he had not been sentenced to any specific term of imprisonment. Messer was told he would remain in prison till he purged his contempt of court.[22] A few months later Messer asked the Home Secretary what was the nature of the offence or offences alleged to have been committed by MacMichael, who had been sent to prison for an undefined term at Luton on 14 January that year, why he was sent to three different prisons and whether his release on 14 March terminated his period of punishment. Sir Samuel Hoare replied that the contempt of court consisted in his having failed to make out and file with the Official Receiver in Bankruptcy a statement of his affairs in the prescribed form and having avoided examination in respect of his affairs. He was committed to Bedford Prison, but was later, for medical reasons, transferred to a London prison; he was accordingly transferred on 21 January to Wandsworth Prison and on the following day to Brixton Prison where he remained until his release by order of the Judge on 4 March. His release terminated his punishment for the contempt of court for which he was committed to prison.[23]

In MacMichael's papers is a transcript of proceedings at the Magistrates' Court in Sutton on 5 August 1943, Rex v MacMichael. The prisoner was undefended and charged with fly posting on private property in Sutton, obstructing a police officer (by ignoring an instruction to desist) and failing to produce an identity card. He pleaded not guilty to damage but pleaded justification in answer to the charge of posting bills without consent. Counsel for the prosecution stated that the illegality of the posting rather than his or the court's opinion of the contents of the posters was the issue: he was not influenced by their contents. (MacMichael wrote 'nonsense' and 'rot' in the margins of the transcript.) The police constable reported that he had

[22] *Parliamentary Debates* (*Commons*) 1937, vol. 319, columns 1760 and 1761.

[23] *Parliamentary Debates* (*Commons*) 1937, vol. 324, column 675.

'a brown suitcase containing posters, which was open on the pavement, and a pot of paste'. MacMichael replied:

> I was entirely justified, because if Parliament holds two enquiries into a form of cruelty, and a local authority supports that form of cruelty, it is the duty of the citizen to protest . . . If I was attacking crime, and I was obstructing you in what you call your duty, you were protecting crime? . . . if any police officer is on the side of cruelty it is quite wrong.

Of 'Sir' Robert Fossett's circus, the subject of the posters, he said:

> Where you have someone sailing under false colours, using a title to which they have no right, and under cover of that title carrying on the nefarious practice of cruelty, they are deceiving the public and getting money by fraud.

MacMichael thought that the importance of the contents of the poster justified its posting, but the court did not. He argued with the chairman and said he would appeal to a higher court. (He noted in the margin of the transcript: 'I maintain that everything that followed is out of order, the bench being judge in its own cause'.) He then refused to cooperate in the rest of the court hearing. He was fined £1 for posting, 10 shillings for obstruction and £5 for failing to produce an identity card.

On 23 September 1958, MacMichael received a summons for non-payment of tax, and on 17 April 1959, he disputed a summons to Marlborough Street Magistrates' Court about non-payment of tax because he believed the nation was in debt to him, presumably for his voluntary police work. He would attend the following Monday, 'but my presence is needed in the High Court in the 'Nino' case' (about a performing dog at the Palladium in 1955):

> Like myself, he has made Legal History, for he has attracted the attention of six Lords Justices of Appeal, two High Court Judges, four Magistrates, six Masters and 'a great many more of lesser degree'.

He was later threatened with arrest. On 26 October 1959, MacMichael returned an income tax summons to Marlborough Street Magistrates' Court because it was signed by someone who had attended a promotional circus lunch. When threatened with arrest, he included in his reply an elegiac verse: 'Perturbabantur Constantinopolitani/Innumerabilibus solicitudinibus' ['The people of Constantinople must have been thrown into confusion with countless worries']. He was later imprisoned at Brixton and then attempted to tell the Governor the imprisonment was unlawful. On 30 September 1960, he refused a summons from Marlborough Street Magistrates' Court because of his outstanding claim on public funds for false imprisonment, arrests as well as his own police work. He was imprisoned in January 1961 for 19 days for contempt of court. Writing on 28 August 1961 to the Collector of Taxes about an application for summons, he referred to this false imprisonment in January 1961:

> It was not I who was the contemnor. It is altogether too reminiscent of Virgil's – 'Hos ego versiculos feci – tulit alter "honores"' ['I composed these lines – another has taken the credit'] – as coming, of course, from the lips of the real culprit.

On 26 February 1962 he informed the Collector of Taxes that he would not attend court that day because he would not admit the legality of any income tax proceedings until he got compensation per capita of the nation for his false imprisonment. He also contested in court rates demands from Westminster City Council, for example, in a letter to Bow Street Magistrates' Court of 12 September 1962. He felt that he should not have to pay rates to pursue his just cause, especially because (as he said in a letter of 20 January 1963) the League had incurred costs in opposing London County Council's illegal staging of a 'cruel performing animal show' in Battersea Park in summer 1962. But he was forced to pay up when a bailiff was sent. The rates applied to the League's two offices on the first floor (north) in Buckingham Chambers at 11 Buckingham Street.

On 24 December 1931, MacMichael wrote to Dick Sheppard, 'the radio parson', who had signed a PADL petition against wild animal turns, asking if he could mention the subject in his Sunday broadcast: '"Be, buy and employ British" blare forth from press, pulpit etc., but not as regards animal turns practically all of which come from abroad'.[24] He complained on 18 September 1936 to Sir John Reith about coverage in the *Radio Times* of a talk for children about the rodeo. He also asked for air time for the PADL, as so far only its opponents had been favoured. If not,

> One day you will have to render an account for your stewardship of the microphone . . . The power of the Radio was not given to be used in the service of the Devil.

MacMichael sent his wireless licence fee to the Head Postmaster at Hitchin on 9 December 1938, informing him of the sustained pro-cruelty performing animals propaganda on the wireless and the BBC's failure to give him airtime. And when the trainer Marcus La Touche appeared on the radio on 4 December 1942 to say in a children's programme that cruelty in training was not necessary, MacMichael protested to the BBC, and through the Postmaster General unsuccessfully asked for a broadcast debate with him (PADL 1945b). On 7 May 1963 he asked magistrates to take action following his call to the BBC to take a cockfight off Cliff Michelmore's television programme, referring to the Town Police Clauses Act 1847. Two days later he complained to magistrates about their lack of action and told them a story of similar inaction: he had entered the ring at a miniature circus at Gamages in the 1930s to remonstrate with a young woman with a performing dog, when a strongman ejected him—the audience thought it was a funny part of the act. The magistrate on that occasion said nothing was wrong.

He notified Marlborough Street Magistrates' Court on 12 March 1965 of his intention to apply for a summons against Mrs Alma Mae Rizza and to produce for inspection in court her eight performing dogs used in the performance of 'Excess Baggage', televised for 'Sunday Night at the Palladium' on 21 February 1965. In a memorandum of 3 April 1964 to Bernard Delfont and Stewart Cruickshank, he had quoted Bernard Levin in *The Daily Mail* of 1 April 1964 on the 'Excess Baggage'

[24] Canon H.R.L. (Dick) Sheppard, a pacifist, Vicar of St Martin-in-the-Fields and Dean of Canterbury, became especially well known for his sermons broadcast by the BBC from 1924.

act in 'Round About Piccadilly' in their production at the Prince of Wales Theatre: 'A dog act even more revolting than most dog acts'. MacMichael had threatened audible protest if 'Excess Baggage' appeared, and if he were hindered or molested, he would apply for a warrant for Delfont's, Cruickshank's and any others' arrest.

Miss Kathleen Haacke, Head of the Programme Correspondence Section of the BBC, received a blistering rebuke on 30 March 1967 about her claim that there was no cruelty to chicks in a conjuring act on the Rolf Harris Show:

> You yourself have been at this game of whitewashing diabolical cruelty for so many years that the attributes of imagination, compassion and even truthfulness itself are ... now foreign to your make-up.

MacMichael said he would have slapped producer Stewart Morris's face. On 6 August 1967 he wrote to Kenneth Robinson, of BBC's *Points of View*, asking if there were no members of staff who would go on strike against the 'pro-cruelty policy of the BBC'. He referred not just to performing animals, but also to a documentary showing animals being unnecessarily used in tobacco tests.

## 6.2 Performing Animals in Film

The use of animals in film from the early twentieth century served both dramatic entertainment and advertising, and animals also began later to play a sometimes contrived part in wildlife and educational documentaries. The likelihood of unpleasant exploitation was from the start made more certain by the commercial imperative: 'Early silent films showed confined crocodiles being slaughtered by gunfire, horses plunging over precipices or being felled by tripwires at full gallop, and lions and tigers in mortal combat' (Ryder 2000, p. 133). The new cinema audiences expressed conflicting needs of excitement spiced with violence and humane treatment of animal participants. John Galsworthy wanted to see 'the animal show' for stage and movie abolished. He said it was 'too ironical altogether that our love of animals should make us tolerate, and even enjoy, what our common sense, when we let it loose, tells us must in the main spell misery for the creatures we profess to love' (quoted in *Literary Digest* 1920). But 'watching motion pictures quickly became one of the nation's favorite pastimes, and as Hollywood brought more and more stories to the screen, some of them inevitably involved animals' (Beers 2006, p. 104). Soon, the charges of cruelty began to appear as frequently as those directed at the circus, music hall and vaudeville: 'The cruelties that have evidently been practised by many people in the production of moving pictures are of a similar nature to those connected with the training of trick animals for the stage and the circus, and which ultimately in all civilized countries will be abolished' (*Literary Digest* 1920, quoting Dr Francis H. Rowley, president of the Massachusetts SPCA).

The RSPCA reminded readers of its *Animal World* that some cinema films featuring animals should also be rigorously boycotted according to the Jack

London Club pledge; however, it was stated that 'No unpleasant notoriety or aggressive propaganda is desired' (*Animal World* 1921, p. v). In 1921, the Jack London Club of Boston had urged the public also to leave when 'pictures of these [trained animal] acts are thrown on the movie screen'. It added that 'keeping animal life caged, and making it perform for profits is an inheritance from man's jungle state. Like all evils, it exists only by reason of the people who patronize it. They, not the trainers, are the responsible ones' (Joys 2011, p. 286). During the 1920s, the film and cinema industry overtook the music hall as a focus of attention, alongside the circus, for the PADL. But when the select committee met in 1921 and 1922, there had been almost no reference to the use of animals in film production. In the decades to come, 'Twentieth-century cinema and television took advantage of the capacity of trained animal performers from the circus in order to create the romance of animals in a natural setting' and in order to serve other purposes (Tait 2012, pp. 166–167). Unfortunately, this included staged fights and killings.

Francis Rowley saw on the cinema screen a leopard or a lioness writhing in agony after being shot (*Literary Digest* 1920). In the film *The big cage* (1933), a tiger was killed in a staged fight with a lion. 'In one report the lion killed the tiger so quickly, the cameramen could not get a good shot of the action. [Clyde] Beatty contended there "wasn't supposed to be any killing—just a fight." But he had to use three tigers before the scene could be properly photographed' (Joys 1983, p. 130; Joys 2011, p. 362). The 8-min lion and tiger battle in *The big cage* was much appreciated by the editorial staff of New York's *Daily News*, who described it as the most exciting they had seen in current movies: 'Nowadays we can see these magnificent fights, which cost the Roman emperors so much, for 50 cents. That certainly looks like progress. Let's have more and better animal fights in the movies'. Despite the disapproval of 'animalarians', it was animal nature to fight, and 'there isn't a more terrific and satisfying spectacle than a fight' (*Daily News* 1933, cited by Joys 2011, pp. 360–361).

In arranging staged fights for the film *Wild cargo* (1934, directed by Armand Denis), 'Frank Buck placed disparate species in small enclosures and then provoked them to fight for the filmmakers' (Joys 2011, pp. 269–270). According to Derek Bouse, this was the 'worst kind of voyeuristic animal pornography'. Denis recalled that Buck only had a penchant for sensationalism and was totally insensitive to animals. When he asked Denis what he thought about a possible fight to the death between a tiger and an orangutan, Denis argued it would be unrealistic. Buck, however, insisted that when he was around the two species would fight. Bouse also noted that the film pioneer, Muybridge, had sacrificed a tied-up buffalo to a tiger, and even Raymond Ditmars (curator at the New York Zoological Park) had staged fights, as did the British educational series, *Secrets of nature*, in which the insistent theme was that the rule of the animal world was to kill or be killed. And Disney's post-World War II, *True life adventure* series, could, like Buck's films, be sensational, condescending and cruel (Bouse 2000, pp. 54–56, cited by Joys 2011, pp. 269–270). But in spite of the continuing instances of cruelty in the production of a variety of film genres, there was enough criticism to affect the outlook of the film companies and to bring about the 'industry's new tendency to look at animals through the eyes of the SPCA and St Francis of Assisi' (Furnas 1933, cited by Joys 2011, p. 361). In December 1933, MacMichael wrote to the wife of a relative who was about to become governor of the British colony of Tanganyika and hoped he would keep an eye on animal film producers at work there.[25]

Burt suggests that in films 'the suspension of belief that is normally in play with regard to humans on screen does not work in the same way for animals': the audience is less prepared to believe that cruelty or suffering is faked (Burt 2002, pp. 137 and 138). This was particularly true of children, and one argument in the critics' armoury was the effect of such films on them. Francis Rowley complained that children were considered a valuable part of the market and 'what the effect of such scenes can be upon growing children in the way of deadening their finer sensibilities only those know who are familiar with the forces that shape the life of the average child' (*Literary Digest* 1920). According to the PADL, films involving or suggesting animal cruelty would 'either arouse bloodlust or ... cause indifference, especially in children, to the suffering of defenceless animals' (PADL 1933, cited by Burt 2002, p. 138; see also PADL 1932).

In films other than those attempting to dramatise the violence of natural (and unnatural) confrontations, animals fulfilled an incidental role that could provoke equally strong criticism, and the reassurances of the film companies were not usually acceptable. Francis Rowley admitted that skilled camera work could substitute for sequences that would be objectionable if real, but had as evidence 'the advertisement of a certain moving-picture producer which states that the "approximate valuation of animals actually slain during the production of the film was $95,000"'. In an earlier case, men had been convicted and fined in a New York State court for compelling a horse to make a death leap over a precipice, and more recently in Portland, Maine, there was a screening in which a span of horses was dragged by a log down a long chute on the mountainside, appearing at the bottom fatally injured, if not dead (*Literary Digest* 1920). (Maine subsequently passed laws protecting animals in movies, as did a few other states as early as 1921.) In similar vein, the film *Jesse James* (1939) included a sequence when a horse was driven over a cliff to its death, but this time a major public outcry had effect, resulting in negotiations between the film studios and the American Humane Association (AHA), and agreement by the Motion Picture Association of America in 1940 to acknowledge the AHA as the official watchdog for animals in the film industry, granting it the right to set guidelines and to monitor animals' treatment during film production. Approved films subsequently bore a 'No Animals Were Harmed' disclaimer at the end of movies (Beers 2006, p. 104).[26] The treatment of animals

---

[25] Sir Harold Alfred MacMichael (1882–1969) was governor of Tanganyika between 1934 and 1938.

[26] The AHA today describes its role: 'Our Film & TV Unit has exclusive on-set jurisdiction from the Screen Actors Guild to supervise the use of animals in movies, TV shows, commercials and short films. Our Certified Animal Safety Representatives™ work around the world – not just in Hollywood – to ensure that productions meet our strict Guidelines for the Safe Use of Animals in Filmed Media' (AHA 2014).

other than horses had also caused concern. Rowley referred to cruelty to a dog with its tail fixed in a knothole in a board fence that was then knocked out, to be dragged around by the animal. Skilled camera work had not substituted for terror and pain in this case, either. Indeed, it had been necessary for the National Board of Censors to issue warnings to film-makers that no films where cruelty was involved would be passed. As for the training of animals to be active rather than passive participants in filmed sequences, the trainers were keen to reassure the press and public of their methods. One trainer of movie animals, 'Curley' Stecher, said that 'The animal who works through fear is a pathetic spectacle and is useless before the camera', and according to director Norman Dawn, 'Persistence, patience, and kindness are the necessary qualities in a movie-animal trainer, whether his subjects be lions or dogs, elephants or chipmunks' (*Literary Digest* 1920).

Ryder notes that it is easier to stop new abuses than to end established ones, and the RSPCA had done well to approach the newly appointed film censor in 1913, in order to secure some protection for animals in film production (Ryder 2000, p. 133). The British Board of Film Censors (BBFC) was founded in that year, and in its first annual report of 1914, 'cruelty to animals' was the first of 22 grounds for cuts or rejection. Criteria for excision were very soon expanded (Burt 2002, p. 133, citing Robertson 1993, p. 6, and Robertson 1985, pp. 20, 21, and 44–46). MacMichael appears to have developed a good opinion about the BBFC's scrutiny of films, and on 28 August 1931 he wrote to the BBFC in appreciation of the censor's rejection of some 'Barkies' advertising films because of apparent cruelty to dogs in their production. In the early 1930s, MacMichael was in correspondence with the BBFC about suspected cruelty in films such as *Trader Horn*[27] and *Hills of Kentucky* (1927, silent), which featured Rin-Tin-Tin. This dog was a star of cinema, radio and television from the 1920s to the 1950s and was represented by several German Shepherd dogs. The first one was seen by the film mogul Darryl F. Zanuck performing at a dog show, and appeared in his first film in 1923, later to complete 26 productions for Vitaphone and Warner Brothers, saving the company from bankruptcy. The original Rin-Tin-Tin had up to 18 trained stand-ins, earned $6,000 per week in 1926 and—to the accompaniment of classical music to aid digestion—dined each day on choice tenderloin steak prepared by a private chef. His trainer, Lee Duncan, maintained that he never used a whip, only kindness in training (Malamud 2007, p. 7; Stevens 2014). On 5 September 1931, the managing director of Kodak in London replied to MacMichael saying that there was no cruelty to Alsatian dogs in *Hills of Kentucky*, which had been passed by the censor. He offered a private view and an explanation of how effects were produced: dummy dogs were used. On 7 September 1931 MacMichael took up the offer, saying that the BBFC had promised that any complaint from the PADL would receive sympathetic consideration. MacMichael also thanked Mr Milder, managing director of Warner Brothers in London, on 24 September 1931 for their meeting that day and

[27] *Trader Horn* (1931), an adventure film made in Africa, featured the killing of animals and animal fights and was censored.

for his suggestion of his writing to the parent firm in the United States advising the issue of certificates confirming that no cruelty had occurred in the production of films. Otherwise, the censor could only rely on his own judgement and only after a film had been made. The existence of certificates could act as a deterrent. MacMichael's relations with United Artists were not so cordial. On 4 July 1938 he wrote to their solicitors who threatened to bring an action against the PADL over a leaflet on cruelty in the 'Goldwyn's Follies' film. MacMichael was defiant. In the New Year number of *The Cinema*, MacMichael said he had asked Samuel Goldwyn to refrain from cruelty in the production of animal films, but 'Silence is Goldwyn'.

MacMichael had received a letter from D. Jeffrey Williams, secretary of the Pit Ponies' Protection Society, on 9 December 1931, congratulating him for organising a conference at Caxton Hall that day on cruel films. At the conference it was suggested that film producers should allow accredited representatives from animal welfare societies to be present during film production (Burt 2002, p. 139, citing BBFC 1931, p. 27). This was the first public challenge to them, and he thought it might be possible to get the cooperation of men like (Richard or Cherry) Kearton (the pioneer wildlife photographers). On 10 December 1931, MacMichael wrote to Edward Shortt, president of the BBFC, about this 'Conference of Delegates from Humane, Educational and other Societies and Organisations' convened to discuss what action should be taken to prevent the exhibition of cruel films in Britain. The Resolution had been passed:

> Having carefully considered the communications which have been issued on the subject of 'Cruelty on the Screen' by the British Board of Film Censors, Mr Robert Wagner, Editor of the Film Magazine, 'Script', and the Performing and Captive Animals' Defence League, [this Conference] is convinced that gross cruelty and brutality have been practised upon animals in the production of some of the films at present being shown on the Screen in this country,[28] and that an interview with the Film Censor and his Board be arranged with a view to preventing the exhibition of cruel films.

On 6 January 1932, Shortt confirmed that the BBFC was vigilant against film cruelty and had been since it was founded in 1913; therefore, a meeting was not necessary. A standing committee was meanwhile formed from the conference under the chairmanship of Robert H. Spurrier, Honorary Treasurer of the PADL.

The RSPCA was also concerned with the problem, and its solution occasioned further disagreements with MacMichael. He wrote to the Prime Minister, Ramsay MacDonald, on 23 February 1934 about its proposed Cinematograph Films (Animals) Bill, describing it as a spurious measure and worthless, like the RSPCA Incorporation Act of 1932 which robbed members of their rights. The PADL had a genuine bill known to promoters of the spurious one. He enclosed detailed criticism of Gower's bill and a copy of the League's proposed 'Cinematograph (Protection of Animals) Bill'. The first dealt solely with what appeared on the screen and not with

[28] *Reynolds Illustrated News* had helped to expose film cruelty.

cruelty inflicted in the production.[29] A meeting would be held on 1 March 1934 followed by a deputation to the House of Commons, and the Government's views were requested. The reply given was that it was a matter for the Home Secretary, and the letter had been forwarded to his Private Secretary. MacMichael then protested, asserting it was the responsibility of the Prime Minister, but if he was too weak to combat vested interests, he should admit it. Acts of legalised film cruelty were increasing rapidly. The deputation was still planned. MacMichael met the Home Secretary's Private Secretary but remained determined to attract the attention of the Prime Minister. On 28 February 1934 the Prime Minister's Private Secretary said again that it was the Home Secretary's area. The next day MacMichael replied to him that the RSPCA's bill was an affront to the House and to the nation and had nothing whatever to do with the Home Secretary's Department. MacMichael's request for the deputation predated that of the Hunger Marchers at the same time: 'It is highly significant that he should be asked in 1 week to hear the case of starving human beings and cruelly exploited animals'. The deputation was still on the cards unless the Prime Minister set up a committee of scrutiny of Gower's bill. There was a telephone conversation as well on 1 March between MacMichael and the Private Secretary, H.G. Vincent. In the end, the Prime Minister could not receive the deputation, and there was no time for the bill on 1 March. On 5 March 1934 MacMichael informed the Private Secretary that the Prime Minister had made a mistake in not receiving a deputation or sending a message about any scrutiny—the Prime Minister should still publicly denounce the bill.

That summer, MacMichael made known in *The Spectator* his feelings about Gower's bill. He accused it of containing provisions which would have to be conceded, leaving a worthless policy which would not stop the cruelty but would deceive the public:

> Under it the real offenders, i.e., animal film producers (quite 80–90 % of whom live abroad and cannot therefore be brought to book) and innocent exhibitors will be liable to fine and/or imprisonment in respect of animal films,
>
> 1. Where suffering has been caused.
> 2. Where suffering may have been caused.
> 3. Which depict suffering.
>
> Both (1) and (2) would have to be deleted owing to legitimate pressure from the trade, and for this very obvious reason. In a film under (1) the cruelty, which would not be obvious, would have to be proved and we should be faced with the monstrous situation of hundreds of absolutely innocent cinema managers and their assistants being fined and/or imprisoned for showing a film which had been passed by Censor and had been exhibited by them months beforehand as being above reproach! Under (2) the same people would be at the mercy of anyone who could show that suffering might have been caused, and this could apply to almost, if not quite, all animal films.

[29] This shortcoming had been corrected by the time the Cinematograph Films (Animals) Act was passed in 1937.

> What, then, is left of the Bill? Films depicting cruelty. This sounds all right, but is it? On the contrary, it is all wrong, for if a man took a photograph of someone illtreating an animal, the effects of a steel-toothed trap, &c., for the purpose of drawing public attention to an existing form of cruelty, instead of being commended by the magistrate he would be liable to fine or imprisonment. But the chief defect of (3) is that it would encourage unscrupulous producers to be as cruel as they wished provided the suffering was not apparent on the screen: Is this not hypocrisy? Matters are made worse by the fact that the policy embodied in the genuine Cinematograph (Protection of Animals) Bill was first in the field. This fact is well known to the headquarters of the RSPCA and it is only lack of funds that prevents our presenting it to Parliament. Perhaps some influential Member who thinks that the time has come for Parliament to pass some genuine measure of animal protection will sponsor this Bill (MacMichael 1934).

Gower replied by describing MacMichael's attitude as pitiable, and 'the circumstance that he describes the [bill] to which he refers as "spurious" and "hypocritical" stamps his criticism with its true value ... We are indifferent to the petty and carping criticism to which at times we are subjected' (Gower 1934).

The need for the Cinematograph Films (Animals) Bill was fully debated in the House of Commons before its enactment in 1937.[30] Gower explained the need for legislation because the film censor was not a government official but appointed by the trade: he was neither entitled nor able to decide which films should be shown and which not, as by law the last word rested with the local licensing authority. Alfred Denville, member for Newcastle upon Tyne Central, spoke for the entertainment industry, including the film side and the theatrical side, as well as the Circus Proprietors' Guild of Great Britain, when giving full support for the bill. Cruelty created aversion and bad feeling among patrons. However Thomas Groves, member for West Ham Stratford, was concerned that it might be used to impose further restrictions on 'the legitimate entertainment industry' and the exhibition of animals in circuses. The Under-Secretary of State for the Home Department, Geoffrey Lloyd, member for Birmingham Ladywood, recalled that a few years previously, there had been a great wave of animal films and the censors had approached them soundly. They set up a panel of experts, nominated by various societies, to advise them. The panel included representatives of the Royal Veterinary College, the Natural History Museum, the RSPCA, the Society for the Preservation of the Fauna of the Empire, the Zoological Society and the London University Animal Welfare Society. Lloyd said that he understood the panel had done a considerable amount of work in advising the BBFC. However, in reply to Lloyd's advocacy of the status quo, Gower pointed out the great dissatisfaction that existed on the part of many members of the panel, particularly of the representative of the RSPCA.

MacMichael was to remain as frustrated with the lack of strong legislation in this area as he was in the matter of animal performance in the circus. In 1939 E.K. Robinson, the secretary of ODFL, wrote an article in *Sight and Sound* criticising the ineffectuality of the 1937 Act, as well as describing continuing

[30] *Parliamentary Debates* (*Commons*) 1937, vol. 322, columns 529–556.

brutalities in training methods, calling for a committee for regulation and trainer supervision (Burt 2002, p. 139, citing Robinson 1939). In the United States, criticism of cruelty in film production still persisted after the Second World War in spite of the presence of agents from the AHA at film shoots. In *Every which way but loose* (1978) the orangutan, Clyde, was allegedly beaten so severely by his trainer that he died a month later. There were supposed to be no trip wires, and softened earth was used for falls by trained horses and riders, but problems faced by horses seemed sometimes to be that in westerns the actors did not know how to ride (Schumach 1959). Joys refers to a report transmitted by WNYC-TV in New York showing how some moviemakers still used cruel means to capture a desired effect: for example, in two productions of 1981, Michael Cimino's western *Heaven's gate* involved fatal injuries from explosions, and Warren Beatty's *Reds* used trip wires to get spectacular falls, both films resulting in the deaths of a number of horses. The trip wires for *Reds* were apparently used in Spain, beyond the oversight of the AHA and without the knowledge of Beatty as director (Joys 2011, p. 439, citing Corry 1986).

Another target for criticism was Ralph Helfer, who trained wild animals for Hollywood, film and television from the 1950s, by what he called 'affection training'. Critics have alleged that this involved declawing, defanging and drugs, but Helfer said such procedures would only be undertaken for medical reasons. At his Gentle Jungle Affection Training School, he was charged by the USDA under the Animal Welfare Act but cleared. Helfer's popular writing is based on the semi-autobiographical memoirs of one later in his career involved in training methods of the 'good' kind, and comparisons are made with the bad and dangerous practices which are allegedly replaced (Helfer 2007).

Since the time of MacMichael's successful criticism of the 'Barkies' advertising films in 1931, objections to the use of animals in filmed advertising have usually been related to perceived assaults on their dignity. The 'Brooke Bond Chimps' are an example. Their use dated back to the 1950s, when Twycross Zoo rented them out for use in the advertisement of tea.[31] Since that time, many complaints that the advertisements were demeaning to the animals were resisted (Hancocks 2007, p. 104). They were dressed up as humans and performed well-known human activities and roles ranging from ironing to James Bond, and, like many historical circus animals in human garb, they were intended to be amusing, as risible substitutes for the real thing. In this their handlers succeeded: they became the UK's favourite advertising characters. When in the 1970s they were left out of advertising as a result of objections, sales of the tea dropped, and they were brought back after 18 months until their last use in broadcast advertising in 2002. It has been tempting for other concerns to use chimpanzees for advertising, by making use of the

[31] David Hancocks has asked: 'Is the silence from zoos about circus conditions due to uneasiness about casting stones from within glass houses?' (Hancocks 2007, p. 114). Zoos have also supplied surplus animals to circuses as part of trading activities with them, and accommodation standards in zoos have been subject to criticism, like those in circuses.

opportunity they present to parody human behaviour. This was the case with the Portman Group, which used chimpanzees in the role of misbehaving drunks in order to illustrate the dangers of binge drinking. The International Fund for Animal Welfare complained: 'The clear implication is that chimps are stupid, aggressive and figures of ridicule. In fact, they are highly intelligent creatures that suffer because of their similarities to humans' (McKie 2004).

## 6.3 Animal Performance and Applied Science

Since Mitchell and Hediger, a limited number of scientists have studied the behavioural components of animal performance either to advance it for commercial reasons or to assess its implications for the participating animals. During the Second World War, Keller and Marian Breland began to study a range of domestic animals with a view to applying the science of animal behaviour to commercial purposes. Since then, McGreevy and Boakes (2007) have investigated the processes of training for entertainment, advising on its best development by analysing selected case studies and at the same time paying some attention to the welfare implications of the training process itself. By the time of their study, other investigations had been carried out in order not to explain the best procedure but primarily to attempt scientific appraisals of what remained a contentious activity. The work of Kiley-Worthington (1990) was intended to provide an objective assessment of the welfare implications of the use of animals in the British circus, and she gave attention to the problem of confinement that was less relevant in McGreevy and Boakes and in the Brelands' publications. This interpretive as opposed to commercially oriented work was soon followed by others that either supported Kiley-Worthington's findings and recommendations or contradicted them, for example, Harris et al. (2006). Scientists had now become fully involved in the controversy, one stimulated by their own disagreements and their association with opposing factions. Against a background of continuing animal activism alongside trade countermeasures, some renewed political intervention came about, in the case of Britain leading to draft legislation to prohibit the use of wild animals in circuses by December 2015.

Keller and Marian Breland had worked under B.F. Skinner during the Second World War, when, after the Japanese attack on Pearl Harbor, they assisted in the development of a secret and eventually unrealised plan to train pigeons to guide missiles, using training methods based on operant conditioning. The Brelands later applied their skills as behavioural technologists to the post-war commercial training of domesticated and 'docile' animals for advertising and entertainment. Their business ventures included a roadside tourist attraction called the IQ Zoo in Hot Springs, Arkansas, which also functioned as a training laboratory between 1955 and 1990. This was part of the Brelands' Animal Behavior Enterprises (ABE) which over the course of 47 years trained over 15,000 animals of approximately 150 species and employed up to 40 staff. As publishing academics who decided to employ their scientific understanding of animal behaviour to practical business purposes, the Brelands therefore represented a new and unique kind of animal trainer. Alongside their entertainment ventures, they developed mass-produced operant-behaviour exhibits and demonstrations for use in advertising. When supplying these for customers, instruction manuals would be included with the packaged animal act, and they trained trainers—in 1955 they wrote perhaps the first manual for training dolphins using operant conditioning (Drumm 2009; Bailey and Gillaspy 2005; Gillaspy and Bihm 2002). They also introduced the use of a secondary reinforcer—a whistle or a click—as a stimulus for, or a time bridge leading to, the upcoming delivery of a primary reinforcer (food), enabling the control of behaviour when food was not immediately available (Bailey and Gillaspy 2005). And a clicker was conveniently transferable between trainers and other presenters.

By 1944, the Brelands were investigating the species-typical behaviours of a large variety of familiar animals, cataloguing their natural behaviour repertoires for bespoke conditioning.[32] They subsequently identified what they termed 'instinctive drift', when attempts at conditioning might be upset by a resumption of instinctive behaviour in the animal. This led them to the heretical belief that operant conditioning theory should accommodate some ethological interpretations and the role in different species of instinctive behaviour, evolutionary history and ecological niche (Breland and Breland 1961, Drumm 2009). Furthermore, sought-after behaviours might not be elicited unless they complied with their precise instinctive uses and associations (e.g. for example, the Brelands' goat would bleat from pain but not for food), and if identified, some instinctive behaviours could be exploited or offered as explanations of behaviour. For example, the instinctive hierarchical herd behaviour of the horse leads it to respond to very subtle cues when under training (as with Morocco or Clever Hans). In their studies of species-specific characteristics, the Brelands attributed qualities to animals that can be compared with the anthropomorphic observations of earlier trainers and commentators when assessing suitability and difficulty in training. For example, domestic cats are less reactive to social signals than dogs, which are by nature more hierarchical, cooperative and communicative, as pack animals responsive to a 'top dog'.[33] In the big cat act, there is no affection for the trainer: the close encounter is a dangerous balance between flight

[32] This remains recognised as a necessity for effective training. Speaking of the training potential of the domestic cat, Bradshaw et al. note "Species-specific behaviour patterns have a clear role to play in providing the behaviour to be shaped, in directing the attention of the cat towards the task to be performed, and in providing an assessment of the value of the reward for correct performance…thus it is much easier to train a cat to obtain a food reward by using a normal component of hunting behaviour" (Bradshaw et al. 2012, p. 54).

[33] Miklósi et al. (2005) suggest that the lack of frequent and early glances at humans shown in their studies of cats may be a sign of relative independence from humans compared with dogs and may also contribute to differences in trainability between domestic cats and dogs. However, Crowell-Davis (2007) argues that domestic cats are neither solitary nor asocial, and the Brelands discuss domestic cat sociability with humans as encouraged by the neotenising effects of domestication (Breland and Breland 1966, pp. 91–3).

and fight reaction (Breland and Breland 1966, p. 94). Therefore of big cats, trainers have preferred the uniquely social lions over such as tigers, which have earned anthropocentric and anthropomorphic criticism for uncooperativeness, sulkiness and 'treachery', leading to subjugation rather than training. Van Amburgh was quoted as saying:

> The tiger is like a reckless, good-for-nothing, drunken rascal, who spends his time carelessly at taverns, and fights in a moment. Tigers have all bad spiteful tempers. The lion is not so irascible, he is slower and cooler, but there is not the generous feeling about him which he has been cracked up for. The leopards are like cats—playful, but easily provoked (*Times* 1838).

In *Haney's*, the lion is described as more dependable and less treacherous than the tiger and always preferred by tamers: burning torches and heated irons are used more frequently for tigers than for lions (Anon. 1869, pp. 124, 126). Harvey Sutherland supplies a more ironic comment on human assessment of the characteristics of this uncooperative animal:

> Tigers are more dangerous than lions. They are treacherous, which is the uncomplimentary epithet we apply to persons and beasts that do not tell us when they conclude that they have stood all the nonsense from us that they propose to. They are stupid, too, which is another uncomplimentary name we have for persons and beasts that have too much self-respect to act the fool for our amusement. I fancy we shall never know what sillies menagerie animals think we are until the tables shall have been turned and a lot of men captured by them and put on exhibition (Sutherland 1902).

Ellen Velvin supplies anthropomorphic descriptions of other animals: 'Jaguars are more noted for their deceit, craftiness, and treachery than all the other treacherous members of the cat tribe ... Baboons are ugly-tempered, vicious, and treacherous animals' (Velvin 1911), and Hornaday insisted that lions are 'the best of all carnivorous performing animals, because of their courage, serenity, self-confidence and absence of jumpy nerves'. He considered leopards the worst, polar bears next, 'with big chimpanzees as a sure third. Beware of all three'. Lions were 'sanguine, courageous, confident, reposeful and very reliable', as compared to the 'nervous, suspicious, treacherous and uncertain tiger' (Hornaday 1922, pp. 242–245). Courtney Ryley Cooper dismissed rhinoceroses and camels as 'evil-tempered, selfish, mean and vengeful', showing the worst of human traits (Bond 1924).

The Brelands proposed that in higher animals, including humans and dolphins, neotenisation eases conditioning and enables varieties of learned behaviour, because the adult has reached maturity without the completion of what would have been the full repertoire of rigid instinctive behaviour in the primitive adult forebear, and neotenised animals typically explore and play. However, this later more 'generalised', versatile and adaptive behaviour still leads back to and serves the primal needs of food, safety and sex. Intelligent and versatile generalised behaviour is also particularly found in omnivores like rats, bears and pigs, animals able to transfer learning from one situation to another. Other animals like chickens (but unlike parrots) do not generalise and instead always make predictable discriminations, which are more suitable for training simple, precise, repetitive performances and which the generalisers are less bound by (Breland and Breland 1966, pp. 77–80, 88, 94–95, 106–107).

These interpretations were put to profitable use at the IQ Zoo and at new dolphin and whale shows in the late 1950s. The IQ Zoo as a roadside tourist attraction benefited from expanding highway systems and increased vacation travelling in the United States after the war (Drumm 2009). Here,

> Chickens walked tightropes, dispensed souvenirs to paying customers, danced to music from jukeboxes, and played baseball. Rabbits kissed their (plastic) girlfriends, rode fire trucks, sounded sirens, and rolled wheels of fortune for lucky customers. Ducks banged out rhythms on drums and played pianos, raccoons played basketball, and parrots rode bicycles (Gillaspy and Bihm 2002).

Other novelty attractions by the roadside and at fairs continued after the demise of the IQ Zoo, sometimes with a single operator attempting to introduce a token educational component to the spectacle. For example, on offer in 1998 were a traditional American high-diving mule act, racing ponies jockeyed by monkeys in a 'Banana Derby', dancing buffalos, motor cycling baboons, a water-skiing squirrel and racing skunks, some of these displays threatened by activists, court actions and state bans. The skunk show was preceded by a brief talk about how the animals needed a balanced diet of meat and vegetables, and the owner of the skunks said, 'With our educational programme, we've been very good about not upsetting anybody'. But the owner of the mules admitted, 'There's nothing educational about mule-diving', and he said he would continue to fight animal rights activists through the courts (Broughton 1998).

Advertisements for the IQ Zoo itself gave the assurance that animals were trained by reward and that there was no punishment, although extreme hunger and electric shocks were sometimes used in shaping behaviour during the Brelands' experimental work. In order to give the establishment educational credibility, there was also information at the IQ Zoo about training methods (Breland and Breland 1966; Drumm 2009). The Brelands had written in 1951:

> Probably the biggest applications exist in the entertainment world. Here we can take over the formal animal training involved in the standard animal act for stage, circus, and movies, and do it faster, cheaper, better, and in multiple units. It is possible to create new acts, whole new circuses, in fact, using unusual animals and unusual acts, and again do it cheaply, quickly, and in numbers limited only by time and production facilities. Television offers unusual opportunities. We can invade the field of night-club entertainment with novel small animals. We can sell or rent trained animal units to hospitals, doctors' offices, waiting rooms of various sorts, or even to private individuals, supplying instructions on care and maintenance (Breland and Breland 1951).

Advertising was also an important market for ABE, which could supply packaged acts designed to run virtually automatically, as show window advertising in which live animals take the place of puppets and robots. Chickens were trained to advertise products at feed stores, trade shows and county and state fairs:

> One hen played a 5-note tune on a small piano, another performed a 'tap dance' in costume and shoes, while a third 'laid' wooden eggs from a nest box; the eggs rolled down a trough into a basket, the audience would call out any number of eggs desired, up to eight, and the

> hen would lay that number, non-stop ... We are now in a position to outstrip old-time professional animal trainers in speed and economy of training ... The success of these acts led to the development of a trained pig show, 'Priscilla the Fastidious Pig,' whose routine included turning on the radio, eating breakfast at a table, picking up the dirty clothes and putting them in a hamper, running the vacuum cleaner around, picking out her favorite feed from those of her competitors, and taking part in a quiz program, answering 'Yes' or 'No' to questions put by the audience, by lighting up the appropriate signs (Breland and Breland 1951).

McGreevy and Boakes echo many of the points made by the Brelands and recognise the role of instinct and species differences. They discuss training—whatever the purpose—as the modification (through an altered environment or experience of positive or negative reinforcement applied by the trainer) of that instinctive behaviour genetically inherited by the chosen animal and peculiar to its species. Early, fundamental and preliminary modifications will include socialisation with humans and habituation, leading to a reduction of fear, as well as positive or negative sensitisation as in the circus use of the spiked bullhook during elephant training and control. (Elephants are unusually sensitive to tactile stimuli; therefore, subtle use is possible, for the whole or part of the animal.)[34] But solitary, unsocial predators do not respond well to coercion. The large cat has to be constantly reminded to 'respect' humans, although generally, 'fear of the trainer, whether the trainer is perceived as a potential predator or a more dominant member of the same species, is perhaps the factor most likely to impede learning' (no doubt a problem with the traditional subjugation of certain wild animals). Socialisation of tigers with humans is achievable from 3 weeks, taking advantage of the critical but limited socialisation period, but is not guaranteed. 'Being solitary hunters, cats rely very little on cues from others when acquiring food. This may account for the relatively minor significance of verbal commands in the training of all felidae when compared to, say, dogs'. McGreevy and Boakes also agree with the Brelands that omnivores are more likely to possess a wide range of opportunistic behaviours and so are more easily trained (McGreevy and Boakes 2007, pp. xii–xiii, 2–4, 6, 15, 121, 161, 188).

The authors advise that good training requires the understanding of an animal's innate responses, temperament, learning abilities and motivational requirements, and 'for the trainer it is important to note that tasks relying on spatial memory may be relatively easy for an animal to acquire, while those relying on perceptual memory are likely to be very difficult'. Taking advantage of an animal's natural abilities, an elephant, for example, can manipulate a paintbrush with dexterity as it would use a switch or twig to deter flies. Secondly, accurate timing is required in the administration of an unconditioned or conditioned stimulus, signal, response, reward or denial of reward. Finally, clarity and consistency in the administration of these elements is important. They note that even mild punishment can destroy an animal's motivation to offer novel responses, and 'only the unenlightened use

[34] Helfer describes proper use of the bullhook for elephants: 'The sharpness of the point is critical: you should be able to hang it from your index finger without drawing blood' (Helfer 2007, p. 114).

physical punishment in the traditional penal sense, because it amounts to abuse'; and learned helplessness caused by aversive stimuli and impoverished environments produces an emotional, cognitive and learning deficit, made longer-lasting by reminders and being variable among species, with the probability that younger animals are more affected (McGreevy and Boakes 2007, pp. 84–86, 97, 196, 280–281).

The relativity of likes and dislikes in laboratory animals presented with choices to make through operant conditioning may indicate if and when animals enjoy being trained: 'Some kinds of training may become enjoyable for their own sake and not just as an escape from boredom. This probably depends on the extent to which the trained behaviour becomes a form of play for the animal'. If they wanted, performing birds could fly off, but they find the tricks rewarding and have been taught later to return to their cage by its learned association with food, and 'acrobatics and chains of complex behaviours . . . can contribute to the maintenance of athletic and mental fitness in captive animals' such as sea lions. Turning to the charge that some tricks seem undignified, McGreevy and Boakes state that animals do not have the self-consciousness that could allow them to feel embarrassed or humiliated, but that people's perception of animals is undoubtedly affected, either by admiring their learning abilities or by seeing them as 'dumb' victims. But it is important to note that instead of having to work for food, presentation of it in a bowl may be as far removed from a natural foraging context as riding a tricycle or balancing a ball on the nose (McGreevy and Boakes 2007, pp. 89, 159, 162, 196–197).

Together with the Universities Federation for Animal Welfare which funded a research assistant, the RSPCA funded an ethologist, Marthe Kiley-Worthington, to carry out in 1988 an independent scientific study of circus animals in Britain in comparison with those in zoos and in the wild. She embarked on an 18-month investigation of 16 circuses involving 3,000 h of observation, in cooperation with individual proprietors and the ACP. There were approximately 20 circuses operating in Britain at the time, of which about 14 were members of the ACP. Each circus with animals had from three to eight animal acts. Only 513 animals lived in the circuses visited in 1988, of which 292 were traditionally 'wild' ones. Kiley-Worthington's investigation resulted in a report in 1989 which the RSPCA decided not to publish but which was then incorporated in a book that the ACP helped to fund. She believed that conditions for animals in the British circus were comparatively acceptable (a relative and personal judgement) and that although improvements were necessary, a ban was not justifiable: this clearly had good promotional potential for the circus industry. Kiley-Worthington's conclusions had not been what the RSPCA expected, and it felt that the data she had collected did not support those conclusions, especially concerning confinement, as discussed above. In the book published the following year, by which time some bad feeling had arisen, she referred to the RSPCA as being devoted to the total banning of circuses. This was a somewhat provocative charge, reminiscent of that long made with such irony by advocates of animal performance in the circus, who accused their opponents of being 'anti animal'.

In spite of its title, *Animals in Circuses and Zoos. Chiron's World?*, the book concentrates on circuses rather than zoos, and it is argued that before banning, we must be sure that pain and suffering are greater in circuses than in other animal enterprises: society accepted millions of intensively reared food animals, and there were only hundreds of circus animals. Following her research she argued that there was no significantly greater distress shown in circus animals than in the same species in zoos or domestic systems: 'It is ... irrational to take a stand against circuses on grounds that the animals in circuses necessarily suffer, unless [objectors] are to take the same stand against zoos, stables, race horses, kennels, pets and all other animal-keeping systems'. (Of course, many did and do, and they do not feel that creation of detailed hierarchies of suffering or cost-benefit analysis is practicable or necessary.) She concluded that there was little evidence of physical maltreatment and cruelty in the husbandry, training or performing of circus animals and that whilst there were improvements that must be made, circuses did not by their nature cause suffering and distress in animals: she believed that funding and agitation should not be directed at prohibition but rather at welfare improvements. Meanwhile she noted circus people's 'intense willingness to listen to criticisms and comment', according to her free movement on premises that were visited with the cooperation of ACP, either by prior arrangement or unannounced (Kiley-Worthington 1990, pp. 13–14, 61, 93, 134, 218, 220, 226–231).

Kiley-Worthington reported that she found no evidence of prolonged or acute distress or high levels of fear or anxiety during the training of any of the circus animals she witnessed.[35] On the contrary, she said, 'Training or educating the animal can act as a form of occupational therapy for the animal in captivity' and a major criticism of British circuses was a lack of training and good trainers, with little of the variety in acts that would make them beneficial.[36] (There is little discussion of types of act in the book.) Large, dangerous animals were not negatively reinforced because when scared or hurt, they would attack in defensive threat or flee: 'The problem to the educator is not one of disciplining and dominating, but one of motivating', and she claimed 'All training sessions used positive reinforcement in the form of food rewards and praise with the voice'. She does not mention the use of the elephant goad. However, she acknowledged that 'Some animals did fight the restraining (training) ropes and leap about: for example, the camels and

[35] She had also reported: 'I saw no evidence in circuses of the use of such techniques as electric shock probes; immobilizing drugs; drugs to reduce anxiety or increase performance, or analgesics used to combat occupational disorders. All of these techniques are used for farm animals, zoo animals, pet and sport animals. There was no evidence either of using other techniques which might come into this category, such as burning, other forms of electric shock – except for fencing – or peculiar or abnormal physical restraint ensuring, for example, that the animals maintained particular postures, as recently reported for show jumpers. Lead reins, side reins, sometimes unnecessarily tight, and occasionally muzzles were used' (Kiley-Worthington 1989).

[36] She suggested that to improve the education of trainers, a training school should be established with the support of organisations including the RSPCA, and an annual award should be given to a circus with the best-trained and most appropriate animal act—this could be financed by the RSPCA and Universities Federation for Animal Welfare (Kiley-Worthington 1989).

some horses when learning to kneel and lie down. In some cases, the use of ropes was unnecessarily rough'. She adds that 'If the whip is one of the reasons for disapproving and trying to ban circuses, current practices in horse training should also be looked at carefully', but she found that the general emphasis was on calming an uncooperative animal rather than exciting it with whips. Meanwhile, 'giving different consideration to wild as opposed to domestic animals must depend more on whether they have been captive bred or wild caught and at what age, and their other individual past experiences, rather than their status as a species of being "wild" or "domestic"'.[37] And of the question whether training and acts were 'natural' or 'unnatural', Kiley-Worthington points out that dogs on leads or cows being milked is not natural, and anyway, she asks, is all unnatural trained behaviour wrong? For example, training elephants to balance on revolving spheres requires time, patience and skill: this is an admirable achievement, she says, depending not on negative reinforcement, just on practice, and it earns respect for both the animal and trainer.[38] Tait has since added: 'The controversy over trained animal acts means that the extraordinary achievements of animal performers [meaning animals] are disregarded – itself a form of speciesism' (Tait 2012, p. 11). But such references to achievement are anthropocentric: who benefits from them?[39] Referring to longevity and health, Kiley-Worthington found from her research that circus animals are usually retired for sentimental reasons rather than slaughtered, so their average age is high when compared with other animal enterprises. Their working lives and fitness are also extended as long as possible because of the investment of time and expense in training (Kiley-Worthington 1990, pp. 53, 115, 118–120, 122–124, 129, 132–134, 140–141, 159).

A little after Kiley-Worthington's report, William Johnson published an invective against animal performance, past and present, and recent attempts to excuse it on educational or conservation grounds. The author concentrates on ordeals in the capture, transport, training and exhibition of wild animals such as the elephant, lion,

[37] Kiley-Worthington (1989) had also stated that it would be wrong always to give wild animals greater consideration because 'as a general rule the fundamental behaviour—social organisation, relationships between individuals, communication systems etc.—has changed little, if at all, during the course of domestication', making domesticated animals equally vulnerable to unfavourable conditions.

[38] The Amboseli Trust for Elephants would disagree: 'They are required to perform behaviors never seen in nature. In short they are treated as commodities, as objects to provide entertainment for humans. The circus experience has nothing to do with the reality of elephant life and behaviour... The totally unnatural existence of captive elephants in a circus, which includes significant physical and emotional suffering, is a travesty. To allow this practice to continue is unjustified and unethical' (Amboseli Trust 2008).

[39] Tait continues: 'In arguments against speciesism, qualities attributed to humans are also phenomenologically observable in nonhumans – as with animal acts. Could the wild animal act perpetuating speciesism paradoxically also induce feelings to undermine it?' (Tait 2012, p. 145). This would be another conundrum in human-animal relations, like the use of animals in psychological experimentation, when the more the animal is like us, the greater the relevance of the experiment, but the greater too its ethical cost (Wilson 2002).

tiger and, especially, the dolphin. A large part of the book assesses the dealership in marine mammals, their high death rate within the entertainment industry since the Second World War, and the ineffectiveness of international conventions, official agencies and the law. Johnson is critical of the British government-sponsored investigation of Margaret Klinowska, whose terms of reference were (1) to advise whether the educational, research or breeding benefits of dolphinaria and similar establishments in the UK are, or can be, of sufficient value to justify the import and display of live cetaceans and (2) to suggest the standards that should be applied to dolphinaria if they are to be justified on educational, research or breeding grounds (or some combination of these purposes) (Klinowska and Brown 1986). Johnson charged Klinowska with ignoring her own evidence within her conclusions, and with lacking impartiality. Kiley-Worthington faced similar criticism, including the observation that she compared circus animal behaviour not with that in the wild but in zoos and safari parks, suggesting that wild captive animals should be downgraded to the domestic category and be taught a 'moral sense' of what is right and wrong (Johnson 1990, pp. 230, 233, 237, 320–323).

John Webster recommended that the welfare implications of training animals to perform 'unnatural' acts needed to be assessed according to (1) the natural affinity of the species with man, (2) the extent to which the training can be achieved by reward and without inducing a chronic sense of anxiety and (3) the quality of life for the circus animal when not 'working'. Dogs may be taught circus tricks only by reward because they are motivated to please their owners. Horses, elephants, seals and some other animals would require fairly benign application of the principle of 'carrot and stick'. But with the large cats, training must be based on the principle of absolute dominance and submission (Webster 1994, p. 226). As for quality of life when not working, this inevitably centres on the experience of confinement. According to Morgan and Tromborg, 'perhaps the greatest stressor in the lives of captive animals is their perceived or actual inability to control most aspects of their surroundings', including their social partners or their mates, the space between themselves and other conspecifics or humans, kinds of food or when it is made available and environmental stimuli such as quantity, quality and timing of lights, sounds, odours or temperatures. Captive environments may then promote helplessness and the loss of adaptive coping strategies. For conservation of bio- and behavioural diversity, such passivity is an undesirable consequence of captivity. Stress in captivity may also be caused by predictability or a lack of it. 'It may be essential to introduce animals to a certain amount of unpredictability. After all, novelty, unpredictability, and variability are all common features of the natural environment'. A problem is that 'many animals respond to stressors by exhibiting diametrically opposite suites of behaviors ... an examination of the symptoms of stress will not generally help us to identify the source of the provocation' (Morgan and Tromborg 2007).

It has been suggested that in the circus, stress from the monotony of confinement can be relieved by performance. The benefits of performance alone for tigers have been questioned, and at the same time it has been suggested that pacing could reflect general excitement at the possible relief from boredom rather than boredom itself:

'The number of performances per day had little influence on [stereotypic] pacing overall . . . pacing was most likely due to external stimulation surrounding the home cage and 'anticipation' of certain events' (Krawczel et al. 2005). Not surprisingly, for tigers 'exercise pens are an important form of enrichment and . . . access for periods longer than 40 minutes should be considered' (Nevill and Friend 2006). For elephants, the picket line is the common and traditional method of restraint in circuses, by chaining one front and the diagonal rear leg to parallel cables or chains for most of the day. Performance as a means of relief has therefore been found to be highly reinforcing, and excited stereotypic behaviour would occur before any special relieving activity such as feeding, performance and watering. The elephants 'were acting like children jumping up and down anticipating something they wanted to do'. It might be suggested that this was evidence of a frustrating and unsatisfactory way of life for the animals, but the author, Ted Friend, decides not to criticise it on the basis that conditions could be worse elsewhere: 'After traveling with this circus and studying its elephants on numerous occasions over 3 years, this author's experience concurs with the conclusions of Kiley-Worthington (1990), that the physical and psychological welfare of circus elephants is not as a rule inferior to that of other animal husbandry systems in zoos, stables, kennels or farms' (Friend 1999). A study of the same year indicated unsurprisingly that accommodation in pens significantly reduced the stereotypical behaviour occasioned by picketing (Friend and Parker 1999).

During consultation for the UK Animal Welfare Bill (enacted in 2006), Animal Defenders International advised that it had conducted an 18-month study on the use of animals in circuses, published as *The Ugliest Show on Earth* in 1998. Observations and recordings were made in 13 UK travelling circuses and winter quarters, and five foreign circuses displaying animals, amounting to over 7,000 h of observations and over 800 h of videotape. The study looked at daily routines, husbandry, animal health, accommodation, exercise, training, psychological and physical effects, abuse, animal supply and international regulations. It also provided the Department of Environment, Food and Rural Affairs (DEFRA) with 'Animal circuses – time for a ban' (2001) and the more comprehensive document, 'Animals in travelling circuses: the science on suffering' (2006). Animal Defenders International found that horses and ponies spent up to 96 % of their day tethered; the large cats between 75 and 99 % of their time in small cages; and elephants spent between 70 and 98 % of their time chained to the ground (United Kingdom Parliament 2004).

The views of Ted Friend and Marthe Kiley-Worthington, who were both later to serve on the UK government's Circus Working Group (CWG), were not supported by the evidence and arguments presented by a team of scientists from Bristol University, commissioned by the RSPCA to contribute to parliamentary deliberations in 2006. Along with others, Harris et al. made a submission to the CWG on 4 December 2006, having reviewed, as was required, the peer-reviewed scientific literature to assess the welfare of wild animals in circuses (Harris et al. 2006).[40]

[40] See also Iossa et al. (2009): 'We conclude that the species of non-domesticated animals commonly kept in circuses appear the least suited to a circus life'.

They examined (1) whether captive animals are different from their wild counterparts, (2) the living and travelling conditions of animals in circuses, (3) the behaviour and health of animals in circuses, (4) welfare related to training and performance, (5) the number and origin of animals involved in the circus industry and (6) the suitability of species for circus life. They proposed that no circus wild animal could be considered domesticated because this was a process that required selection for tameness over thousands of years. Restriction of natural, inherited, highly motivated behavioural repertoires in domesticated animals was known to cause stress, so the problem would be worse for wild animals: 'We argue that animals in circuses face particular challenges unique to the circus environment with which their evolutionary history has ill-equipped them to cope'. Stressful restriction of natural behaviour and environment is worse than in zoos and is 'deleterious for a captive animal's cognitive development, normal social development and, later in life, reproduction and health'. Health problems included spread of disease, tuberculosis, bad feeding practices, lack of exercise, obesity, arthritis and lameness; joint problems as well as hernias were thought to result from circus elephants repeatedly assuming unnatural positions during performance, and 'in circuses, the lack of mud baths, which wild elephants use for skin care, commonly causes severe skin problems'. They also found that 'Evidence suggests that both handrearing and performing have deleterious effects on an animal's subsequent reproductive ability and health'. They stated that 'Animal welfare may simply be defined as an animal's ability to cope with its environment'. Factors preventing this were the stressful effects of mixing unsocial or rival animals or different species, or of isolating social animals; travel and confinement, resulting in more stereotypies than in zoos; and the stressful effect of the audience, notably on large felines. They added that

> There has been no scientific evaluation of how much training is carried out in circuses, or the methods used, or whether the time spent training and the associated stimulation are effective substitutes for involvement in more natural activities. Whilst training based on positive reinforcement has the potential to alleviate some of the negative aspects of a circus life, poor training has the potential to negatively affect both short- and long-term welfare ... Some circuses have no training and, in general, animals are in their cages most of the time.

In the early years of the twenty-first century, it was confirmed that the large majority of MPs and general public wanted to see an end to the use of wild animals in circuses. The trade interests have at the same time attempted to defend their activities by forming public-facing organisations such as the Circus Zoological Society and the Classical Circus Association, emphasising the high standards of practice their members say they apply in order to ensure welfare.[41] But in March

[41] The Circus Zoological Society, founded in 2003, 'supports those circuses who uphold the long tradition of featuring animals, both domestic and exotic, in their shows and strives to disseminate verifiable and factual information about the health and occupation of the animals in those circuses which show a caring attitude'. The Classical Circus Association aims to 'promote and advocate the concept and practice of the art form of Classical Circus with Animals' and to 'ensure CCA members who work with animals practise, promote and maintain the highest possible animal welfare standards'.

2006, Ben Bradshaw, the then minister with responsibility for animal welfare, announced in the House of Commons that under the Animal Welfare Act 2006, the government intended to ban certain wild animals from travelling circuses. A few months later the CWG was set up to investigate a possible partial ban in travelling circuses and to examine the existing welfare implications of transport, accommodation and behavioural needs, but not those of training. The Group invited the external submission of scientific evidence to its Academic Panel, prompting the submission of Harris et al. (2006). However, this submission was not referred to in the report of the CWG in October 2007, 'Wild animals in travelling circuses', issued by the chairman, Mike Radford. The report recommended no immediate changes to the legislation because its Academic Panel believed it had not been supplied with scientific evidence of the specified welfare considerations being in circuses either better or worse than in other captive environments. Radford reported that further primary legislation would therefore be needed for any ban, that this must now become a political issue, and as things stood, current bans by local authorities were questionable. At the time, there were less than 50 wild animals in UK circuses, but the matter was important, he said, as the UK's decision was expected to be influential beyond its borders. In the meantime, Radford recommended strengthening the enforcement of the existing law, development of codes of practice and future revision of the Performing Animals (Regulation) Act of 1925.

Political attention was reactivated in 2011 following the exposure, and accompanying press coverage, of the abuse of the UK's last remaining circus elephant, Anne, at Bobby Roberts Circus. In May the government announced its intention to implement a new licensing system rather than a ban, referring to the difficulties experienced in Austria when a ban was challenged under European Union law. The licensing proposal was met with anger, and in June 2011, there was a successful parliamentary motion for a ban by backbench MPs. In July 2012 legislation for a licensing scheme was laid before parliament, together with an announcement of the preparation of a bill for a ban. In due course, in April 2013, the draft Wild Animals in Circuses Bill was published for scrutiny. The DEFRA select committee published its report in July, recommending a partial ban only affecting big cats and elephants in travelling circuses (none of these animals actually then remained in English circuses). The government rejected this recommendation in the face of continued support for a full ban on wild animals in English circuses by a large majority of MPs as well as the RSPCA, British Veterinary Association, Born Free Foundation, CAPS and others, and in October 2013 the government confirmed that legislation would ensure that all wild animals would be out of travelling circuses by December 2015. Meanwhile, the use of domesticated animals would not be affected, and their protection would depend on the provisions of the 2006 Animal Welfare Act and on the provisions of any legislation replacing the 1925 Act.[42]

[42] Voluntary guidelines for the industry were produced by the RSPCA's Performing Animals Advisory Service in 2012 and are intended to ensure welfare rather than to address ethical issues. They are described as relevant to any animal suppliers, trainers, transporters, handlers or

# References

## *(Articles in periodicals are anonymous or untitled unless otherwise shown)*

AHA (American Humane Association) (2014) http://www.americanhumane.org/animals/programs/no-animals-were-harmed/. Accessed 1 Apr 2014

Amboseli Trust for Elephants (2008) Statement on circus elephants. http://www.elephanttrust.org/article/1 March. Accessed 3 Apr 2014

Animal Defenders International (1998) The ugliest show on earth. Animal Defenders International, London

Animal Defenders International (2001) Animal circuses – time for a ban. Parliamentary briefing. Animal Defenders International, London

Animal Defenders International (2006) Animals in travelling circuses: the science on suffering. Animal Defenders International, London

The Animal World (1921) January

Anon. (1869) Haney's art of training animals. A practical guide for amateur or professional trainers. Giving full instructions for breaking, taming and teaching all kinds of animals including an improved method of horse breaking, management of farm animals, training of sporting dogs; serpent charming, care and tuition of talking, singing and performing birds; and detailed instructions for teaching all circus tricks, and many other wonderful feats. Jesse Haney, New York

The Argus (2008) 23 April

Bailey RE, Gillaspy JA (2005) Operant psychology goes to the fair: Marian and Keller Breland in the Popular Press (1947–1966) Behav Analyst 28(2):143–159

BBFC (British Board of Film Censors) (1931) Performing and Captive Animals' Defence League – preliminary conference on cruel films. Verbatim reports (held by the British Film Institute) 9 Dec

Beers DL (2006) For the prevention of cruelty. The history and legacy of animal rights activism in the United States. Swallow Press, Athens

Bond FF (1924) Modern miracles of the big circus menagerie: *Lions'n' tigers'n' everything.* A review. The New York Times 5 Oct, p X12

Bouse D (2000) Wildlife films. University of Pennsylvania Press, Philadelphia

Bradshaw JWS, Casey RA, Brown SL (2012) The behaviour of the domestic cat. CABI, Wallingford

Breland K, Breland M (1951) A field of applied animal psychology. American Psychologist 6:202–204

Breland K, Breland M (1961) The misbehavior of organisms. American Psychologist 16:681–684

Breland K, Breland M (1966) Animal behavior. Macmillan, New York

Broughton PD (1998) Mule-diving for the high jump. The Spectator 21 Aug, p 16

Burt J (2002) Animals in film. Reaktion, London

Corry J (1986) Cruel camera. The New York Times 24 Mar, p C18

Crowell-Davis SL (2007) Cat behaviour: social organization, communication and development. In: Rochlitz I (ed) The welfare of cats. Springer, Dordrecht

---

production staff who will have influence on animal use, with further information and advice available from http://performinganimals.rspca.org.uk and the RSPCA Performing Animals Advisory Service. The use of Animal Welfare Risk Assessments is advised, and templates are made available (RSPCA 2012).

The Daily News (1933) Great movie animal fights. Editorial 5 May, p 18
Drumm P (2009) Applied animal psychology at an American roadside attraction: Animal Behavior Enterprises and the IQ Zoo of Hot Springs, Arkansas. American Journal of Psychology 122 (4):537–545
Friend TH (1999) Behavior of picketed circus elephants. Applied Animal Behaviour Science 62:73–88
Friend TH, Parker ML (1999) The effect of penning versus picketing on stereotypic behavior of circus elephants. Applied Animal Behaviour Science 64:213–225
Furnas JC (1933) The be-kind-to-animals trend in recent Hollywood offerings. The New York Herald Tribune 7 May, p 3
Gillaspy JA, Bihm EM (2002) Obituary: Marian Breland Bailey (1920–2001). Am Psychol 57 (4):292–293
Gower R (1934) Letter. The Spectator 20 Jul, pp 88–89
Hancocks D (2007) Zoo animals as entertainment exhibitions. In: Malamud R (ed) A cultural history of animals in the modern age. Berg, Oxford and New York
Harris S, Iossa G, Soulsbury CD (2006) A review of the welfare of wild animals in circuses. School of Biological Sciences, University of Bristol and RSPCA, Horsham
Helfer R (2007) The beauty of the beasts. Tales of Hollywood's wild animal stars, Harper, New York
Hornaday WT (1922) The minds and manners of wild animals. A book of personal observations. Charles Scribner's Sons, New York
The Hull Evening News (1929) "Noah's Ark" in a gale. Arrival of wild animals at Hull delayed. Good sailors. 9 December
Iossa G, Soulsbury CD, Harris S (2009) Are wild animals suited to a travelling circus life? Animal Welfare 18:129–140
Johnson W (1990) The rose-tinted menagerie. Heretic Books, London
Joys JC (1983) The wild animal trainer in America. Pruett, Boulder
Joys JC (2011) The wild things. Dissertation, Bowling Green State University
The Kensington News and West London Times (1929) The circus at Olympia. Feats of daring: no wild animals. 21 December
Kiley-Worthington M (1989) The training of circus animals. In: Animal training. A review and commentary on current practice. Proceedings of a symposium organized by the Universities Federation for Animal Welfare, 26–27 Sept, pp 65–81
Kiley-Worthington M (1990) Animals in circuses and zoos. Chiron's world? Little Eco-Farms Publishing, Basildon
Klinowska M, Brown S (1986) A review of dolphinaria. Department of the Environment, London
Krawczel PD, Friend TH, Windom A (2005) Stereotypic behavior of circus tigers: effects of performance. Applied Animal Behaviour Science 95:189–198
The Literary Digest (1920) Cruelty charged in training trick animals for stage and movie. 25 Sept, pp 102–110
Lowther HC (1930) Animal turns. Letter. The Spectator 6 Dec, pp 882–883
Lowther HC (1931) Performing animals. Letter. The Spectator 17 Jan, p 80
MacMichael ET (1930) Letter. The Spectator 13 Dec, pp 939–940
MacMichael ET (1931a) Performing animals. Letter. The Spectator 24 Jan, p 115
MacMichael ET (1931b) Cruelty on the films. Letter. The Spectator 11 Apr, p 586
MacMichael ET (1934) Letter. The Spectator 6 Jul, p 18
MacMichael Papers (uncatalogued and unpublished material relating to the director of the Performing Animals' Defence League between 1929 and 1968, kindly lent to the present author by Edmund MacMichael's daughter, Mrs Esther Denham)
Malamud R (ed) (2007) A cultural history of animals in the modern age. Berg, Oxford and New York
The Manchester Guardian (1929) Lions spring a surprise. Permit required for an "Extra". Another vanishes. 11 December

McGreevy PD, Boakes RA (2007) Carrots and sticks. Principles of animal training. Cambridge University Press, Cambridge
McKie R (2004) Unnatural behaviour: activists condemn ads with boozing chimps. The Observer 30 May
Miklósi A, Pongracz P, Lakatos G, Topal J, Csanyi V (2005) A comparative study of the use of visual communicative signals in interactions between dogs and humans and cats and humans. Journal of Comparative Psychology 119:179–186
Morgan KN, Tromborg CT (2007) Sources of stress in captivity. Applied Animal Behaviour Science 102:262–302
Nevill CH, Friend TH (2006) A preliminary study on the effects of limited access to an exercise pen on stereotypic pacing in circus tigers. Applied Animal Behaviour Science 101:355–361
Ouseley GJR (1904) The gospel of the holy twelve, known also as The gospel of the perfect life. Order of At-one-ment and United Templars' Society, Paris
PADL (Performing Animals' Defence League) (1932) Cruel films. Pamphlet
PADL (Performing Animals' Defence League) (1933) The scandal of animal films. Pamphlet
PADL (Performing Animals' Defence League) (1944) List of subscriptions and donations and balance sheet for 1944. Richard Vincent Hughes Collection, Theatre Museum THM/131
PADL (Performing Animals' Defence League) (1945a) Newsletter no. 26 Aug
PADL (Performing Animals' Defence League) (1945b) Newsletter no. 27 Winter
The Performer (1929) Olympia circus opens. 24 December
Robertson JC (1985) The British Board of Film Censors: film censorship in Britain 1896–1950. Croom Helm, London
Robertson JC (1993) The hidden cinema: British film censorship in action 1913–1975. Routledge, London
Robinson EK (1939) Wild animals in the films. Sight and Sound 8:8–10
RSPCA (Royal Society for the Prevention of Cruelty to Animals) Archives, Horsham
RSPCA (2012) Guidelines for the welfare of performing animals. RSPCA, Horsham
Rutherford L (1986) Managers in a small way: the professionalisation of variety artistes, 1860–1914. In: Bailey P (ed) Music Hall. The business of pleasure. Open University Press, Milton Keynes, pp 93–117
Ryder RD (2000) Animal revolution. Changing attitudes towards speciesism. Blackwell, Oxford
Schumach M (1959) Human watchdog eyes film beasts: association agent guards welfare of animals. The New York Times 5 May, p 40
Stevens M (2014) Biography of Rin-Tin-Tin in Internet movie database. www.imdb.com. Accessed 31 Mar 2014
Sutherland H (1902) Training wild animals. Ainslee's Magazine 8, 6 Jan, pp 518–525
Tait P (2012) Wild and dangerous performances: animals, emotions, circus. Palgrave Macmillan, Basingstoke
Thorley D, Brown D (2008) Kathleen Hanchett-Stamford wins £2.3 m and gives it away. The Times 22 March
The Times (1838) Van Amburgh, the lion tamer. 19 Sept, p 6
The Times (1927) 29 Dec, p 6
The Times (1928a) 7 Dec, p 12
The Times (1928b) 14 Dec, p 10
The Times (1929) News in brief. 28 Oct, p 11
The Times (1931) 3 Jan, p 12
United Kingdom Parliament (various dates) Parliamentary Debates (Commons). HMSO, London
United Kingdom Parliament (various dates) Parliamentary Debates (Lords). HMSO, London
United Kingdom Parliament (2004) Select Committee on Environment, Food and Rural Affairs Written Evidence. Memorandum submitted by Animal Defenders International and the National Anti-Vivisection Society. Annex A: Circuses and Performing Animals, 6 Oct, as published in Animal Defenders International (1998) The ugliest show on Earth. http://www.publications.parliament.uk/pa/cm200405/cmselect/cmenvfru/52/52we136.htm

VAF (Variety Artistes' Federation) Archives, box VAF 5, held by Equity (British Actors Equity Association), London
Velvin E (1911) Critical moments with wild animals, second paper. McClure's Magazine 37:155–163
Ward E (1931) Lord Lonsdale and the RSPCA. Letter. The Spectator 31 Jan, pp 152–153
Webster J (1994) Animal welfare: a cool eye towards Eden. Blackwell Scientific, Oxford
Westacott E (1962) Spotlights on performing animals. Being extracts of evidence given before the Select Committee of 1921 and 1922. Selected by E Westacott, with reference to existing conditions. CW Daniel, Rochford
Wilmut R (1989) Kindly leave the stage! The story of Variety 1919–1960. Methuen, London
Wilson DAH (2002) Animal psychology and ethology in Britain and the emergence of professional concern for the concept of ethical cost. Studies in History and Philosophy of Biological and Biomedical Sciences 33:235–261
Yeats-Brown F (1930) Letter. The Spectator 3 Oct, pp 438–439

# Chapter 7
# Conclusion

**Abstract** Reasons are proposed for the sudden occurrence of a major performing animals controversy in Britain very soon after the First World War. The controversy can also be compared with that over vivisection and the level of public involvement in it. The role of the press and of political interest is noted, alongside charges of social-class discrimination and the reasons for national prejudices. After the weak legislation of 1925, a few outspoken campaigners ensured that opposition to animal performance would remain in the public eye, and growing concerns about the confinement of animals began to have more widespread effect. Many began to prefer to see and understand animals in their natural environment and to respect their independent interests.

Animal performance has been a neglected area of human-animal studies: it has certainly caused much public argument for over 100 years, and doubts over its acceptability as a form of popular entertainment have revived in recent decades.[1] The first controversy that grew up around it in Britain deserves attention if only because its very timing is so interesting. It might be thought that British society in 1921 had more pressing subjects competing for attention, such as reflection on the horrors of the recent war or concern for the national economy, unemployment and strikes. But the aftermath of the First World War might well have provided a suitable context for a public examination of a subject like this. The early 1920s was clearly a time of great social upheaval, doubt and change that finally marked a watershed from the Victorian age. Women had new access to employment and the professions and (if they were over 30) to the vote, and the media was becoming a 'mass media', with modern presses and the advent of radio broadcasts. New forms

[1] Davis complains: 'More than in any previous period, popular culture today continues to favor and perpetuate negative images and narratives of trainers—a recurrent theme in the endlessly rerun *Simpsons,* as well as in the news media, and talk shows. Hooking into powerful memes, the politics of animal rights has infected the discourse of animal welfare, resulting in mostly negative consequences for both animals and those who work them.' Although he rightly highlights contemporary popular criticism, this resentful claim of 'infection' and (for animals) 'negative consequences' is not argued or substantiated in his article (Davis 2007).



D.A.H. Wilson, *The Welfare of Performing Animals*, Animal Welfare 15,
DOI 10.1007/978-3-662-45834-1_7

of modern entertainment like jazz and revue reflected a different way of looking at the world: as with fashions, there was less of the fussy and contrived and more that was expressive and honest. In these exciting, uncertain and challenging circumstances, a new examination of a familiar topic with accessible and easily understood moral implications might have fitted the mood of the time very well. Once the controversy was under way, it was easy to identify the defenders of animal performance with a pre-war conservative outlook and the opponents with the new, outspoken, free-thinking and radical elements of contemplative post-war Britain. The occurrence of the performing animals controversy is another example of the fundamental effect the First World War had on British society, especially as part of a renewed collective recognition of public morality. Nevertheless, during the controversy, defenders of animal performance often expressed a concern at the threat of overregulation in public life.

Sporadic public attention to the question of animal performance from the late nineteenth century to 1914, assisted by some press coverage, had led to concentrated opposition with the establishment of the Performing Animals Committee and then the post-war appearance of the Jack London Club, inspired by London's *Michael, Brother of Jerry*. The subsequent emergence of the performing animals controversy after the First World War followed a pattern that has since become increasingly familiar. After a new pressure group was formed, an articulation of public feeling resulted in press interest; then the issue became attractive to politicians, who saw it as fertile ground if not as a cause of genuine personal concern, and its consolidation became secured. The vivisection controversy 50 years previously had also fed off widespread interest, but in the 1920s, the role of the press as mass media was more effective and had greater influence in setting the agenda of parliamentary interests, so that the sustenance of the performing animals controversy was not solely dependent on the continued commitment of a limited number of individuals in Parliament. In the case of animal performance, each agent in the controversy contributed to its development until a temporary denouement was reached in 1925. However, the substance and significance of the issue was confirmed by interest in it that has continued until the present day, and after the legislation of 1925, the uncompromising, high-profile, direct-action tactics of Edmund MacMichael ensured that it remained publicly visible and that parliamentarians like Danesfort, Freeman and David Kenworthy would continue to draw attention to it in the following decades.

The select committee investigation of 1921–1922 was well organized and fairly conducted, but largely because of a context of professional secrecy, it was unable to reveal conclusive evidence, and its findings were therefore somewhat tentative, resulting later in half-hearted legislation that represented a compromise at the time but promised continued agitation. The issue became centred not so much on what could be evidenced as about what society thought. And in 2011, backbench MPs in the British parliament successfully insisted on prohibition measures even though they had recently been informed (probably incorrectly) that there was no scientific evidence to warrant them. As with vivisection, this controversy had turned into an example of mass public awareness of an aspect of the human-animal relationship.[2] There were, however, at first no further major legislative developments after 1925 because the majority of MPs had become disaffected by the continuing irreconcilability of extremely opposed assertions by a few about the nature of animal performance. Furthermore, they knew that considerable time and effort had already been invested in the issue between 1921 and 1925 and felt that little more could be added.

The restricted effect of the legislation of 1925, like that for vivisection of 1876, ensured continued maintenance of the original campaigns of opposition. In the case of opposition to animal performance, a new form of activist group took shape in the PADL. It was led by those from the well-educated upper and middle classes, but unlike opposition to vivisection, it dealt with an aspect of mass public entertainment about which ordinary working people were also able to express an informed point of view. This point of view was of economic concern to the animal trainers, exhibitors and promoters, whereas in the vivisection controversy, ordinary members of the public were able only to express simple philosophical opinions on procedures whose detail they usually did not understand and never encountered—their views were therefore more easily dismissed as inexpert, as they often are today.

In parallel with the earlier vivisection controversy, the role of professional secrecy and the difficulty of securing reliable evidence during the period of official enquiry meant that external criticisms could be described as unfounded and the result of prejudice, a morbid imagination or 'crankitis'. Any criticisms from within the profession might be dismissed as grudges. The opponents of animal performance were also accused of being killjoys and busybodies and employing in their societies staff who had to earn their keep by finding problems where they did not exist. Charges of hypocrisy could also be made when opponents were known to indulge in field sports or to attend exploitative events such as the Grand National horse race. The supposed legitimacy of the sports of the privileged compared with criticism of the entertainment of working people provided defenders of animal performance with the means to claim that their activity—of economic and alleged educational benefit, too—was being made subject to social or class prejudice.

The campaign against animal performance was given added intensity in the early 1920s by an appeal to criticize the behaviour of trainers from the ranks of the 'alien enemy'. Germany had supplied the most impressive and frequent animal turns before the war, but after its conclusion, this source suffered xenophobic and economic boycott as well as charges of cruelty designed to deflect criticism from domestic training methods. The sincerity of such criticism on the part of British trainers could be called into question because while German methods were described simultaneously as the most effective and the most cruel, at the same

[2] Like vivisection, animal performance can still inspire two reactions: firstly, deciding case by case and often with some difficulty that some is at least innocuous and some bad or, secondly, deciding (more easily) that all is bad, because it is philosophically unacceptable.

time the select committee was assured that training by kindness was the only successful method, one adhered to in Britain (and, of course, advocated by Hagenbeck in Germany before the war). Meanwhile, both sides were ready to criticize any imported acts, the professionals in order to protect domestic employment and the opponents because humane standards could not be controlled or guaranteed in foreign sources.

Although, as with vivisection, organized opponents of animal performance were to be found at first mainly among the middle and upper classes, especially writers and those educated in the arts, animal performance in the circus and music hall began to attract the attention also of those to whom such entertainment was actually directed: a broad section of society able more easily to pass an opinion on activities less esoteric and incomprehensible than those of the laboratories of animal science. However, in both cases, the professional interests managed to weaken legislation so that it acquired a legitimizing function based on registration, licensing and limited inspection. It could be argued by its proponents that vivisection, unlike animal performance, was 'essential'; and unlike the Cruelty to Animals Act of 1876 (and its successor of 1986), the Performing Animals (Regulation) Act was not fully 'enabling'. A vivisection licence was granted at national level and once issued protected the work of the scientist; and a local authority could not be involved or oppose it. Although local authorities were not entitled to deny applications for licences for animal performance unless the applicant had been convicted for cruelty, they have been free to respond to continuing public pressure since 1925 by imposing bans on performing animal acts on their land.

Although the select committee had struggled to acquire hard evidence on training methods to serve its enquiry, verifiable accounts of conditions of transport and accommodation were easier to come by, and these were more often the basis of successful prosecutions concerning conduct in the animal performance sector of the entertainments industry. The purpose of the enquiry itself was to assess the presence or absence of cruelty in training and exhibition, but as the investigation proceeded, the nature of transport and accommodation began to win almost as much attention. The performing animals controversy in the immediate post-war period then served to bring to the forefront this new and developing area of concern about animal welfare, which had relevance as well to many other situations in which animals were managed for human purposes. Travelling menageries came to be regarded as unacceptable, and the last of them ceased operations in 1932: the collection was sold to London Zoo, and thereafter, there were no more proper examples of this form of entertainment. Further discussion on the nature and effect of animal captivity was then directed towards zoos and afterwards linked with concerns about 'factory farming'.

In the decades since the eruption of the performing animals controversy, the treatment of animals has been made increasingly accountable, in theory if not in practice. Animal performance has also come to be seen as outdated as an attraction unless it is presented convincingly as unpressured and natural, and so the exhibition of wild animals out of their element in the circus arena in Britain has dwindled through loss of favour. The public has found more interesting and sympathetic ways of enjoying animal behaviour, as through the wildlife documentaries on television that have helped to effect a better understanding of—and a changing attitude to—the vulnerable natural world. This effect of the broadcast media had been paralleled in earlier times by the results of the gradual emergence of ideas of human-animal kinship facilitated by Darwinian theory (and Darwin would have influenced those well-educated classes from which criticism of animal performance soon came). During the twentieth century, an urge to dominate nature gave way to a concern to defend it; the gun more often gave way to the camera; animals were more easily associated with their habitats than with exhibition cages; and the traditional bravery of the wild animal trainer was often newly criticized as stupid or cowardly. Animals, with better recognized interests if not rights, were now worthy of respect, as well as being sources of curiosity: their activities and behaviour in the natural environment became more fascinating than in an artificial one.

## Reference

Davis CB (2007) Cultural evolution and performance genres: memetics in theatre history and performance studies. Theat J 59:595–614

# Index




D.A.H. Wilson, *The Welfare of Performing Animals*, Animal Welfare 15,
DOI 10.1007/978-3-662-45834-1

C

MIX
Papier aus verantwortungsvollen Quellen
Paper from responsible sources
FSC® C105338

If you have any concerns about our products,
you can contact us on
**ProductSafety@springernature.com**

In case Publisher is established outside the EU,
the EU authorized representative is:
**Springer Nature Customer Service Center GmbH**
**Europaplatz 3, 69115 Heidelberg, Germany**

Printed by Libri Plureos GmbH
in Hamburg, Germany